Salyut – The First Space Station
Triumph and Tragedy

Grujica S. Ivanovich

Salyut – The First Space Station

Triumph and Tragedy

 Springer

Published in association with
Praxis Publishing
Chichester, UK

Grujica S. Ivanovich
Principal Engineer Distribution Planning
Ergon Energy
Toowoomba
Queensland
Australia

SPRINGER–PRAXIS BOOKS IN SPACE EXPLORATION
SUBJECT *ADVISORY EDITOR*: John Mason B.Sc., M.Sc., Ph.D.

ISBN 978-0-387-73585-6 Springer Berlin Heidelberg New York

Springer is a part of Springer Science + Business Media (*springer.com*)

Library of Congress Control Number: 200794102

Cover design: Jim Wilkie
Editor: David M. Harland
Typesetting: BookEns Ltd, Royston, Herts., UK

Printed in Germany on acid-free paper

This book is dedicated to the heroic crew
of the Soyuz 11 spacecraft

Viktor I. Patsayev (1933–1971)
Georgiy T. Dobrovolskiy (1928–1971)
Vladislav N. Volkov (1935–1971)

Viktor Patsayev, Georgiy Dobrovolskiy and Vladislav Volkov

and to the thousands of designers, technicians and workers who
participated in the development of the pioneering Salyut space
station, and thereby in the beginning of a new era in the
history of mankind.

Contents

List of Illustrations

Front cover

A technically accurate painting by artists Andrey Sokolov and Aleksey Leonov of a Soyuz spacecraft about to dock with the Salyut space station.

Rear cover

The main picture shows cosmonauts Patsayev (left), Dobrovolskiy and Volkov. The top two pictures on the right show Dobrovolskiy (left) and Volkov on board the Salyut space station in space; the Soyuz 11 descent module immediately following its landing; and an attempt to resuscitate Dobrovolskiy.

Dedication

Chapter 1

Foreword

"THE STAR KETS"

In 1936 the book *The Star KETs*[1] was published by the well-known Soviet science-fiction writer Aleksandr Belyayev. The main events in this work took place aboard an enormous "exo-atmospheric laboratory".

This was an entire city in near-Earth orbit. By the will of the author, the designers of the space station equipped it with a rocket base for receiving vehicles from the Earth, as well as with a gigantic greenhouse to provide the inhabitants of the station with oxygen and fresh food, and also with numerous living, support and scientific compartments offering comfortable conditions for the crew and to enable them to work "for the benefit of humanity".

Any inhabitant of the Earth could see this "man-made star". It was sufficient to go outside at night and glance upward. Aleksandr Belyayev believed that it would not take much time – 40 to 50 years – before a real "Star KETs" would grace the sky.

The prophecies of this visionary did indeed come true. Although this will be seen by some as sheer coincidence, the world's first "exo-atmospheric laboratory" by the name Salyut was launched by the Soviet Union on 19 April 1971, 35 years after the appearance of Belyayev's book, and within the interval predicted by its author. Only five days later, the first spacecraft with cosmonauts on board docked with the station, and one and a half months later the first crew began to work on board it.

The launch of Salyut was the logical culmination of work that began almost ten years earlier at the design bureaus of Sergey Korolev and Vladimir Chelomey, in which the first sketches of the civilian DOS[2] and military Almaz stations were made. It had been a difficult path, filled with sleepless nights, agonising bitter reflections, disappointments and . . . unexpected flashes of inspiration. But at the end of this path there was VICTORY!

The book which you are holding in your hands relates the development of the first

---

[1] KETs – Konstantin Eduardovich Tsiolkovskiy
[2] DOS – Long-duration Orbital Station

Soviet orbital stations. About how they were conceived. About the people who made them. About the difficulties that had to be overcome. About the cosmonauts who worked in near-Earth orbit. In other words, about that which made possible the development of the International Space Station that nowadays warms the hearts of earthlings as it crosses the night sky and resembles, albeit remotely, Belyayev's "Star KETs".

Aleksandr Zheleznyakov
Member-Correspondent of the Russian Academy of Cosmonautics Named After K.E. Tsiolkovskiy
Adviser of the President of RKK Energiya

"ЗВЕЗДЫ КЭЦ"

В 1936 году увидела свет книга известного советского писателя-фантаста Александра Беляева "Звезда КЭЦ". Основные события в этом произведении разворачивались на борту огромной "заатмосферной лаборатории".

Это был целый город на околоземной орбите. Волей автора создатели орбитальной станции оснастили ее и ракетодромом для приема ракет с Земли, и гигантской оранжереей, обеспечивающей обитателей станции кислородом и свежими продуктами питания, и множеством жилых, служебных и научных отсеков, позволяющим космонавтам вести комфортную жизнь и создающим необходимые условия для работы "на благо человечества".

Любой житель Земли мог увидеть эту "рукотворную звезду". Достаточно было выйти ночью на улицу и взглянуть вверх. Александр Беляев считал, что пройдет совсем немного времени, лет 40-50, и не выдуманная, а настоящая "звезда КЭЦ" зажжется на небе.

И пророчества фантаста сбылись. Кому-то это может показаться случайным совпадением, но первая в мире "заатмосферная лаборатория" под именем "Салют" была запущена в Советском Союзе 19 апреля 1971 года, спустя 35 лет после появления книги Александра Беляева. Именно в те сроки, о которых говорил писатель. Уже через пять дней к ней причалил корабль с космонавтами на борту. А через полтора месяца к работе на станции приступил первый экипаж.

Запуск "Салюта" стал логическим завершением работы, которая началась почти десятью годами раньше, когда в конструкторских бюро Сергея Королева и Владимира Челомея были сделаны первые наброски будущих станций – гражданской ДОС и военного "Алмаза". Это был трудный путь, наполненный бессонными ночами, мучительными размышлениями, разочарованиями и … неожиданными озарениями. Но в конце этого пути была ПОБЕДА!

Книга, которую вы держите в своих руках, рассказывает о разработке первых советских орбитальных станциях. О том, как зарождалась идея этих уникальных для своего времени комплексов. О людях, которые их делали. О трудностях, которые пришлось при этом преодолеть. О космонавтах, которые работали на околоземной орбите. То есть, о том, что сделало возможным появление Международной космической станции, сегодня "греющей" своим светом землян и, хотя и отдаленно, но напоминающей беляевскую "Звезду КЭЦ".

Александр Железняков,
член-корреспондент
Российской академии космонавтики им. К.Э. Циолковского

Author's preface

The mission of the Soyuz 11 crew who lived on board the first Salyut space station is remembered by the phrase *triumph* and *tragedy*.

Triumph stands for the successful designing, testing and launching of the world's first space station in an unbelievably short period of time. In fact, it was done in less than 16 months. It also stands for the ability of the Soyuz 11 crew to dock and enter the station after the preceding crew had been prevented from doing so. And then it stands for their ability to conduct a broad programme of scientific research on board the station. Finally, it stands for their perseverance in conditions that were far from the norm to establish a new world record for the duration of a space mission.

Tragedy stands for the fact that with only a few minutes remaining from returning to their motherland, they were overwhelmed by an emergency which, within just a few seconds, claimed their lives. It stands for the shock of the recovery team which, on opening the capsule, found their inert bodies. It stands for the trauma suffered by their families and colleagues, and indeed the entire nation. And it stands for how, on reflection, the loss of this brave crew ought never to have happened.

In a less than a year and a half after the worst tragedy in Soviet cosmonautics, the book *Salyut in Orbit* was published. The first time that I laid my hands on it was in the mid-1980s, on a visit to the Russian Home of Culture in Belgrade, Serbia. What caught my interest was that a book intended as a memorial to the fallen cosmonauts should contain a wealth of information describing the first Salyut space station, its apparatus and the experiments that were conducted by the unlucky crew. It contains the cosmonauts' diaries, and even some of their conversations with the controllers on Earth – it was astonishing that such a book was allowed to be published during the Soviet era. However, it seemed incomplete because it said little of the tragic end of the mission. Why did cosmonauts Georgiy Dobrovolskiy, Vladislav Volkov and Viktor Patsayev lose their lives? Was it a design error in the Soyuz spacecraft? Was it the result of an error by the cosmonauts? Was it utterly inexplicable? The official story was that a ventilation valve inadvertently opened in space and the cosmonauts died when the air suddenly escaped from the cabin. However, in my inner self, I felt that there had to be more to the worst tragedy of the Soviet space programme. I was also fascinated by the fact that even in the latest books, published without official censorship, the reasons for the loss of the Soyuz 11 crew were still not explained. It was as if there was simply no desire to uncover the details of such a traumatic event. This stirred within me the challenge of finding out what, directly or indirectly, led to the loss of this heroic crew.

As my analysis of the material progressed, slowly the veil of mystery began to lift. I realised that the story of the valve prematurely opening was just a part of the story, and also that if the people who prepared the spacecraft had adhered to the stipulated procedure, then the cosmonauts would have survived the flight because the opening of this valve would not have caused the air to escape!

For a long time while working on my research, I was under the impression that no one else had looked deeply into this subject, but I was wrong. Just before I finished this book, I managed to get in contact with Viktor Patsayev's daughter and son. It was only then that I found out that their mother, Vera Patsayeva, had over a period of many years gathered material and interviewed the designers, engineers and other specialists who worked on the preparation of the Soyuz 11 mission. In fact, she was the driving force behind the publication of *Salyut in Orbit*. Her daughter, Svetlana, kindly sent me some of Vera Patsayeva's material. This corroborated the results of my own analysis. I am grateful to Svetlana for allowing me to included in my book an extract from Vera Patsayeva's notes.

As I worked on this book, I came to develop an emotional bond with its heroes – Dobrovolskiy, Volkov and Patsayev. I understood that in an odd way they had been murdered twice. The first time was when they became the first human beings to die in the vacuum of space. But when the truth about the cause of their loss was hidden, they were effectively murdered a second time. This is my attempt to shed light on how these inspiring men lived and died. The urns containing their ashes have rested in the wall of the Kremlin for over 36 years. Now it is time for the truth to be told.

Grujica S. Ivanovich
Toowoomba, Australia
27 September 2007

Note on transliteration

I have used a modified version of the standard for English translations of Russian names and toponyms, as they are often phonetically inappropriate.

For example:

- Baykonur, instead Baikonur
- Dobrovolskiy, instead Dobrovolski
- Sergey, instead Sergei.

However, because they have been used so widely, I have retained Korolev (which is more correctly, Karalyof) and Kamanin (Kamanyin).

I have noticed that some authors use Russian titles in Latin, and some even combine Latin and English.

For example:

Semyonov, Y.P., ed, *Raketnaya-kosmicheskaya korporatsiya Energiya* named after S.P. Korolev, 1996.

In the Bibliography, I have added the English translation beneath each Russian title; viz:

Ракетно-космическая корпорация "Энергия" им. С. П. Королева/Под. ред. Ю. П. Семенова, 1996

Semyonov Y.P., ed, Rocket and Space Corporation Energiya named after S.P. Korolev, 1996

However, it should be noted that not all of these books are translated into English.

Note on illustrations

I have illustrated this book with as many unique or rare pictures as possible, some of which have never been published before. In some cases, reflecting their historic importance, I have used pictures that are of poor quality, but I hope that they do not detract from your enjoyment of the book. For permission to reproduce illustrations appearing in this book, please correspond directly with the owners, as specified in the individual captions. Uncredited pictures belong to the author.

Acknowledgements

Writing this book involved extensive research, but it is a logical continuation of my interest in space flight which was sparked by the television series *Star Trek* when I was only 11 years of age. The idea for this book arose when the Serbian magazine *Astronomija* (Astronomy) published a series which I wrote detailing the disasters of the space programme, one of which was an account of the Soyuz 11 tragedy.

Seeing in *Spaceflight* magazine of the British Interplanetary Society a short letter from Praxis Publishing encouraging new authors interested in space to join them, on 10 October 2006 I sent them my first email offering the story of the greatest tragedy in Soviet cosmonautics. To my great delight, they accepted. In the ensuing months, I read all the material available to me on the Salyut space station, ranging from the early releases in 1971 to the most recent books published in Russia, England and America. It would have been very difficult to write this book without the generous assistance and support of enthusiasts in Australia, Russia, Serbia, England, Scotland, Ireland, America, Israel, Spain and Sweden – some of whom have spent decades probing the secrets of the Soviet space programme – and I thank them all from the bottom of my heart. In particular, I am grateful to:

- My love **Natasha** and our little angels **Tijana Sara** and **Dushan** – for their understanding, support, strength, tolerance and endurance during these long months;
- **David Harland** – for his comprehensive preparation of the manuscript and illustrations;
- **Vadim Anosov** – for continuous support, and for sharing his knowledge, interest and endless enthusiasm for cosmonautics;
- **Marina Dobrovolskiy** – for memories of her heroic father;
- **Aleksandr Zheleznyakov** – for kindly contributing the foreword;
- **Svetlana Patsayeva** – for sensitive words about her exceptional father, unselfish assistance, and for exclusive access to the materials pertaining to the Soyuz 11 tragedy collected by her mother, Vera Patsayeva, over many years;
- **Brian Harvey** – for archive materials of the Salyut space station, and for reviewing an early draft of the manuscript;
- **Rex Hall** – for providing photographs;

- **Dmitriy Patsayev** – for sharing memories about his father, and also for professional comments;
- **Clive Horwood** – for continuous support and belief in the project;
- **Ivana Lukic** – for reviewing my English, providing translations and advice, and for encouragement to work on this project;
- **Leon Rosenblum** – for information regarding the tracking ships;
- **Aleksandar Zorkic** – for continuous support, encouragement and help;
- **Sven Grahn** – for Salyut radio-tracking data;
- **Dmitriy Payson** – for help in establishing contact with Marina Dobrovolskiy;
- **Mark Wade** – for providing diagrams;
- **Asif Siddiqi** – for his support and assistance;
- **Peter Pesavento** – for providing photographs;
- **Slobodan Zlokolica** – for archive materials of the Soyuz 11 mission from the National Serbian Library.

During the long and silent nights that I studied the material about the first Salyut space station, glances at my rested and blessed parents **Stale** and **Mila** provided me an additional strength. They wholeheartedly supported my love of the heavens. Ten years ago, they proudly assisted the presentation of my first book in Serbia. I know how proud they would have been to see this book too.

Again, to all concerned, I kindly thank you, and bow to the immensity of space! After all, "we are all made of stars".

1

From Almaz to Salyut

EARLY DAYS

Special Design Bureau 1, OKB-1,[1] is situated some 25 km northeast of the centre of Moscow in Podlipok, Kaliningrad (renamed Korolev in 1997), and it played a key role in the Soviet manned space programme: it designed the first satellites, the first lunar and interplanetary probes, and the Vostok spacecraft that carried the first man into orbit. In the years that followed those early achievements, it defined the major strands of the manned space programme.

The leader of OKB-1, and the main driving force of Soviet cosmonautics, was the legendary Chief Designer Sergey Pavlovich Korolev. After Korolev's death during what had been expected to be routine surgery in January 1966, he was succeeded by his deputy Vasiliy Pavlovich Mishin, a rocket engineer who had worked closely with Korolev since 1945. Mishin promptly reorganised the work force of more than 60,000 employees, and on 6 March 1966, at the direction of the Ministry of General Machine Building (MOM), and no doubt in an effort to confuse spies, OKB-1 was renamed the Central Design Bureau of Experimental Machine Building (TsKBEM).

Mishin inherited from Korolev the task of completing the development of the new manned spacecraft named Soyuz ('Союз', meaning 'Union'), and using this for the L1 programme in which two cosmonauts were to fly in a very high orbit that looped around the far side of the Moon before returning to Earth. But for Mishin the most important task was the development of the giant N1 rocket for the L3 programme to land a Soviet cosmonaut on the lunar surface.

The development of the Soyuz proved to be more difficult than expected, with a series of unmanned test flights revealing a variety of problems, but in April 1967 it was decided to proceed with the first manned test in which one spacecraft would be launched into orbit with a single cosmonaut and a second spacecraft with a

---

[1] The 'O' in the abbreviation OKB-1 is the word *ossobeniy*, which can also be translated as 'particular' or 'experimental'.

The TsKBEM building at Kaliningrad, Moscow.

The founder of the Soviet space programme, Sergey Korolev (left) and his successor Vasiliy Mishin, who was Chief Designer of the TsKBEM from 1966 to 1974.

The Soyuz spacecraft was the workhorse of the Soviet manned space programme. On the left is the orbital module with the active docking probe, then the descent module with the crew cabin, and finally the propulsion module containing the main engine and solar panels.

Leonid Brezhnyev with the crews of the Soyuz 4/5 joint mission.

crew of three would follow the next day. The two spacecraft were to rendezvous and dock, and two of the cosmonauts were to cross from one vehicle to the other by spacewalking. However, Soyuz 1, flown by Vladimir Komarov, ran into difficulties immediately on entering orbit. First, one of two solar panels failed to deploy and this resulted in problems with the star sensor, which made it difficult for the vehicle to maintain the desired orientation in space. The State Commission at the Baykonur cosmodrome in Kazakhstan cancelled the launch of Soyuz 2. After overcoming numerous technical problems, Komarov finally succeeded in orientating his craft and made the de-orbit burn. Unfortunately, the parachute failed to deploy and the descent module hit the ground at great speed and Komarov perished.

When flights were resumed in October 1968, Soyuz 2 was launched unmanned. Georgiy Beregovoy, launched the next day in Soyuz 3, performed a rendezvous, but could not achieve a docking.

When two manned Soyuz spacecraft were finally able to dock in January 1969, Yevgeniy Khrunov and Aleksey Yeliseyev performed a spacewalk to transfer from Soyuz 5 to Soyuz 4, then returned to Earth with Vladimir Shatalov. When Boris Volynov attempted to land in Soyuz 5 the next day, the propulsion module failed to separate from the descent module, causing the vehicle to start its re-entry with the hatch – as opposed to the heat shield – facing in the direction of flight. Fortunately, the connections between two modules were severed by the heat before the descent module suffered damage, and the capsule rotated into the safe orientation. However, the off-nominal re-entry caused the capsule to descend 600 km from the planned recovery point and the impact was so violent that Volynov suffered several broken front teeth.

The docking of two manned spacecraft was one of the rare Soviet achievements during the race to the Moon. But the success of Apollo 8 in performing 10 orbits around the Moon in December 1968 had rendered politically pointless the simpler circumlunar mission for which the L1 version of Korolev's spacecraft had been designed.

When Apollo 11 landed on the Moon in July 1969, the Americans won the race to the Moon, and the mood of the Kremlin was further diminished by two failures of the N1 rocket. In an attempt to once again impress the Soviet nation, and indeed the world, it was decided that the next mission should included three manned spacecraft with a total of seven cosmonauts.

Accordingly on successive days in October 1969 Georgiy Shonin and Valeriy Kubasov were launched on Soyuz 6, Anatoliy Filipchenko, Vladislav Volkov and Viktor Gorbatko were launched on Soyuz 7, and Vladimir Shatalov and Aleksey Yeliseyev – both of whom were veterans from the successful Soyuz 4/5 docking – were launched on Soyuz 8. Once all three spacecraft had rendezvoused in space, the crew of Soyuz 6 were to film Soyuz 8 docking with Soyuz 7. This time, however, it was not intended that any cosmonauts should make a spacewalk. Unfortunately, the Igla automatic rendezvous system onboard Soyuz 8 malfunctioned, and despite four manual attempts Shatalov was unable to complete the approach. Pursuing their own programme, Shonin and Kubasov performed the first vacuum-welding operation in space, then returned to Earth, followed in turn by Soyuz 7 and 8 over the next two days. As much as the Kremlin and TASS, the official news agency, had portrayed this 'group flight' as another achievement of Soviet cosmonautics, Mishin and his engineers were disappointed.

Mishin's dilemma was that because the Soyuz was to be a 'universal' spacecraft, delays in perfecting it were holding up the programmes that were to exploit it, some of which, including the N1-L3 lunar landing programme, were already years behind schedule.

CHELOMEY AND THE KREMLIN

Mishin's TsKBEM was not the only design bureau in the USSR involved in the development of manned spacecraft. In Moscow's eastern suburb of Reutov, 30 km south of Kaliningrad, was the headquarters of OKB-52, which in 1966 changed its name to the Central Design Bureau of Machine Building (TsKBM). It was led by Vladimir Nikolayevich Chelomey. Although there was only one letter different in the titles of the two bureaus, namely the 'E', Chelomey, having a staff of only 8,000 employees, had much more modest capabilities. However, because Chelomey had good relations with the military, having developed a number of cruise missiles, and because one of his engineers was the son of Nikitha Khrushchov, in the early 1960s his bureau was the main competitor to OKB-1.

In 1963 Chelomey conceived the idea to develop a military Orbital Piloted Station (OPS) equipped with cameras to monitor the US and NATO military facilities. The project was named Almaz ('Diamond'), this name being in keeping with the practice of naming his products after precious stones. When designers from the Central Scientific-Research Institute for Machine Building (TsNIIMash) visited OKB-52 in the spring of 1964 they were shown the mockup of the station and its return capsule. It was to be launched by the powerful UR-500 Proton rocket that Chelomey was developing.[2] However, the Ministry of Defence was unwilling to finance the project. Undeterred, Chelomey sought the behind-the-scenes support of his military contacts.

In the meantime, after the assassination of John F. Kennedy, Lyndon B. Johnson became the American president. On 10 December 1963 he cancelled the US Air Force project to build a small winged 'space plane' named Dyna-Soar, and it was announced that plans would be drawn up for a new military space programme: the Manned Orbital Laboratory (MOL). This was to monitor the activities of Soviet military forces and observe rocket launching sites, airfields and naval bases. Since methods for rendezvousing and docking in space had yet to be developed, the plan was to launch the MOL with the crew of two military astronauts riding on top in a modified form of the Gemini spacecraft which NASA was at that time developing. The mission would last a month, and the MOL would be abandoned when the crew departed.

The capabilities of the MOL prompted the Kremlin to back Chelomey's proposal, and the project was given to OKB-52's Branch No. 1 at Fili, in the heart of Moscow, which had developed the Proton launch vehicle. The manager was Branch No. 1's Chief Designer, Viktor Bugayskiy. On 12 October 1964, the day that Chelomey announced the start of work, the first Voskhod spacecraft was launched for a 1-day flight with a crew of three cosmonauts. While they were in space, Khrushchov was overthrown – and Chelomey lost his main supporter. The situation was particularly dire because, as Khrushchov's favourite, Chelomey had gained many enemies. Not only was the new Kremlin leader, Leonid Brezhnyev, not an ally, the new Prime Minister, Aleksey Kosygin, was very rude to Chelomey during their first telephone conversation regarding the future of the UR-200 rocket programme. In fact, neither Brezhnyev nor Kosygin shared Khrushchov's enthusiasm for manned space flights.

Another man of special importance was Dmitriy Ustinov. Since 1946 he had been responsible for the development of the Strategic Rocket Forces. He was known as 'Uncle Mitya' to the leaders of the design bureaus. His influence declined in the Khrushchov years, but his position was reinforced by the arrival of Brezhnyev, and in March 1965, in a major restructuring of the Soviet rocket and space programmes, he became the Secretary of the Central Committee of the Soviet Communist Party responsible for defence and space.

---

[2] The UR-500 rocket first flew in July 1965, and became known as the 'Proton' after its first scientific payloads.

Vladimir Chelomey (left), the Secretary for Defence and Space Dmitriy Ustinov and 'Space Minister' Sergey Afanasyev.

Despite the scepticism of Brezhnyev, Kosygin and Ustinov, Chelomey still had the strong support of the generals in the Soviet Air Force and the Strategic Rocket Forces. He also had the support of Mstislav Keldysh. As a long-time companion of Korolev and proponent of using rockets and satellites to facilitate scientific studies, Keldysh was one of the most eminent figures in the rocket and space programme. In fact, he had played a key role in the establishment of OKB-52 in 1955. To mark his contribution to the management of the pioneering manned space flight by cosmonaut Yuriy Gagarin in April 1961, Keldysh had been appointed President of the Soviet Academy of Sciences. In Brezhnyev's government, the Ministry of General Machine Building was the public name for the secret rocket and space industry – the bland name was to mask the significance of its work. In March 1965, Kosygin nominated Sergey Afanasyev as the first 'Space Minister'. On 25 August 1965 President Johnson gave the formal go-ahead for the MOL project, which was to make its first flight by the end of 1968. Two months later, on 27 October, Afanasyev signed the order for Almaz. The preliminary paperwork was drawn up in 1966 and, based on regulations signed by the Council of Ministers and the Central Committee of the Soviet Communist Party, on 14 August 1967 the technical requirements and timescale were specified.

ALMAZ

The Almaz orbital complex had four major segments:

- the manned spacecraft which formed the re-entry vehicle (VA);
- the working compartment;
- the compartment with the apparatus for taking long-focus photographs; and
- the propulsion module.

As with the American MOL, in its original design the Almaz was to be launched with its crew riding in a spacecraft on top. This eliminated the task of developing a rendezvous and docking system. However, further analysis led to a revision of this concept. In particular, because the presence of the heavy manned spacecraft would reduce the mass of the space station, and hence the amount of scientific and military equipment that it could carry, in 1967 the State Commission endorsed a two-launch option in which the space station and the manned spacecraft would be autonomous vehicles. This would not only enable the station to grow in mass to exploit the 20-tonne payload capacity of the Proton, it would also allow the station to be operated by a series of crews. Furthermore, because the crew was to be launched by a Proton, it was decided to mate the re-entry vehicle to a Functional-Cargo Block (FGB)[3] to produce the 20-tonne Transport and Supply Ship (TKS)[4] which would dock with the station to deliver a crew together with the cargo required for their tour of duty. The crews would be exchanged at intervals of two or three months, and the station would have an operational life of up to two years, being unoccupied only during the short intervals between one crew departing and the next one arriving. This revision would make more efficient use of the hardware than the original plan.

On board Almaz, the crew would use equipment that could be precisely aimed to study military targets on Earth, including camouflaged and mobile ones. In addition, it would be possible to undertake scientific and ecological monitoring, including the early detection of bushfires and the spread of pollution by rivers to the oceans. The equipment was state of the art for that time. The primary optical instrument was a photographic camera that used a mirror with a diameter of almost 2 metres and a focal length of 10 metres.[5] In fact, the design was so complex that it took 3 months to negotiate with Zenith, the Krasnodar firm assigned the task of manufacturing it, precisely how the system was to operate. When the design was judged too complex to be built within the specified 18-month period, it was decided to produce a simpler apparatus, which was named Agat ('Agate').

The crew would work around the clock. During one shift, two cosmonauts would work while the third rested. One of the two active cosmonauts would work full-time, with the other providing assistance during breaks from the physical exercise regime. They would rotate shifts every 8 hours. One of the serious issues was logistics, not only to sustain the crew but also to operate the camera, which would require a lot of film. In effect, the long-term use of the Almaz was dependent on the cargo capacity of the TKS. In order to maximise the operational life of the station, the docked TKS was to be responsible for controlling the attitude of the orbital complex.

The station and the re-entry vehicle of the TKS were developed at TsKBM under Chelomey's leadership, and the FGB was designed by Branch No. 1 in Fili, which was often referred to as TsKBM(F).

In fact, this was not Chelomey's first attempt to develop a spacecraft for manned use. In the mid-1960s he conceived the LK-1 for a circumlunar mission. This was to be a Gemini-shaped spacecraft that would be launched by Proton and carry two cosmonauts on a trajectory around the back of the Moon and straight back to Earth. But Chelomey had seen this merely as a precursor to a programme to beat the

---

[3] *Funktsionalno-Gruzovoy Blok.*
[4] *Transportniy Korabl Snabzheniya.*
[5] In effect, it was a Hubble Space Telescope designed for observing the surface of the Earth.

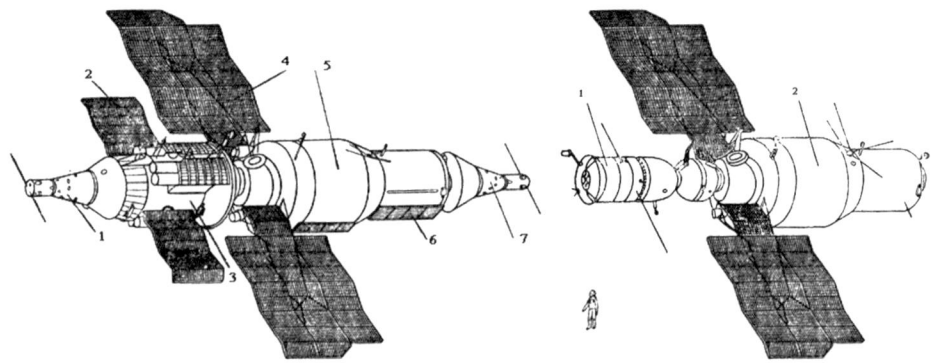

Two versions of the Almaz station. Left: the original project with the TKS resupply ship showing the station (5); the functional cargo block of the TKS (3); re-entry vehicles (1, TKS and 7, station – original concept only); solar panels (2, TKS and 4, station); and a radio-locator (6). Right: the version that used a Soyuz ferry showing the Soyuz (1); and the station (2). (Copyright Igor Afanasyev)

Americans to a lunar landing. Although during Khrushchov's time the official effort for this goal was Korolev's development of the N1 rocket for the programme that became known as N1-L3, Chelomey sought funds for a massive new rocket for his LK-700 programme in which a spacecraft of his own design would make the lunar landing. When the LK-1 was cancelled in a favour of using the Proton to launch the L1 spacecraft designed by Korolev-Mishin, and work on the LK-700 ceased, Almaz became Chelomey's main project and he incorporated in it all the lessons which he had learned from developing military and scientific satellites, the Proton rocket, and the preliminary design work on the LK-1 and LK-700 projects.

At launch, the mass of the Almaz station was 18.9 tonnes. It was 11.61 metres in length, had a maximum diameter of 4.15 metres, and a usable volume of the order of 90 cubic metres. The hermetic section was in the form of a stepped cylinder, with the crew compartment in 'front' of, and adjoining, the wider working compartment. At the rear of the working compartment was an unpressurised section housing the propulsion system, through which ran a small transfer tunnel leading to the passive portion of the docking system.

The crew compartment was 3.8 metres in length and 2.9 metres in diameter. A variety of apparatus was mounted on its exterior, including the antennas for the Igla rendezvous system, solar orientation sensors, a television camera, a laser device and an infrared sensor. At lift-off, this section was protected by an aerodynamic shroud that was jettisoned once the vehicle was above the atmosphere. This compartment had the OD-4 optical port and the POU-II apparatus to take panoramic images with a resolution of 8 metres, the Kolos-5D water tanks and a mechanism to measure the body mass of the weightless cosmonauts. In order to minimise the use of propellant, and thereby maximise the operational life of the station, this compartment housed a system that used electrically driven gyroscopes to control the orientation of the craft. In effect, this compartment was a 'room' in which the cosmonauts could take meals, do physical exercise using a treadmill, perform medical examinations and rest while off duty. There was a small table with a food warmer. Around the table were small chairs and food stores. Above the table were the controls for the station's guidance system. Beneath the table there were removable panels providing access to medical equipment, medicines, clothes, the cosmonauts' personal items, a tape player with audiocassettes and a radio receiver.

The narrow cylinder of the crew compartment was attached by a 1.2-metre-long conical frustum to the working compartment. This was 4.15 metres in diameter and 4.1 metres in length, and it contained the station's primary apparatus. The protective covers for the windows and external equipment were to be discarded in orbit. There were 14 cameras and optical devices. The rear part of the compartment was almost fully occupied by the Agat-1 apparatus and the OPS guidance system. The core of the Agat-1 was an optical telescope for monitoring objects on land, at sea and in the air. It used a telescope in a hermetic conical section that viewed through an aperture in the 'floor', with the imagers installed on top, almost at ceiling level. There was equipment to process images, and to enable the crew to study them. Important data could be coded and sent to the Flight Control Centre (TsUP) by the Biryuza radio transmitter, which used antennas located at the rear of the station. If a more detailed analysis was required, the imagery could be returned to Earth by a special capsule that was accessible from the transfer compartment.[6] A cosmonaut could place film and video into it using a mechanical manipulator. Once released, the capsule would automatically perform re-entry and land by parachute. With mass of 360 kg, it could accommodate 120 kg of film or 2 km of recording tape. The transfer compartment also contained two EVA suits, and could be hermetically isolated from the working compartment to serve as an airlock to enable cosmonauts to work outside the station.

The propulsion system had two rocket engines, each of which had a thrust of 400 kilogram-second, four correction engines with a thrust of 40 kilogram-second, and 28 smaller engines with thrusts of 20 kilogram-second and 1.2 kilogram-second to provide respectively 'rough' and 'fine' control over the station's stabilisation. Most of the engines were positioned around the axial transfer compartment. An unfolding solar panel was mounted on each side of the transfer compartment. With a total area of 52 square metres, their solar transducers were capable of providing a maximum electrical output of 3.12 kVA.[7]

The intention was to build the entire Almaz system, comprising the Proton rocket, the OPS station and the TKS spacecraft, in the M.V. Khrunichev Machine Building Plant (ZIKh) in Fili, which was then under Chelomey's control, and to initiate flight operations in 1969. But because the systems were required to operate reliably for up to two years with minimal maintenance this made Almaz extremely sophisticated for its time, and despite their early work on the LK-1 the TsKBM engineers did not have the experience of systems for manned spacecraft that had been gained by their

---

[6] This KSI capsule had the designator 11F76.
[7] A kilovolt-ampere (kVA) is equivalent to a kilowatt (kW).

TsKBEM rivals. As a result, the programme soon fell behind schedule, making the first operational flight in 1969 impracticable. Although by 1970 the cores of ten stations had been assembled – eight for testing and training, and two for flight – the real challenge for Chelomey's designers was the TKS spacecraft.

In the meantime, in September 1966 the 'Almaz Group' of military cosmonauts started to train at the Cosmonaut Training Centre (TsPK) in Zvyozdniy Gorodok (Star Town) near Moscow to operate the first Soviet space station. By the end of 1971 there were 28 cosmonauts in training for Almaz, making this the largest group ever formed for one space programme.

OKB-1'S SPACE STATIONS

Owing to the protracted delays with the design and development of the TKS, it was decided to start Almaz operations using the Soyuz spacecraft as a crew ferry. The manner in which this decision was made is interesting. When the Almaz programme began in 1964, OKB-1 was involved in so many projects that it was overcommitted. In addition to adapting the Vostok capsule for the Voskhod missions, developing lunar and interplanetary probes, and developing several versions of the new Soyuz spacecraft – including the Soyuz-P and Soyuz-R for military missions – Korolev's designers were developing the N1 launch vehicle. When the Americans announced their intention to develop the MOL, Korolev transferred the military Soyuz projects to OKB-1's Branch No. 3 in Kybishev (now Samara), which had developed the R-7 missile that was used to launch the early Sputniks and, with an additional stage, the Vostok spacecraft. Chief Designer Dmitriy Kozlov, who had led Branch No. 3 since 1959, eagerly accepted the transferred projects. The objective of the Soyuz-P was to rendezvous with an American military satellite in order to inspect and, if required, destroy it.[8] However, it was decided that to have a crew fly such a mission would be too risky, and in 1965 the project was cancelled in a favour of an unmanned satellite interceptor (IS) proposed by Chelomey.

This left Branch No. 3 of OKB-1 with only the Soyuz-R.[9] For our story, this is an important project since it was actually the first space station ever to be endorsed by the Soviet government – although admittedly it was of modest scope in comparison to Almaz. The order was signed by Defence Minister Marshal Rodion Malinovskiy on 18 June 1964, six months after the announcement by the Americans of their intention to develop the MOL, and it was included in the 5-year development plan drawn up for the Soviet military space programme covering the period 1964 to 1969. Representatives of the Ministry of Defence, MOM and the Academy of Sciences conducted a major technical and scientific assessment of the project in early 1965, and accepted that it was viable. The mission was to involve two separately launched unmanned spacecraft, both of which were based on the Soyuz design. Once docked,

---

[8] In the case of Soyuz-P, the 'P' stood for *Perehvatchik*, meaning 'interceptor'.
[9] The 'R' stood for *Razvedchik*, meaning 'intelligence gatherer'.

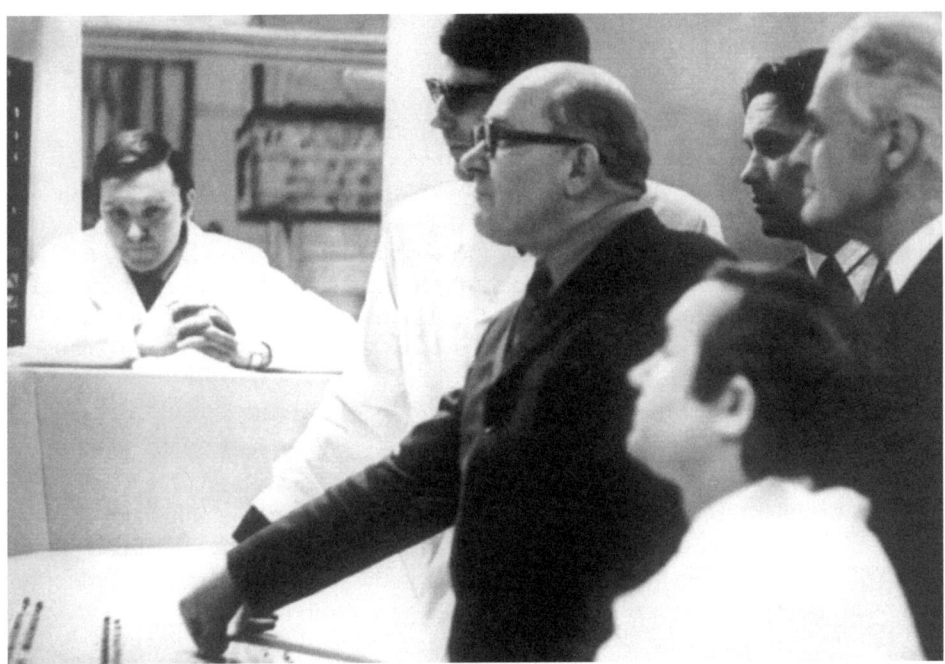

Dmitriy Kozlov (centre), the Chief Designer of Branch No. 3 of TsKBEM in Kybishev (now Samara).

they would form a small space station with total mass of 13 tonnes, a length of 15 metres and a habitable volume of about 31 cubic metres. In documents drawn up by OKB-1 Branch No. 3, this two-part facility had the technical code 11F71. A third Soyuz (11F72) would be launched with two cosmonauts. After docking, they would move into the station through a hermetic transfer tunnel to pursue a programme of military observations and experiments. In December 1965 Kozlov visited General Nikolay Kamanin who as Deputy Chief of the Soviet Air Force was in charge of the manned space programme to develop a joint plan to make use of the Soyuz-R station, for which military cosmonauts were already in training at the TsPK.

However, when the Americans began to fly Gemini missions in 1965 the Kremlin, fearful that this spacecraft would be used to conduct satellite interceptions, realised that even if the work at OKB-1 and OKB-52 progressed as planned, their respective spacecraft would not become available until 1968. In August 1965, therefore, the Kremlin ordered Kozlov to urgently develop a new military spacecraft which would be able to be introduced before the end of 1966.[10] The project was named 'Zvezda' ('Star'), but was also known as the Soyuz-VI.[11] So, having lost the Soyuz-P, Kozlov once again had two manned spacecraft for military use.

---

[10] Such a rapid time scale would prove to be impracticable, owing to the slow pace of the development of the Soyuz spacecraft on which Zvezda was based.

[11] The designation 'VI' stood for *Voenniy Issledovatel* (Военый исследовател), meaning Military Researcher.

The Soviet space programme suffered a setback in January 1966 when Korolev died. The Science and Technical Committee of the Ministry of Defence conducted a detailed review of the two long-term military projects, and decided to terminate the Soyuz-R. The 11F71 code was reassigned to Almaz. Then, reluctant to wait for Chelomey's TKS, the committee recommended using the 11F72 Soyuz that Kozlov was developing to ferry crews to the Soyuz-R station. At this point it is necessary to explain that the general designation for Soyuz spacecraft was 7K, with the Soyuz-P being 7K-P, the Soyuz-R being 7K-R, the Soyuz-VI being 7K-VI, and the 11F72 variant being 7K-TK.[12] Kozlov was told to give the technical documentation for the Soyuz-R station to Chelomey to enable the Almaz to be modified to accommodate its ferry craft. The 7K-TK would deliver crews to the Almaz stations until the more capable TKS became available. This made Almaz the first case of a major Soviet manned space programme to integrate work by highly competitive design bureaus. But establishing the necessary coordination of the two teams in order to revise the Almaz design to use the much smaller Soyuz as a crew ferry took time, and in late 1966 the Military-Industrial Commission (VPK), which was an institution created by the Council of Ministers to implement the decisions of Communist Party, issued decree No. 304 accepting the delay in the development of the Soyuz crew ferry for Almaz and calling for tests of Almaz systems in 1968 and the first operational flight in 1969. Hence, by 1967, after much political manoeuvring by politicians, generals and chief designers, Almaz had become the principal Soviet military manned space programme, and the nation's only space station project.

Nevertheless, the Zvezda reconnaissance spacecraft was still under development by Kozlov. After a series of technical problems during the unmanned test flights of the Soyuz, Kozlov directed his engineers to change the configuration for Zvezda. In particular, it was to have a crew of two cosmonauts who would wear pressure suits, whereas the Soyuz was to have a crew of three who would not wear pressure suits – the precedent for this decision by Korolev being the three-man Voskhod mission in 1964. At launch the Zvezda spacecraft would weigh 6.6 tonnes, be 8 metres in length, 2.8 metres in diameter and have a volume of 12 cubic metres. The total mass of a man, his pressure suit, couch and life support system was approximately 400 kg. Early in the development of the Zvezda project it was expected that the capacity of the Soyuz rocket would restrict the spacecraft to a single cosmonaut, but continued redesign of the descent module enabled a second couch to be installed. The Zvezda spacecraft was to be capable of a 1-month mission, which was twice the maximum duration of the American Gemini.

An interesting difference in design between the Soyuz and Zvezda spacecraft was the descent module. In the case of the Soyuz, this was located in line between the orbital module (in front) and the propulsion module (behind). The descent module is actually a command module, with the couches on its broad base facing an array of controls and instrument panels. This arrangement severely limited the visibility. The

---

[12] 'TK' stood for *Transportniy Korabl*, meaning 'transport spacecraft'.

spherical orbital module blocked the view directly ahead. Sitting in the centre couch, the commander had only a 15-degree-wide periscope for rendezvous and docking operations. The cosmonauts seated left and right had side-facing windows, but were unable to see the target vehicle. As Zvezda was not required to dock with a satellite, Kozlov optimised visibility to enable the crew to conduct a visual inspection of the target. In particular, he moved the descent module to the forward end of the vehicle and inserted the orbital module between it and the propulsion module. Although this arrangement had obvious advantages in comparison to the Soyuz, the design had the important disadvantage that the access hatch to the orbital module was through the heat shield of the descent module, and there was some concern that the hatch would not withstand the thermal stress of re-entry.[13] Although dynamic tests conducted at Branch No. 3 of OKB-1 showed that the hatch was safe, there were lingering doubts. Another issue was Zvezda's power system. Instead of using either chemical storage batteries or solar panels, it was to have a pair of radioisotope thermal generators that would use thermocouples to transform radiogenic heat into electricity. Although these were to be located at the rear of the propulsion module, the potential exposure of the crew to radiation from this system was a matter of some concern. Finally, in view of the fact that the mission was overtly military, a rapid-firing Nudelyman gun was added for protection from an American satellite-killer.[14,15] The entire spacecraft, with its gun, entered ground testing in 1967 and, despite the safety concerns, in late July the Central Committee and Council of Ministers endorsed Zvezda with the first launch being scheduled in 1968 as a prelude to in-orbit military operations in 1969.

In the meantime six military cosmonauts began to train at the TsPK in September 1966 for the Zvezda programme. They were later joined by two new cosmonauts. The group was led by veteran cosmonaut Pavel Popovich. The first two potential crews were soon selected. However, the project was derailed in October 1967 when Vasiliy Mishin, now Chief Designer of the TsKBEM and Kozlov's boss, intervened. The root of the issue was that Mishin took exception to the degree of independence that Korolev had given Kozlov. By machinations and political intrigues exploiting his MOM and VPK contacts, in January 1968 Mishin, with the support of Minister Afanasyev, directed Kozlov to cancel Zvezda. When General Kamanin heard of this he supported Kozlov, but was unable to have the cancellation order reversed.

As a substitute for Zvezda, Mishin suggested the Orbital Research Station (OIS) Soyuz-VI (11F730), and in May 1968 sent a technical specification to the Ministry of Defence. It was based on the Soyuz-R, would fly at an altitude of 250 km at an inclination of 51.6 degrees to the equator, and have solar panels for power and a passive docking system incorporating a hermetic tunnel. The crew would arrive in a

---

[13] The designers of the American MOL made the same compromise, placing the hatch in the heat shield of the Gemini spacecraft. An unmanned test flight demonstrated that the hatch could survive re-entry, but no manned mission was ever flown.

[14] This special-purpose gun was designed by Aleksandr Emmanuelovich Nudelyman.

[15] In fact, the Americans did not have such a capability.

Soyuz 7K-S (11F732)[16] and spend typically a month on board using about 1,000 kg of equipment supplied by the Academy of Sciences and the Ministry of Defence. In addition, the plan called for an unmanned cargo craft 7K-SG (11F735) to supply the occupied station with food, water, air and additional equipment.[17,18] Kozlov was not excluded – Branch No. 3 of the TsKBEM joined this project in June 1968, but in a subsidiary rather than a leading role.

However, at that time the TsKBEM was fully committed both to redesigning the Soyuz following the loss of Vladimir Komarov on its first manned mission, and to the development of the L1 and L3 lunar programmes. The low priority assigned to the OIS is indicated by the fact that the TsPK never even received a simulator for it, and when the Zvezda cosmonauts were transferred to the OIS they were given only theoretical, physical and survival training. By the end of 1969 it had been decided to start a much more ambitious project, and in February 1970 Afanasyev cancelled the OIS and the cosmonauts were reassigned to the Almaz space station.

There were also considerably more ambitious space stations projects initiated by OKB-1/TsKBEM. Notably, in Korolev's time there was the Multirole Space Base Station (MKBS) that was to be launched by the N1 rocket. Work on this station had to be halted, awaiting the introduction of the giant rocket. When the N1 tests finally began in early 1969, Mishin appointed Vitaliy Bezverby, an expert in the ballistics of space vehicles, to manage the MKBS project. The core of the station was to be a cylinder 20 metres in length and 6 metres in diameter. Some 60 metres away, and connected by three long supports, there was to be a nuclear power unit and a plasma electric engine, increasing the length to 100 metres. The total mass of between 220

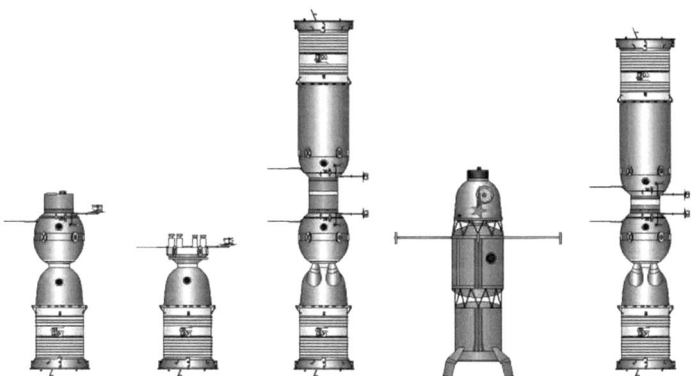

Early planned Soyuz variants: the Earth orbital version (left), Soyuz-P, Soyuz-R space station, Zvezda, and Soyuz-VI miniature space station. (Courtesy Mark Wade)

---

[16] The 'S' stood for *Snabzheniya*, meaning 'supply', so the role of this spacecraft was to transport a crew and their immediate supplies.

[17] The 'G' stood for *Gruzovoy*, meaning 'cargo', so the 'SG' model was a cargo transporter.

[18] A decade later, a modified form of the 7K-SG was launched as Progress 1 to resupply Salyut 6.

and 250 tonnes was to have included 80-88 tonnes of modules and 15-20 tonnes of scientific equipment. It would operate at an altitude in the range 400-450 km and at inclinations of either 51.6 or 91 degrees – the latter being a polar orbit which would enable it to survey the entire globe on a daily basis. It was to have a main crew of six cosmonauts, and an operational life of at least a decade. Two modules providing a volume of 30 cubic metres were to be spun in the manner of a centrifuge to simulate a gravity of 0.8 g. The 200-kW output of the nuclear power unit was to be supplemented by 14 kW from solar panels having a total area of 140 square metres. The station was to have eight docking ports to enable it to serve as a 'space port' for a variety of types of spacecraft, some of which would be unmanned – after being serviced by the station's crew, an unmanned spacecraft would depart to conduct an automated programme of military reconnaissance. The MKBS was to be equipped to protect itself. In fact, its design included numerous concepts similar to those that were envisaged for the 'Star Wars' programme which was initiated in the 1980s by President Ronald Reagan. Although the MKBS remained a paper study, many years later some of its elements were included in the design of the Mir space station.

In all the time that the TsKBEM was concentrating on the redesign of the Soyuz, the L1 project and the development of the N1-L3, Chelomey progressively worked to reduce the degree to which Almaz was dependent on Mishin's bureau. In 1969 he rejected the Soyuz 7K-TK as the crew ferry in favour of the TKS, which was being developed under the leadership of Yakov Nodelyman. By the time the draft design was completed in 1969, the TKS had grown in length to 13 metres, had a volume of 50 cubic metres and a mass of almost 22 tonnes – making it heavier than the Almaz station itself! Although eager to be free of the TsKBEM, Chelomey borrowed some aspects of the Zvezda spacecraft; in particular modifying a quick-firing Nudelyman-Richter NR-23 cannon used by the Tu-22 bomber.[19] This had a maximum range of

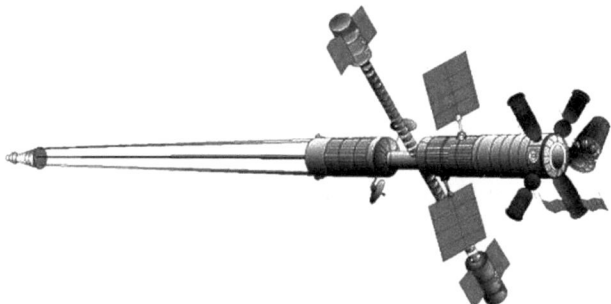

The giant MKBS space station designed by the TsKBEM, showing four Soyuz type spacecraft docked at the main compartment (far right), an artificial-gravity module at each end of the perpendicular boom, and a nuclear power module supported by three long pylons. (Courtesy Mark Wade)

[19] Note that whereas the large Almaz was able to accommodate a straightforward conversion of an aircraft cannon, Nudelyman had to develop a much more compact weapon for the smaller Soyuz-VI.

3 km, fired 0.2-kg projectiles at a speed of 690 metres per second and had a rate of 950 rounds per minute. Since the gun would be in a fixed position, it would be necessary to align the station to aim the gun at a target, and the correction engines were to maintain the station's stability while the gun was firing. It was expected that the gun would be able to hit and destroy a target within five seconds.

The great irony was that while the Soviet Union was working on all these military projects, the development of the American MOL had fallen behind schedule, and in 1969 this suffered the same fate as the Dyna-Soar 'space plane' by being cancelled shortly before its preliminary test flight. Nevertheless, the Americans had not given up on the idea of a space station.

THE CONSPIRACY

In contrast to the low priority assigned to the military space station projects at the overcommitted TsKBEM, the Ministry of Defence encouraged the development of Chelomey's Almaz. Although this project suffered protracted delays, by 1969 it was the only *real* Soviet space station project. It is true that there were ideas for joint endeavours in space station development between Mishin's team in Kaliningrad and Chelomey's in Reutov, but owing to the poor relationship between the two Chief Designers no one at the TsKBEM wished to approach Mishin officially to propose a formal collaboration. Even the Kremlin recognised that rivalry between the bureaus had seriously damaged the Soviet space programme. Once, even the mighty Ustinov said that Mishin and Chelomey behaved just as if the bureaus were their personal "principalities". Although the Kremlin could have ordered strategic integration, in practice it did little to force the bureaus to collaborate.

In the meantime, the Americans had been very busy. In August 1965 NASA had assigned a group of experts the task of defining a programme of long-term scientific research in Earth orbit using Apollo hardware. This drew up a phased programme that would lead to a scientific space station. Over the years most of these projects were dismissed, but in the spring of 1969 it was announced that a 'Sky Laboratory' (Skylab) would be launched in 1972. It would be a 90-tonne giant, and with a length of 36 metres and a diameter of 6 metres it would have a volume of 400 cubic metres – four times that of Almaz. The pace of the American space programme renewed the Kremlin's concern. It was clear that the USSR had lost the race to land a man on the Moon, work on the Almaz space station was seriously behind schedule, and all that the TsKBEM had to offer was the Soyuz spacecraft. The Soviet response to the American plan would therefore have to be quick and efficient.

In August 1969 a group of designers led by Boris Raushenbakh, the TsKBEM Department Chief responsible for the development of spacecraft guidance systems, put it to Boris Chertok, their boss, that a propellant tank of the Soyuz rocket should be converted into a space station. It was estimated that this could be done within a year, and could be launched before Almaz – and before Skylab, of course. As the Chief Designer, Mishin was the top man. His First Deputy was Sergey Okhapkin, who was in charge of the development of rocket systems, including the N1 launcher. Next was Konstantin Bushuyev, the Deputy Chief Designer for the development of unmanned and manned spacecraft, including the Soyuz spacecraft and its L1 and L3 variants. Deputy Chief Designer Chertok was the fourth man, and his responsibility was the development of guidance, control and electrical systems for launchers and spacecraft. Chertok was one of the pioneers of Soviet rocketry, having worked with Korolev and the other leading Soviet rocket designers in analysing the design of the V-2 rockets that were confiscated from the Nazis. Raushenbakh had left Keldysh's tutelage to join Chertok's group in the early days of OKB-1, and was one of the few top Soviet spacecraft designers whose name was known in the West. His proposal was to modify a tank to accommodate various systems from a Soyuz spacecraft, and to install solar panels, a docking mechanism and a hermetic tunnel to provide access from a docked Soyuz. The fact that this structure was to be launched by the Proton rocket meant that its mass could be no greater than that of Almaz, but it would be much simpler.[20]

Initially, Chertok hesitated. His main concern was the limitations implicit in the systems developed for the Soyuz spacecraft. An additional issue was that a vehicle having three times the mass of the Soyuz would require more powerful engines to maintain its orbit and to control its orientation. Furthermore, this propulsion system would require to be able to support a mission of many months, rather than a brief Soyuz flight. Chertok consulted his old friend Aleksey Isayev. In 1944 Isayev had been appointed the Chief Designer of OKB-2 (in 1966 renamed Himmash), and had worked with Chertok and Korolev in Germany. He was now the leading designer of rocket engines for both unmanned and manned spacecraft. When Chertok explained the TsKBEM's idea, Isayev said that he had already developed such a propulsion system for Chelomey's Almaz. The logical way to proceed would be to combine the proven systems of the Soyuz spacecraft with those already developed for the Almaz station. Thus was born an idea with dramatic implications for the future of world cosmonautics.

Interestingly, only a small group were involved in originating this project. Taking the lead was Konstantin Feoktistov. As a Department Chief in Bushuyev's group, he was of similar rank to Raushenbakh. He had been involved from the earliest days in the design of the Vostok spacecraft, and in return for leading the modification of that capsule to accommodate three cosmonauts he had been assigned to the crew of the first Voskhod flight in October 1964. The Soyuz spacecraft was very much one of his 'offspring'. On hearing of the proposal to convert a propellant tank into a station, Feoktistov asked: why start with an empty tank? There were several Almaz prototypes standing idle in Chelomey's factory in Fili. It would be better to modify one of these.

However, Chelomey was sure to oppose any attempt to requisition his spacecraft, the Ministry of Defence and Minister Afanasyev would reject any further delays in Almaz development and, of course, Mishin would not appreciate a proposal to use a

---

[20] The main habitable compartment of the Skylab space station was the fuel tank of the second stage of a Saturn IB launch vehicle, so the basis for Raushenbakh's idea is obvious.

The development of the Soyuz spacecraft was led by Department Chief Konstantin Feoktistov.

competitor's hardware in a TsKBEM project. But Feoktistov and Chertok thought differently. Their strategy was to avoid anyone who might raise an objection, and to go straight to Dmitriy Ustinov, who was on the Central Committee of the Kremlin and was in overall control of the Soviet space programme. They were sure he would understand the strategic implications of the idea. However, it was no simple matter to contact Ustinov. Normally such an approach would be made by Mishin, as head of the bureau. But Mishin was in Kyslovodsk, taking his annual leave; and anyway he would object. In Mishin's absence, Bushuyev was one of the few people with the authority to seek a meeting with Ustinov.

Feoktistov recalls: "Several times Bushuyev, Chertok and I reviewed this matter. Chertok, and his engineers who'd worked on the development of guidance systems, supported the idea of moving immediately. But Bushuyev hesitated because Mishin would be against the idea, and we would not have the support of our own bureau."

Someone suggested that Bushuyev should call Ustinov and ask for a meeting, but Bushuyev did not wish to take such an important step without the knowledge of his bureau chief. However, Feoktistov had a reputation for being disobedient, and he proposed that *he* call Ustinov. Intriguingly, although Feoktistov was not a member of Communist Party, he readily arranged a meeting with one of the most influential men in the Central Committee.

Ustinov was aware that even under the most optimistic scenario, Almaz would not be ready until early 1972. If everything went to plan Almaz would beat Skylab, but if the launch were to fail, or if the station were to experience a problem that would prevent a crew from boarding it, then the Soviet Union would again trail behind the Americans. Another issue was that as a military project, the design and operation of the Almaz station should remain a secret. Skylab was a scientific project funded by NASA. If the first Soviet space station could be portrayed as a civilian space project, and it was given lavish coverage in the newspapers, then it would serve to mask the true role of the subsequent Almaz stations – about which much less information would be released. That is, to launch a scientific station first would serve as a *maskirovka*, or deception, designed to hide the real project. Ustinov fully appreciated this point. He invited Chertok, Bushuyev, Feoktistov, Raushenbakh and Okhapkin to his office on 5 December 1969. Also present were Leonid Smirnov, who was Aleksey Kosygin's deputy for space matters and chairman of the VPK since 1963, Afanasyev, Keldysh and some of Ustinov's officials. As Mishin was on vacation it was reasonable that he should not be invited, and Chelomey, being in hospital, was conveniently unavailable.

In advance of the meeting, the TsKBEM people agreed to let Feoktistov talk first. His presentation was very convincing. It would be possible to equip the core of one of Chelomey's stations with the solar panels of the Soyuz spacecraft, together with its guidance and command systems. In approximately a year's time, Feoktistov said, the Soviet Union would have the world's first space station. Chertok then noted that the systems of the Soyuz spacecraft were considered to be reliable because they had been tested during 14 unmanned and manned orbital flights. The development of a docking system incorporating an internal tunnel was underway. Keldysh asked how the construction of such a space station would interfere with the development of the N1-L3 lunar programme. Okhapkin said that the two projects were separate, and the designers involved in the lunar programme would not be needed for the station. Of course, Ustinov knew that both Soviet lunar programmes were under review. After the success of Apollo 8 in December 1968 the L1 circumlunar project launched by a Proton rocket had lost its purpose, and the N1-L3 lunar landing was contingent on successfully introducing the N1 launch vehicle – and after two spectacular failures in January and July 1969 some people were beginning to doubt that this would ever fly. And then, of course, the Americans had already won the race to the Moon.

Ustinov was enthusiastic about the space station conversion, not only because if it worked it would demonstrate that the Soviet Union was ahead of the Americans in this aspect of manned spaceflight, and not only to provide a *maskirovka* for Almaz, but also because Ustinov had never liked how Chelomey had exploited the personal support of Khrushchov and his links with the Kremlin and the Ministry of Defence

Mishin's deputies: Konstantin Bushuyev (left) for satellites and manned spacecraft, and Boris Chertok for control and guidance systems.

Department Chief Boris Raushenbakh (left) worked on guidance systems at the TsKBEM, and Academician Mstislav Keldysh led the scientific programmes for Soviet satellites.

The N1 lunar rocket was Vasiliy Mishin's dream.

to expand his activities into manned spacecraft. Ustinov wanted all such work to be undertaken by a single design bureau. Converting the core of a military Almaz into a civilian space station would not only enable the Soviet Union to once again claim leadership in space, it would also put Chelomey in his place!

The meeting ended with the decision to immediately prepare a project time-scale, and by the end of January 1970 to issue a decree to endorse the plan. Although the TsKBEM rebels were surprised by the ready acceptance of their proposal, they had (to coin a phrase) been 'pushing an open door'. Brezhnyev accepted the importance of space stations for national prestige. In fact, he had referred to them several times in speeches which he made that autumn, and on 22 October, in welcoming home the crews of the 'group flight' of Soyuz 6, 7 and 8, he had asserted that the USSR had a broad space programme which was planned years in advance and would unfold in a logical manner. The strategy was to downplay American successes and not to admit Soviet failures. This was why the USSR was only one of two European states (the other being Albania) not to run 'live' TV coverage of the first manned lunar landing. In order to convey the impression that the Soviet space programme was following a grand plan, Brezhnyev had spoken of "space cosmodromes" from which men would set off on journeys to the planets. Obviously, however, this plan would unfold by a series of ever more ambitious steps, the first of which would be relatively modest. By the end of November 1969 Academicians Keldysh and Boris Petrov had written in newspaper articles that space stations would permit unprecedented monitoring of meteorology, oceanology, ecology and aspects of the economy; they would serve as laboratories to study physics, geophysics, advanced technology and astronomy; they would serve as factories; and later they would test systems needed by the promised interplanetary spaceships.

Space stations and the Kremlin. Kosygin (left) and Brezhnyev (second right) with the crew of Soyuz 9: Sevastyanov and Nikolayev.

DOS IS BORN

Although Mishin and Chelomey were united in their opposition to the plan to create a hybrid Long-Duration Orbital Station (DOS) by using Almaz and Soyuz systems, the Kremlin's directive was firm. Chelomey was satisfied to ensure that this project would not further delay Almaz, but Mishin was furious at what he referred to as the "conspiracy". In one meeting Mishin threatened: "If I hear that anybody else apart from these two – Bushuyev and Feoktistov – occupies himself with this DOS, I will send him to hell." He opposed the DOS effort not only because his staff had gone behind his back to initiate it, but also out of concern that, despite assurances to the contrary, it would jeopardise the N1-L3 programme. Even once it was underway he never really endorsed the project, and at times he openly criticised it.

Not only were the TsKBEM designers eager to develop the hybrid space station, so too were the engineers in Fili who had spent five years designing the systems for Almaz and wished to find out how well they performed in space. In fact, Chelomey himself was not very popular in Fili. Initially, Fili had been an independent design bureau (OKB-23) headed by the famous Chief Designer Vladimir Myasishchev, and between 1951 and 1960 had created the successful M-4 and 3M strategic bombers. While it was designing the M-50 jet bomber and a manned rocket plane, Chelomey, with the support of Khrushchov, but against the will of the Air Force, had drawn the bureau into his own organisation, naming it Branch No. 1. Myasishchev had gone to the Moscow Aviation Institute. The DOS project provided an opportunity for Fili to regain a degree of autonomy, and Viktor Bugayskiy, who was in charge there, was keen to collaborate with his TsKBEM counterparts.

In fact, the first task was to establish a genuine management structure that would integrate the Kaliningrad and Fili design teams. In December 1969, shortly after the meeting with Ustinov, Okhapkin, Bushuyev and Chertok asked Mishin to nominate Yuriy Semyonov as the Leading Designer for the DOS programme. Semyonov had participated in the design of the Soyuz spacecraft and managed the L1 circumlunar programme, whose cancellation was imminent. Semyonov was also a son-in-law of Andrey Kirilenko, the fourth man in the Kremlin's hierarchy. Although it is only a supposition, it is possible that Ustinov played a role in the nomination; the rationale being that someone with Semyonov's connections ought to be able to counter any attempts by either Mishin or Chelomey to undermine the rapid pace set for the DOS development. On 31 December the basic organisational documents were drawn up. In January 1970 Mishin officially appointed Semyonov and three deputies: Dmitriy Slesarev was responsible for modifying the Soyuz for use as a space station ferry;[21] Valeriy Ryumin was responsible for the station's systems; and Viktor Inelaur was responsible for the guidance apparatus. Later, Arvid Pallo was appointed as a fourth deputy. Also, Mishin nominated his own deputies as general managers of the entire programme. Bushuyev, assisted by Feoktistov, was responsible for the development of all aspects of the programme. Under their direct control were Pavel Tsybin, who

---

[21] This was to be the 7K-T ('T' for *Transportniy*, or 'transporter') version of the Soyuz spacecraft.

Yuriy Semyonov led the development of the DOS space station at the TsKBEM.

managed the development of the Soyuz, and Leonid Gorshkov, the designer of the Orbital Block (i.e. the station itself). In addition, Chertok led the guidance group, with Raushenbakh and Igor Yurasov as deputies; Lev Vilnitskiy was responsible for the docking systems; Vladimir Pravetskiy was responsible for life support systems; Oleg Surgachov was responsible for thermal regulation systems; Yakov Tregub and his deputy, Boris Zelenshchikov, were responsible for the testing of all the systems, cosmonaut training and mission control; Gherman Semyonov was to supervise the preparation of the station for shipment to the cosmodrome; and Aleksey Abramov and Vladimir Karashtin were to manage the launch preparations. In Fili, Bugayskiy nominated Vladimir Pallo as his deputy for the DOS project. This was a wise choice, because when Semyonov added Arvid Pallo to his team the two brothers were well placed to coordinate joint activities. All the leading people of the DOS project have been named here because, by managing the activities of thousands of engineers, technicians and others, they defined the basis for not only the Soviet manned space programme but also, in the long term, the world's manned space programme.

On 9 February 1970 the Central Committee and the Council of Ministers issued decree No. 105-41. It was one of the most important decrees in the history of space station development. One of its directives was that all pertinent documentation and all existing hardware, including Almaz cores, be transferred to the DOS programme.

After studying the design documents, Feoktistov drew up the specifications of the station to maximally exploit the capabilities of the Proton launcher: it was to have a maximum diameter of 4.15 metres, a length of 14 metres and an initial mass of 19 tonnes. With a volume of almost 100 cubic metres, which was almost ten times that of the Soyuz, it would be able to accommodate comfortable facilities for the crew, consumables for a long mission and a wide variety of apparatus. One of the design requirements was that most of the built-in apparatus must be accessible to the crew for maintenance, repairs or replacement. In fact, this requirement became one of the greatest design challenges. The complexity of the DOS station is evident from the fact that it had 980 instruments (according to another source 1,300) connected by in excess of 1,000 cables that had a total length of 350 km and a mass of 1.3 tonnes!

The next big decision was the maximum possible operating life of the first station, designated DOS-1. This would depend on the altitude of the orbit, the available fuel and the power supply. Although the upper atmosphere is exceedingly rarefied, if the station were to start off in the range 200–250 km the drag would cause the orbit to decay at an increasing rate, until the station re-entered and was destroyed. It would be necessary to fire the rocket engine periodically to maintain the desired altitude. It was calculated that it would be necessary to use about 3 tonnes of fuel annually to maintain DOS-1 at an altitude of 300 km, 1 tonne at 350 km, and a mere 200 kg at 400 km. A higher orbit was therefore desirable to maximise the operating life of the station. However, the higher the station's altitude, the more fuel the Soyuz would use to make a rendezvous. Furthermore, a higher altitude would expose the crew to more intense space radiation. The next big issue was the total period of occupancy. This would be dependent on the reserves of air, water and food. Since one man would consume about 10 kg of materials per day, it was decided to load the station with sufficient stores to support three men for three months – a period that would be accumulated by a succession of crews. It was on the basis of such analyses that the documentation for the DOS-1 station was drawn up in February 1970.

The first meeting between the TsKBEM and TsKBM experts was in March 1970. Feoktistov presented the technical specifications to the Fili team. Then Semyonov outlined the structure of the programme, its management, and the responsibilities of not only the TsKBEM and the TsKBM but also their subsidiary factories. The M.V. Khrunichev Machine Building Plant (ZIKh), which the TsKBM managed, was to be responsible for building the DOS stations and the Proton rockets that would launch them. The Plant for Experimental Machine Building (ZEM) had been part of the TsKBEM since 1966, and its role would be to test the station's apparatus. Because each institution had its own structure, work philosophy, methodology and standards, the task of coordination was formidable. If prior experience was anything to go by, designing, developing, testing and launching a space station would take at least five years, but the DOS managers set out to do so in a period of approximately one year!

The first challenge was to arrange the transfer of the Almaz cores to the TsKBEM. Several days after the first meeting between the two engineering teams, Semyonov went to see Chelomey in Reutov. It was a difficult and strained meeting. Although Semyonov was armed with the Kremlin's decree, Chelomey accused the TsKBEM of "stealing" his work. Only after a telephone call to Afanasyev was Semyonov able to persuade Chelomey to transfer four Almaz cores.[22]

---

[22] It was one of these cores which, some 13 months later, was successfully launched as the world's first space station.

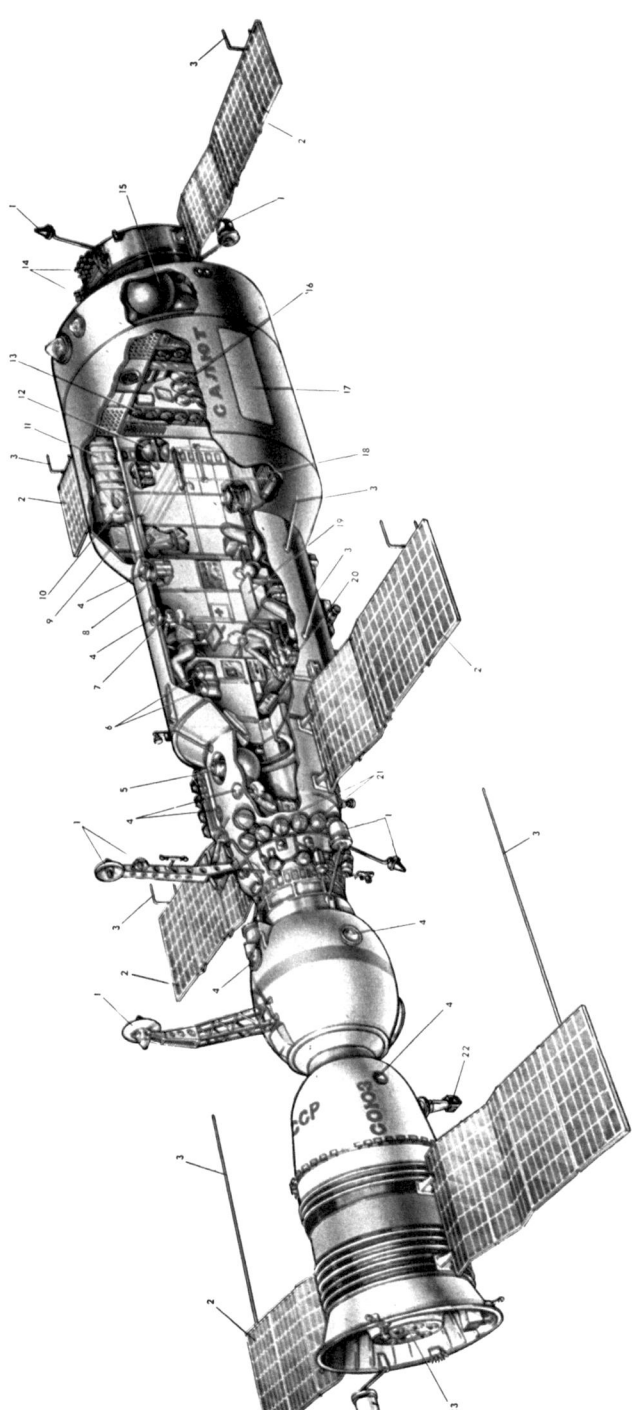

The first DOS space station and a docked Soyuz ferry: (1) rendezvous antennas; (2) solar panels; (3) radio-telemetry antennas; (4) portholes; (5) the Orion astrophysical telescope; (6) the atmospheric regeneration system; (7) a movie camera; (8) a photo camera; (9) biological research equipment; (10) a food refrigeration unit; (11) crew sleeping bags; (12) water tanks; (13) waste collectors; (14) attitude control engines; (15) propellant tanks for the KTDU-66 main engine; (16) the sanitary and hygienic systems; (17) micrometeoroid panel; (18) exercise treadmill (not shown, but it was aft of the large conical housing for scientific equipment viewing through the floor); (19) the crew's work table; (20) the main control panel; (21) oxygen tanks; (22) the periscope visor of the Soyuz descent module; (23) the KTDU-35 main engine of the Soyuz spacecraft. The conical housing for the main scientific equipment is not shown.

The DOS-1 station will be described in detail later, and here it is necessary only to explain how it differed from Almaz. The transfer compartment housing the docking system was at the front of DOS-1, rather than at the rear. Whereas on Almaz there was a hermetic tunnel through the unpressurised propulsion module, in the case of DOS-1 the docking system provided access to a small compartment that had been added to the front of the Almaz structure. On the exterior of this compartment were two solar panels of the type developed for the Soyuz spacecraft. A hatch led to the compartment which combined the Almaz crew and work compartments.[23] As in the case of Almaz, the rear of the main compartment was dominated by a large conical housing, but now the apparatus was for scientific rather than military observations. Another change was that the propulsion system developed for Almaz was discarded, and a system based on that of the Soyuz spacecraft was affixed in its place. This unit carried a second pair of solar panels.

The following DOS-1 systems were taken from Soyuz spacecraft:

- guidance and orientation
- solar panels
- Zarya radio-equipment
- RTS-9 telemetry system
- Rubin radio-control system
- command radio lines
- central post and main control panel
- Igla rendezvous and docking, and
- regenerators for oxygen.

In addition, the system for controlling the complex was taken from the Soyuz, but it was modified to take account of the station's greater mass. The thermal regulation system had also to be upgraded. These were in-house systems to the TsKBEM. The

A model of a Soyuz spacecraft (left) about to dock with the first DOS space station. The conical housing for the main scientific equipment has been 'airbrushed out'.

---

[23] On the original Almaz, this forward hatch would have enabled the crew to enter the station from the capsule mounted on the front at launch.

Two engineers work at the main control panel of the DOS station, with the open hatch
to the transfer compartment in the background.

Sirius system for information analysis was supplied by Sergey Darevskiy's Special
Design Bureau. It was based on the Soyuz command display, and on DOS-1 it was
on the left-hand side of the main control panel, in front of the commander's seat. It
provided the following indicators:

- the pressure in the fuel tanks
- the distance and speed of the station relative to an approaching spacecraft
  during rendezvous and docking
- the voltage and current in the electrical power system
- the environmental parameters inside the station
- onboard clocks, and
- a globe to enable the cosmonauts to readily determine the position of the
  station in relation to terrestrial geography.

The development of the various scientific and medical apparatus also challenged
the designers. Never before had so many scientific instruments been installed in one
spacecraft: this apparatus weighed 1.5 tonnes in total. Most of it was designed and
developed outside the TsKBEM, in coordination with the Academy of Sciences. For
example, the Orion ultraviolet telescope was devised by the Byurakan Observatory
and the OST-1 solar telescope by the Crimean Observatory. For each instrument on

the station, the mission planners had to develop a programme of experiments for the crew to conduct.

Everyone involved in the project worked without holidays in order to build, test and launch the first space station within a period of one year! The project itself, and all the basic systems, were developed by Kaliningrad. Design schemes and system diagrams were prepared by Fili. The manufacturing process was organised by ZEM, where Ryumin and Pallo, Semyonov's deputies, worked alternate shifts around the clock. The station and its mockups (including wooden ones) were fabricated in the Khrunichev Plant. The final testing of the station was planned and conducted by the TsKBEM.

Even more remarkably, this coordinated effort was conducted without the support – and indeed against the wishes – of the leaders of the two design bureaus: Mishin and Chelomey!

In December 1970, after less than a year, Khrunichev completed the construction of the DOS-1 station. It was transferred to the TsKBEM for further testing, and then delivered to the Baykonur cosmodrome in March 1971.

Specific references

1.   Chertok, B.Y., *Rockets and People – The Moon Race, Book 4*. Mashinostrenie, Moscow, 2002, pp. 239–249 (in Russian).
2.   Afanasyev, I.B., Baturin, Y.M. and Belozerskiy, A.G., *The World Manned Cosmonautics*. RTSoft, Moscow, 2005, pp. 224–226 (in Russian).
3.   Afanasyev, I.B., *Unknown Spacecrafts*. Znaniye, 12/1991 (in Russian).
4.   Semyonov, Y.P., ed, *Rocket and Space Corporation Energiya named after S.P. Korolev*. 1996, pp. 264–269 (in Russian).

2

DOS-1 crews

STAR TOWN

Zvyozdniy Gorodok (Star Town), home of the Cosmonaut Training Centre (TsPK) where Soviet military cosmonauts live and train for space missions, is located in a wood of 100-year-old birch trees in the Shchelkovo area about 40 km northeast of Moscow and 10 km east of Kaliningrad.

In 1958 General Nikolay Kamanin became Deputy Chief of the Soviet Air Force. He was responsible for the selection of all military cosmonauts, their training and nomination for space missions. He was also on the military commission that decided to build Zvyozdniy, and when construction started in the early 1960s all decrees relating to its development required his signature.

Kamanin maintained a good association with his boss, Commander of the Air Force Marshal Konstantin Vershinin, but his relationship with Sergey Korolev was often tense. They got on well during the years of the Vostok flights, but in 1963 OKB-1 set out to modify this capsule to carry up to three cosmonauts and this led to a conflict. Kamanin wished the Voskhod cosmonauts to be drawn exclusively from the Air Force, as in the case of Vostok, but Korolev wished to give his engineers the opportunity to fly in order to personally assess their designs. Korolev got his way for the first Voskhod mission, on which Air Force cosmonaut Vladimir Komarov flew as commander, Konstantin Feoktistov flew as engineer, and Boris Yegorov, a physician whose father was a friend of Korolev's, flew to investigate the symptoms of 'space sickness' that were reported by Vostok cosmonaut Gherman Titov.

When Vasiliy Mishin succeeded Korolev upon the latter's death in January 1966, the conflict between Zvyozdniy and Kaliningrad became even more intense. And when in August 1966 the Kremlin granted Mishin permission to recruit civilians for the L1 and L3 lunar programmes, Mishin argued that the TsKBEM (as OKB-1 had by then become) should have its own training facility – a proposal that was resisted by Kamanin. However, as the TsPK grew, Kamanin faced management problems. By the mid-1960s the manned space programme was based on the Soyuz spacecraft whose variants were to support a variety of projects, including autonomous flights, circumlunar and the lunar landing missions, developing techniques for rendezvous and docking, a variety of military tasks, and serving as a ferry for a space station. Appropriate simulators had to be installed at the TsPK, and training procedures and methodologies developed. The installation of the first Soyuz simulator in late 1966 coincided with the arrival of the first cosmonaut-engineers from the TsKBEM. As there were not yet simulators for either the circumlunar L1 or the military Soyuz-VI, the civilians joined the military cosmonauts in training for Soyuz missions. The L3 simulator was an even less likely prospect, in part because Mishin hoped to squeeze the Air Force out of the lunar landing programme and to build the simulator at the TsKBEM. Many of the problems that Kamanin faced were beyond his control. To make matters worse, the death of Yuriy Gagarin while flying a MiG-15 in training in March 1968 reflected poorly on the TsPK. Both Kamanin and General Nikolay Kuznyetsov, who had been appointed as Commander of the Cosmonaut Training Centre in 1963, felt that they were partly to blame for the accident.

Furthermore, Kamanin suffered from the diminishment of his Khrushchov-era allies in the Ministry of Defence and the Air Force. In 1967 Rodion Malinovskiy was replaced as Minister for Defence by Marshal Andrey Grechko, who had not been a supporter of manned space flights. In 1968 the TsPK gained orbital, military, and lunar training facilities, and was expanded to include engineering and medical departments. It was also renamed the Yu. A. Gagarin Test and Research Centre for Space Flight. For almost 11 years Kamanin had worked closely with Vershinin, but Grechko wanted his own man running the Air Force, and in 1969 he replaced Vershinin with General Pavel Kutakhov, who in turn decided to replace Kamanin as soon as possible.

General Nikolay Kamanin, who managed the training of cosmonauts at the TsPK. (From the book *Hidden Space*, courtesy astronaut.ru)

The residence and training building for Soviet cosmonauts at the TsPK located at Zvyozdniy Gorodok ('Star Town') near Moscow.

It was in this intense atmosphere that the crews for the DOS-1 programme were nominated.

THE FIRST CREWS

Soon after decree No. 105-41 was issued in February 1970 directing that work start on the DOS project, Kamanin asked Mishin to immediately assign crews for the first space station, and Mishin directed his subordinates who dealt with the selection and training of cosmonaut-engineers to do so.

One of the first to be nominated was Aleksey Yeliseyev, who had flown two Soyuz missions in 1969: "Deputy Chief Designer Yakov Tregub called and said that he would like to include Nikolay Rukavishnikov and I in the first crew. He also suggested that we familiarise ourselves with all works related to the orbital station and the preparation for its launch. Tregub led the testing of the spacecraft systems, the technical training of the cosmonauts, and managed mission control. His opinion was important, and we thought we had good chances. We were so excited to work on the first orbital station!"

At the end of April 1970 Tregub and Colonel Sergey Anyokhin, who was head of the TsKBEM's cosmonaut group, paid Kamanin a visit and explained that the plan was to build two identical space stations, each of which would be occupied twice. Four crews had been selected. The first two would be assigned to DOS-1; the first flying a 30-day mission and the second a 45-day mission.[1] The third and fourth crews would serve in a backup role for DOS-1 and then become the prime crews for DOS-2.

The nominations were:

- Crew 1: Vladimir Shatalov, Aleksey Yeliseyev and Nikolay Rukavishnikov
- Crew 2: Georgiy Shonin, Valeriy Kubasov and Pyotr Kolodin
- Crew 3: Boris Volynov, Konstantin Feoktistov and Viktor Patsayev
- Crew 4: Yevgeniy Khrunov, Vladislav Volkov and Vitaliy Sevastyanov.

The commander of the first crew, Colonel Shatalov, had been recruited by the TsPK in 1963 as a member of the second group of military cosmonauts. He had flown twice – the first time performing the first docking in space of two manned spacecraft. In 1966 Yeliseyev had become a member of the TsKBEM's first group of cosmonaut-engineers. He was one of three Soviet cosmonauts with experience of spacewalking.[2] In January 1969, after Shatalov had docked Soyuz 4 with Soyuz 5, Yeliseyev and Khrunov had made an external transfer to join him. Also, Shatalov and Yeliseyev had flown together on Soyuz 8 in October 1969. Rukavishnikov was also a member of the first group of cosmonaut-engineers, but had not been able to enter training until early 1967. His assignment on the space station crew was as the research engineer.

Colonel Shonin was to command the second mission to the station. Although he had been recruited in 1960 as a member of the first group of the cosmonauts, he did not make his first flight until October 1969, when he commanded Soyuz 6 and spent five days in space. His engineer on that mission was Kubasov, who, like Yeliseyev, was a member of the TsKBEM's first group of cosmonaut-engineers. Lieutenant-Colonel Kolodin was recruited in 1963 as a member of the second group of military cosmonauts. He had served in a backup role for the 'group flight' of 1969. On the space station crew he would serve as the research engineer.

Colonel Volynov, the commander of the third crew, was a member of the first group of cosmonauts. He commanded Soyuz 5, which served as the passive target for Soyuz 4. Spacecraft designer Feoktistov had flown as the engineer of the Voskhod mission in 1964. Because Volynov had been backup commander for that mission, he knew Feoktistov well. Patsayev, a rookie TsKBEM cosmonaut-engineer, was to be the research engineer.

The fourth crew was to be commanded by Colonel Khrunov, who was a member of the first group of cosmonauts. On his first flight he had launched on Soyuz 5 and, with Yeliseyev, had spacewalked to Soyuz 4 to return to Earth. The flight engineer, Volkov, was a TsKBEM cosmonaut-engineer who had flown on Soyuz 7. Although Sevastyanov was chosen as a member of the first group of cosmonaut-engineers, he had not entered training until early 1967. At the time of his assignment as a space

---

[1] According to Mishin, Ustinov ordered that the first visit to the station should last one month.
[2] The pioneering spacewalker was Aleksey Leonov in 1965.

station research engineer, he was training for the Soyuz 9 'marathon' mission to be flown in June 1970.

Although Mishin and Kamanin had previously argued that the DOS crews should be drawn exclusively from his own side, five of the nominations that the TsKBEM proposed were Air Force and seven were civilians. Kamanin acknowledged that the commanders were military cosmonauts, but wanted to have two military officers on each crew – only the second nominated crew had two military officers; in the others there were two civilians. He also criticised having two veterans on each crew. He particularly objected to having two of the most experienced cosmonauts – Shatalov and Yeliseyev – on the same crew. There were Air Force cosmonauts who had been waiting many years to make their first space flight.

Kamanin also criticised the nomination of Feoktistov. Every time that he had seen Feoktistov's name on a list of candidates for an assignment, he had opposed it. In 1964 he had argued against Korolev's desire to fly Feoktistov on the first Voskhod mission. After the death of Komarov on Soyuz 1 in 1967, Mishin had proposed that since the primary task of the manned mission planned for October 1968 would be to test the modified Soyuz, the best man to fly it would be Feoktistov, but Kamanin had insisted that the renowned test pilot Colonel Beregovoy be assigned. However, Beregovoy failed do dock his Soyuz 3 with the unmanned Soyuz 2 – despite the fact that unmanned Soyuz spacecraft had twice previously achieved automated dockings. Kamanin's hostility to Feoktistov was not limited to crew assignments. In 1969 the Americans had invited the Soviet Union to send two cosmonauts on a goodwill trip to the United States. The TsPK candidate was Beregovoy. When Mishin nominated Feoktistov, Kamanin argued that another military officer, Pavel Belyayev, who had commanded Voskhod 2, should be sent instead. On that occasion, Mishin won. As regards the DOS nomination, the basis of Kamanin's criticism was that Feoktistov's state of health was too poor, he wore glasses and was divorcing for the second time. But the real reason for Kamanin's persistent antipathy might have been that, unlike the other cosmonauts, Feoktistov never joined the Communist Party. In fact, given that Feoktistov had gone behind Mishin's back to get the DOS programme started, it was perhaps surprising that the TsKBEM's Chief Designer had allowed his name to go forward at all!

Volynov's nomination also caused Kamanin a difficulty. Volynov had been one of the strongest candidates in the first group of cosmonauts, but his mother was of Jewish heritage and this had attracted the criticism of the Kremlin's anti-Semites. Ivan Serbin, who was the Chief of the Industries Department, had openly warned Kamanin after the successful Soyuz 4/5 mission that not only must Volynov not be assigned another space flight, he should not even be allowed to travel to abroad. In 1964 Volynov had been on the verge of commanding the historic Voskhod mission, but at the last moment Kamanin, yielding to Korolev's argument to fly Feoktistov, who was on the backup crew, and to criticism of Volynov's appointment by Serbin and others in the Kremlin, had allowed the backup crew to fly. In the spring of 1966 Volynov had gone to Baykonur to command the planned long-duration Voskhod 3 mission, but this was cancelled – although not owing to criticism of Volynov. Now, with Grechko's man Kutakhov running the Air Force, Kamanin knew that his own position was too weak to resist the criticism which Volynov's nomination to a DOS crew would draw. In February 1970 Kamanin had given Volynov the 'low profile' job of commanding the new recruits; now he told him not to expect a nomination to a space flight for at least several years.

Finally, Kamanin was stunned at the nomination of Khrunov. In 1969, while he was a backup commander for the Soyuz 'group flight', Khrunov had been involved in a car accident two months prior to launch and had left the scene without assisting an injured person. As punishment for this irresponsible behaviour, Kamanin had temporarily excluded Khrunov from training for a future space mission.

REVISED APPOINTMENTS

On 6 May 1970 revised crews were nominated. They were:

- Crew 1: Georgiy Shonin, Aleksey Yeliseyev and Nikolay Rukavishnikov
- Crew 2: Aleksey Leonov, Valeriy Kubasov and Pyotr Kolodin
- Crew 3: Vladimir Shatalov, Vladislav Volkov and Viktor Patsayev
- Crew 4: Georgiy Dobrovolskiy, Vitaliy Sevastyanov and Anatoliy Voronov

By this point, relations between Mishin and Kamanin were improving. Mishin had accepted most of Kamanin's criticisms. Splitting Shatalov from Yeliseyev led to Shatalov having an unpleasant conversation with Kamanin at being demoted from commanding the first crew to visit the space station. Mishin and Kamanin agreed a more equitable share of the nominations: with the first and third crews having one TsPK cosmonaut (who was commander) and two TsKBEM cosmonaut-engineers; and with the second and fourth crews having two military cosmonauts and one civilian. In return, Kamanin allowed his cosmonauts to visit Kaliningrad to perform part of their training with their civilian counterparts – thereby relieving the demand on the TsPK's simulators. Since the L1 and L3 lunar programmes were both stalled, Kamanin reassigned all the military cosmonauts nominally in training for such missions to other projects, including DOS.

In the reshuffle Shatalov was moved from the first crew to replace Volynov on the third, with Shonin taking Shatalov's place. Kamanin nominated Colonel Leonov, another veteran from the first group of cosmonauts who had trained to command the first L1 circumlunar mission, to fill Shonin's place. Khrunov was replaced as commander of the fourth crew by Lieutenant-Colonel Dobrovolskiy who, although he had been recruited in 1963 as a member of the second Air Force group, had not yet flown in space. Mishin replaced Feoktistov by advancing Volkov from the fourth crew, then reassigned Sevastyanov's role from research engineer to flight engineer. Kamanin completed this crew with Lieutenant-Colonel Voronov, who was another member of the Air Force's second group who had yet to fly. All the research engineers were rookies, and the first three crews each had two experienced cosmonauts – one from each community. The inexperience of Dobrovolskiy's crew was not considered to be a problem, because they would have the longest time to

Kubasov (left), Volkov and Yeliseyev, flight engineers assigned to the first DOS station, in conversation with an Air Force representative (back to camera).

A rare photo showing some of the original DOS crewmembers at an early stage of training listening to a presentation at the TsKBEM: Shonin (left), Shatalov, Kubasov, an unidentified person and Volkov.

train and would be able to benefit from the lessons learned by their predecessors in operating a space station. On 13 May 1970 Mishin and Kamanin signed a decree which confirmed the crew assignments.

The names of all the cosmonauts have been mentioned intentionally, even those who were not actually able to train for a mission to the DOS-1 station, as this shows how the destinies of these men were influenced by incidents such as Khrunov's car crash, Volynov's Jewish blood, Kamanin's dislike for Feoktistov, and the need to agree a fair balance of assignments between the two communities of cosmonauts.

In late July the Military-Industrial Commission (VPK) met specifically to discuss the progress with the DOS programme. Okhapkin, Mishin's First Deputy, gave the TsKBEM report. In accordance with the original plan, two identical stations were to be built. DOS-1 was to be launched in early 1971 and be visited by two crews who would undertake a variety of scientific experiments and make terrestrial, solar and astronomical observations. Two further crews would visit DOS-2 in 1972. However, Okhapkin reported that as a result of a number of problems the project was about 2 months late. After Soyuz 8 experienced difficulties with its Igla rendezvous system on the 'group flight' in 1969, the design had been revised and the system transferred from the descent module to the orbital module in order to improve its 'field of view'. In addition, since the flight to the station would take only one day and the capsule would return to Earth within hours of departing from the station, the designers had simplified its life support system; but there were delays in testing the revisions. The major change to the Soyuz was the inclusion of a docking system incorporating a 0.8-metre-diameter hatch to enable the cosmonauts to access the station. There had been delays in constructing this new system. Nevertheless, as soon as the schedule allowed, DOS-1 would be launched by one of Chelomey's Proton rockets. After 8 to 10 days, the Soyuz 10 mission would be launched using the first 7K-T crew ferry. If everything went to plan, Shonin, Yeliseyev and Rukavishnikov would spend 30 days

A theory lecture in the early stage of DOS training, showing Shatalov and Patsayev (foreground), and Yeliseyev and Rukavishnikov.

on the station. Twenty-five days after Soyuz 10's return, Soyuz 11 would be launched with Leonov, Kubasov and Kolodin, who would spend up to 45 days in space, with the actual duration being determined by how well the flight progressed. It was therefore hoped that the DOS-1 station would be able to be occupied for 75 days of its expected service life of 80–90 days. However, it was accepted that this would be a pioneering venture. The longest time that American astronauts had spent in space was 14 days, on a Gemini flight in December 1965. Several weeks prior to the VPK meeting, the 18-day flight of Soyuz 9 by cosmonauts Andriyan Nikolayev and Vitaliy Sevastyanov had broken this endurance record. As the days passed, the cosmonauts had become so tired that mission control had used a siren to wake them up. On their return to Earth their heart rates were twice the norm, and for three days neither man had been able to walk. It took them a month to recover fully. In fact, Nikolayev had to retire several months later owing to ongoing heart issues. In view of the experiences of this crew, Kamanin said that missions of 30 or 45 days were unrealistic until more information was gained on how the human body was affected by prolonged exposure to weightlessness, and he argued that the early DOS flights should not exceed 20 days.

The VPK meeting ended without specifying the length of the missions for the DOS-1 station. Leonid Smirnov, the chairman of the commission, ordered that all testing must be done by the end of the year, and that the station must be launched in time to be celebrated by the 24th Congress of the Soviet Communist Party in March 1971.

In June 1970 the engineers of the first and third crews (Yeliseyev, Rukavishnikov, Volkov and Patsayev) started to train at the TsPK, and began to pay regular visits to the Khrunichev factory in Fili to monitor the building the first station.

Yeliseyev recalls: "The construction of the station was rapid. It took only a few days to build a wooden mockup – all its sections and elements were in accordance with the design, but instead of real devices and apparatus it had wooden models. With Rukavishnikov, I went to see the mockup of the first station. Compared to the Soyuz, it looked like a giant – it was more than 10 metres from one end to the other. There was room for several people to work, without hindrances. ... Engineers were working continuously, checking every detail of the documentation. Every revision was tested on the mockup, with a detailed inspection. If the change was acceptable, then it was made to the station. We were involved in testing the positioning of the controls, instrument panels and the apparatus for visual monitoring. We were also consulted on how the crew should work and rest. ... This work was very interesting. However, I must admit that the most impressive thing was to watch the real station be born. ... I had a feeling of being present at the nativity of a secret miracle that the public knew nothing about. However, the whole world would hear about it very soon."

The commanders of the first and the third crews (Shonin and Shatalov) joined in the training on 17 August. The second crew (Leonov, Kubasov and Kolodin) began to train on 18 September. The members of the fourth crew were each busy with his individual tasks. Dobrovolskiy's assignment was to the Contact project, testing the rendezvous and docking techniques for the N1-L3 lunar programme, and he did not

Cosmonauts Sevastyanov (left) and Nikolayev shown on TV from Soyuz 9 during a communication session of their record-breaking mission in 1970.

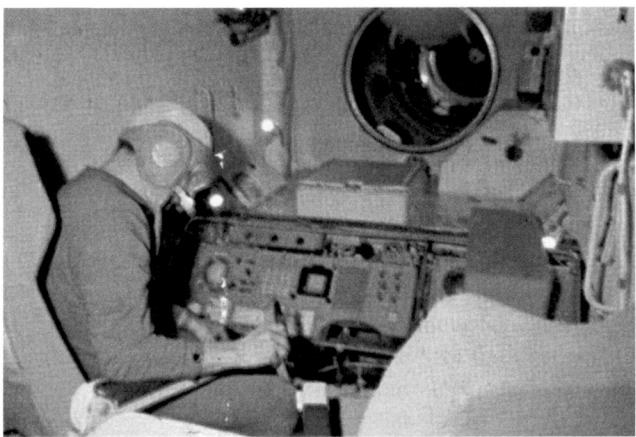

Photos of cosmonauts training in the DOS-1 simulator are extremely rare. Here, an unidentified cosmonaut is in the commander's seat, facing the main control panel.

begin DOS training until January 1971, after the cancellation of the N1-L3. After his Soyuz 9 mission Sevastyanov spent several months recuperating, and so did not start DOS training until October 1970. However, as was usual for cosmonauts who had just made their first flight, he was frequently sent on goodwill visits, both to the member republics of the USSR and to foreign countries. The third crewmember, Voronov, was also initially involved in another project. As a result, the fourth crew did not begin serious training for DOS until January 1971, and expected to have at least 18 months before making their flight.

THE DISMISSAL OF SHONIN

When the DOS programme started in early 1970, it was only one of several manned space projects. In addition to the 12 cosmonauts assigned to the DOS crews there was a group of ten cosmonauts in training at the TsPK for projects involving other versions of the Soyuz. Pavel Popovich led the 20 military cosmonauts in the Almaz group. Valeriy Bykovskiy led the now much reduced lunar group. There was also a team in training for the Spirala 'rocket plane' project. Because the facilities were in constant use, some of the simulators were in need of significant maintenance. The mockup of the DOS-1 station was not able to be installed at the TSPK until October 1970, barely four months before (on the target schedule) the first crew were due to be launched.

Having worked 24 hours a day, the Khrunichev factory managed to deliver three stations to the TsKBEM in December 1970 for testing.

The Ministry of General Machine Building formed the DOS-1 State Commission, drawing its members from the leading people responsible for the design and testing of the spacecraft, training the cosmonauts, launch preparations, mission control and recovery activities. By tradition, at the final meeting of a State Commission prior to a launch, representatives from the TsKBEM, the Baykonur cosmodrome and all the other institutions which participated in the preparation of the mission would assess their readiness. After the decision of the date and time of launch, Kamanin would present the prime and backup crews. Since 1966 the State Commissions for Soyuz flights had been chaired by Major-General Kerim Kerimov, one of the Ministry of General Machine Building's directorate chiefs. He had previously been responsible for developing and operating the facilities of the Strategic Rocket Forces. At the inaugural meeting of the new State Commission in late December 1970, the Chief Designers reported on the status of the programme, and the progress in constructing the two stations and the 7K-T ferries for the Soyuz 10 and Soyuz 11 missions. As a result of problems testing its subsystems, it would not be possible to launch DOS-1 in February as hoped, and its launch was rescheduled for 15 March 1971. There was an argument about the duration of the first missions. Mishin and the medical experts of the Institute for Biomedical Problems (IBMP) in Tushino, which was the most prestigious of the civilian space medicine institutions, wanted the first crew to make a 30-day flight, but Kamanin and the physicians of the Central Air Force Scientific Research Hospital (TsVNIAG) argued for a maximum target of 25 days; the issue was left unresolved.

The DOS-1 station undergoing final system tests.

Shortly after the meeting, General Kutakhov, the Commander in Chief of the Air Force, visited the TsKBEM to inspect progress with the first station. Mishin tried to convince him to overrule Kamanin and support a 30-day flight for the first mission, but Kutakhov diplomatically replied that the matter should be decided by aerospace physicians and those who were responsible for training the cosmonauts. Although Mishin inferred from this that Kutakhov supported Kamanin, this was not the case. Aware that his days were numbered, Kamanin wrote to Kutakhov proposing to retire in favour of his aide, Major-General Leonid Goreglyad. But Kutakhov had his own candidate, a man who had worked under his command many years ago – cosmonaut Vladimir Shatalov. Kamanin thought that the more experienced Goreglyad would be a better choice, at least until Shatalov had matured as a manager.

In general the crew training was efficient, and the cosmonauts divided their time between the facilities in Zvyozdniy and Kaliningrad as necessary. One of the most important military experiments, which was to be done by both of the DOS-1 crews, was to monitor the launch of intercontinental ballistic missiles. In the early days of February 1971, Shonin's and Leonov's crews went to Baykonur for special training. Flying in a Tu-104 aircraft, they observed night launches of ballistic missiles using the Svinetz ('Lead') apparatus which was to be installed on the station. However, on 5 February, soon after their return, Shonin missed a session of important quality control and testing (KIS) training in Kaliningrad, apparently because he was drunk. A furious Mishin called Kamanin and loudly announced his firm decision: "He'll never fly again in my spaceships!" Kamanin called Shonin, and promptly realised how stupid Shonin had been. Kamanin had received reports of Shonin's drinking habits in March 1970 but, having been impressed by Shonin on the Soyuz 6 mission,

Kubasov observes Shonin (right) and Kamanin (left) playing chess.

had taken no action. Now Kamanin realised that for months the TsPK managers had been covering for a drunken cosmonaut. This left Kamanin with no option. Shonin pleaded his case: "Take my Hero's star, strip me of my colonel's rank, but please don't take my spaceflight from me!" Shonin had received the Gold Star of Hero of the Soviet Union for his space flight, and it was the highest honour the nation could award a military officer. Leonov, who served as one of the deputies to Kuznyetsov, the head of the TsPK, approached Kamanin in an attempt to defend Shonin, but it was too late.[3]

Kamanin now suggested the strongest military cosmonaut, Shatalov, to command the first crew. Although this reinstated the crew nominated by the TsKBEM in May 1970, Mishin was against the idea. He sought to exploit Shonin's dismissal to call for assigning an all-civilian crew to the first DOS-1 mission: Yeliseyev would be in command, with Kubasov as flight engineer and Rukavishnikov as research engineer. Of course, Kamanin rejected this. After much argument, Mishin was obliged to accept Shatalov as the commander of the first crew. Dobrovolskiy took Shatalov's place on the third crew, and Lieutenant-Colonel Aleksey Gubaryev, a rookie from the second group of military cosmonauts, was given command of the fourth crew. In this reshuffle, only Leonov's crew remained untouched.

[3] Soon after this, Shonin was admitted to the Burdenko Hospital in Moscow suffering from depression. On being discharged in March 1971 he was urged to undergo a lengthy medical treatment. He recovered, but never flew in space again. He died from a heart attack in April 1997.

Shatalov (foreground) and Volkov training with photo-equipment.

After Shonin's dismissal, Shatalov (second from the left) joined the 'first crew' with 2 months remaining to the launch of the DOS-1 station. Here he, Yeliseyev (left) and Rukavishnikov listen to Deputy Chief Designer Yakov Tregub, the head of manned flight control. (From the book *Life – A Drop in the Sea*, courtesy astronaut.ru)

On 12 February 1971, with the launch of the first space station imminent, the crews were changed for the third time:

- Crew 1: Vladimir Shatalov, Aleksey Yeliseyev and Nikolay Rukavishnikov
- Crew 2: Aleksey Leonov, Valeriy Kubasov and Pyotr Kolodin
- Crew 3: Georgiy Dobrovolskiy, Vladislav Volkov and Viktor Patsayev
- Crew 4: Aleksey Gubaryev, Vitaliy Sevastyanov and Anatoliy Voronov

However, as events transpired, this plan did not last.

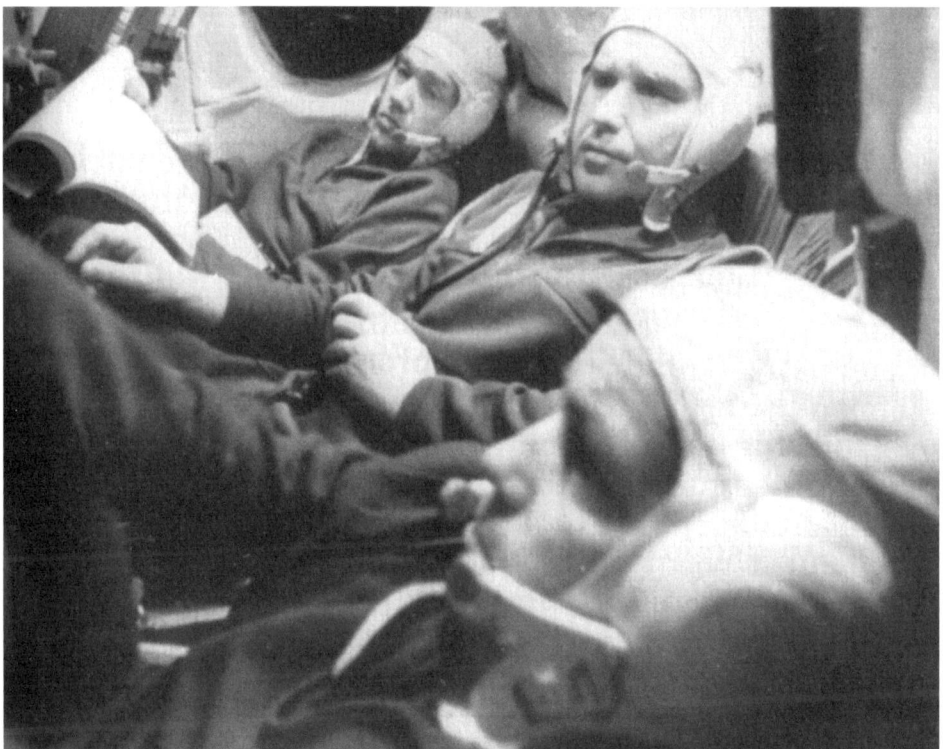

The 'first crew' for DOS-1: Yeliseyev, Shatalov and Rukavishnikov (foreground) in the Soyuz simulator.

Specific references

1. Kamanin, N.P., *Hidden Space, Book 4*. Novosti kosmonavtiki, 2001, pp. 153–160 and 260–262 (in Russian).
2. Yeliseyev, A.S., *Life – A Drop in the Sea*. ID Aviatsiya and kosmonavtika, Moscow, 1998, pp. 70–72 (in Russian).

The 'second crew' for DOS-1: Kubasov (standing), Leonov and Kolodin inside the Soyuz descent module simulator. (From the private collection of Rex Hall)

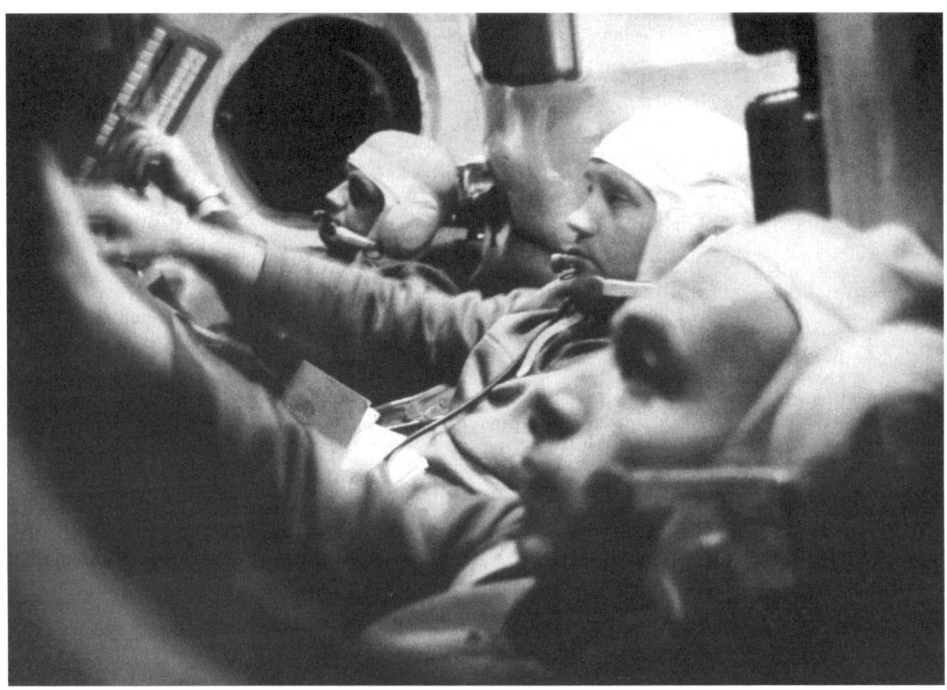

The 'third crew' for DOS-1: Volkov (rear), Dobrovolskiy and Patsayev.

The 'fourth crew' for DOS-1: Gubaryev (left), Voronov and Sevastyanov.

DOS cosmonauts. Sitting: Leonov (left), Yeliseyev, Shatalov, Rukavishnikov and Kubasov. Standing: Kolodin (left), Dobrovolskiy, Volkov and Patsayev. (From the private collection of Rex Hall)

3

Salyut in space

FINAL PREPARATIONS

As the cosmonauts were training for flights to the first space station, on 2 March 1971 the Council of Chief Designers met at the TsKBEM for its first session in relation to the DOS-1 work. The Council had been formed in late 1947 by Sergey Korolev to oversee the technical management of rocket and spacecraft development. It was chaired by Korolev, and originally comprised the six Chief Designers of the primary rocket design bureaus: Valentin Glushko for rocket engines, Nikolay Pilyugin for guidance systems, Viktor Kuznetsov for gyroscopes, Vladimir Barmin for launch equipment and Mikhail Ryazanskiy for radio-control systems. After the death of Korolev in January 1966 Vasiliy Mishin took his place. He now chaired the meeting. The main presentation was by Yevgeniy Shabarov, Bushuyev's deputy for the testing of manned spacecraft, who said that all testing had been successfully completed and DOS-1 and the two Soyuz 7K-T ferries were ready to be sent to Baykonur. The preparations to launch the station – which was to be named Zarya ('Dawn') – were to begin on 9 March with a view to achieving a launch on 15 April. If all went well, the first crew would follow within five days. Mishin criticised the delays in vibration testing the DOS-1 mockup, and asked that this be completed by 29 March. There was also the issue of the warranty on the parachutes of the Soyuz, which would expire on 15 April. The Council also discussed the efficiency of the Igla rendezvous system. But the main concern was that the first phase of testing the station's life support system, which had only recently been completed, had revealed a number of faults and it would not be possible to start the second phase of testing until these issues had been resolved. Interestingly, although there were less than 45 days remaining to the station's planned launch, the question of how long the crews should occupy it remained undecided.

SPACE STATION LAUNCH

Meanwhile, the crews were wrapping up their training programmes in Zvyozdniy and Kaliningrad. On 9 March Shatalov, Yeliseyev and Rukavishnikov spent more than 14 hours in the Soyuz simulator, rehearsing each major phase of the mission – launch, rendezvous and docking, undocking, re-entry and landing – in the process overcoming five simulated anomalies, including the failure of the main engine and an excessive rate of fuel consumption. Leonov's crew had a similar session the next day, and Dobrovolskiy's crew three days later. This completed the formal training. On 16 March they had their final exams, and achieved the best possible score. Three days later, the State Commission confirmed that an attempt would be made to launch the station in the period 15–18 April, but left undecided the duration of the first crew's mission.

Did anyone in the West have a suspicion of the imminence of such an important event in cosmonautics? Of course, Brezhnyev's speech and the newspaper articles by Academicians Keldysh and Petrov had said that the Soviet Union was interested in a space station. Some of this was reported overseas. For example, on 16 March 1971 the *Guardian* in London published an article entitled 'Russia Plans the First Station in Space'. And in an interview with the newspaper *Socialist Industry*, an unidentified Chief Designer (Mishin?) predicted: "It seems to me expedient to build in the near future a station in space near the Earth that would operate for a long time. The flight of Soyuz 9, which lasted 18 days, was an important step in this direction. A time will come when manned Soviet research laboratories will orbit the Earth." Western newspapers speculated that the next long-duration manned mission might coincide with the 24th Congress of the Soviet Communist Party, which was to open on 30 March.

On 20 March the three crews travelled to Baykonur on separate Tu-104 aircraft, and on arrival went to the so-called 'Cosmonaut Hotel'. Over the next three days they performed final rehearsals on board the Soyuz 10 spacecraft in the Assembly-Test Building (MIK) near Pad No. 1; the pad from which first Sputnik and later Gagarin had lifted off. The rocket was being checked elsewhere in the building. Chelomey's preparation facility was near Pad No. 82, from which his Protons were launched. The DOS-1 station and its launcher were being checked out in one building while construction was underway nearby of the building in which the spacecraft for the Almaz programme would be prepared.

The cosmonauts flew back to Moscow in order to attend the meeting of the VPK Military-Industrial Commission on 25 March. Mishin said that it should be possible to launch the station in the period 15–20 April, with Soyuz 10 lifting off 3 days later. He recommended that Shatalov, Yeliseyev and Rukavishnikov's mission should last a month, and that 25 days later Leonov, Kubasov and Kolodin should set off for a mission of up to 45 days. It was decided to assign the first crew a 30-day mission, but to postpone making a decision on the length of the second mission until the state of health of the first crew had been assessed upon their return. However, looking on the optimistic side, the second crew were to train on the presumption that they were to conduct a 45-day mission.

The Soyuz 10 crew before their final training: Rukavishnikov (left), Yeliseyev and Shatalov. (From the private collection of Rex Hall)

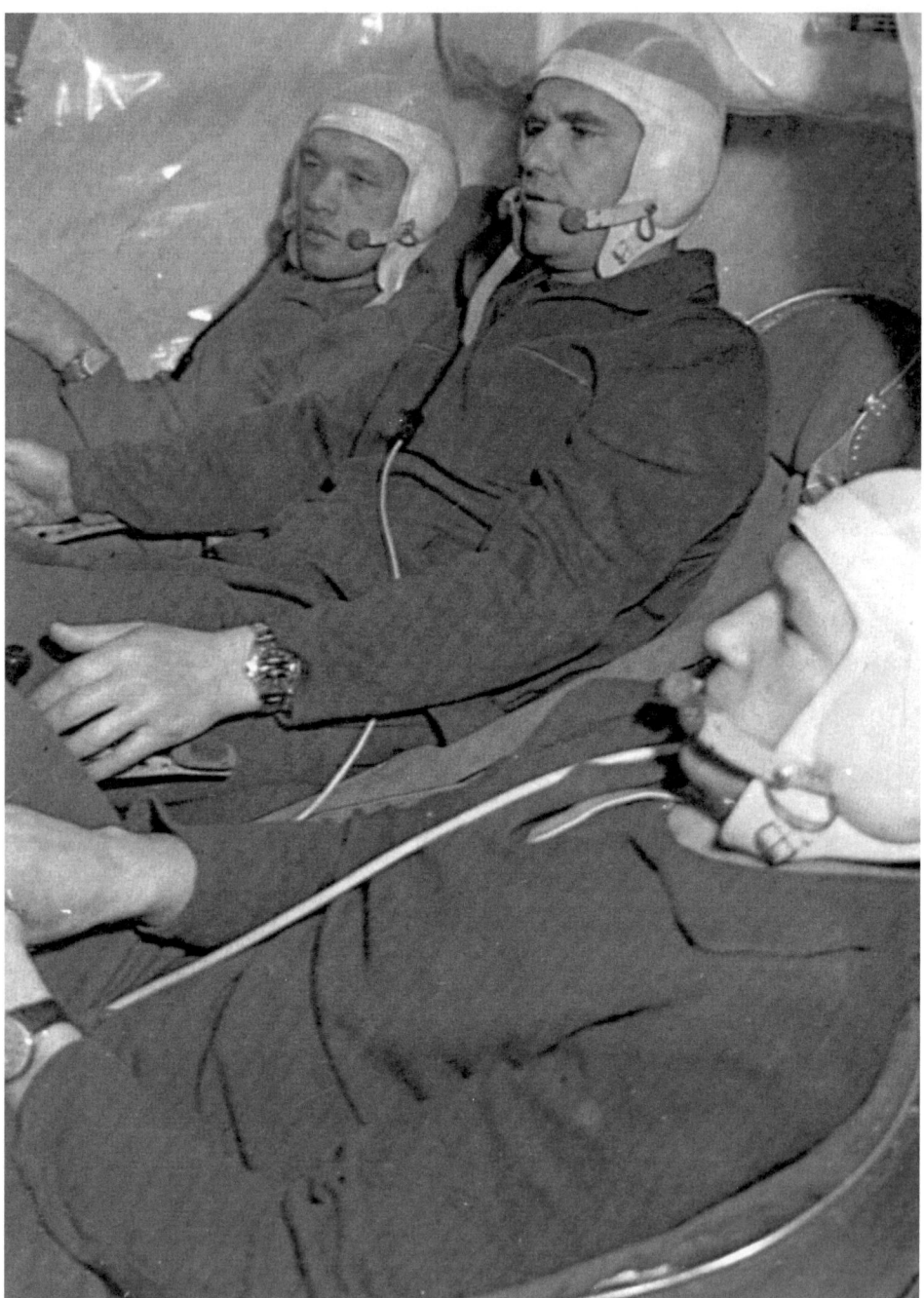

Yeliseyev (rear), Shatalov and Rukavishnikov in the Soyuz simulator.

Shatalov (left), Yeliseyev and Rukavishnikov after 14 hours in the Soyuz simulator.

As 'known' cosmonauts, Shatalov and Yeliseyev were directed to attend the 24th Congress of the Soviet Communist Party. Less than a month before the launch of the first space station, two members of the prime crew spent 6 days as delegates at a political meeting! In accordance with tradition, the cosmonauts visited Red Square and Gagarin's office in Zvyozdniy. Then there was a final meeting with Mishin, Bushuyev, Chertok and Semyonov at the TsKBEM, which was attended by Nikolay Lobanov, who was the Chief Designer of the Scientific-Research and Experimental Institute of the Parachute Landing Service, and Gay Severin, the General Designer of OKB Zvezda which supplied the spacecraft crew couches and the suits and other apparatus required to undertake spacewalks.

On 6 April the three crews returned to Baykonur. They conducted a final round of training on board the space station and the Soyuz 10 spacecraft, and then these were mated with their launchers. On 9 April the State Commission decided to install the Proton on the pad at 7.00 a.m. local time on 15 April, with a view to launching it at 6.40 a.m. (4.40 a.m. Moscow Time) on 19 April. General Kerimov, who chaired the meeting, expressed the Kremlin's dissatisfaction with the name 'Zarya'. It appeared that China was developing a rocket with this name. The Kremlin wanted the station to be renamed. In fact, there was another reason to change the name, as 'Zarya' was the radio call-sign of the mission control centre in communications with a manned spacecraft. Kerimov suggested that the name on the side of the vehicle be repainted. Chertok objected, pointing out that as the Americans would surely not be able to photograph the station in orbit it was unnecessary to repaint the vehicle! After much discussion, someone proposed Salyut ('Салют'; meaning 'Salutation' or 'Greeting')

It is traditional for cosmonauts to visit Gagarin's office at the TsPK in Zvyozdniy prior to departing for Baykonur. Here Volkov (left), Yeliseyev, Dobrovolskiy and Shatalov (foreground) look on as Leonov writes in the visitors' book. A picture of Konstantin Tsiolkovskiy hangs on the wall.

and this was unanimously accepted. As there was no time to repaint the name, the station was mated with the third stage of the Proton bearing the name 'Zarya' on its side.[1]

The final meeting of the State Commission prior to the launch of DOS-1 was held on 18 April. After representatives of the TsKBEM and the TsKBM reported on the status of the Proton and its payload, the preparations for Soyuz 10 were reviewed.

At 4.39 a.m. Moscow Time on Monday, 19 April 1971, Salyut (DOS-7K No. 1) was successfully launched as the world's first space station – it was only 16 months since the decision was taken to initiate the programme!

The initial parameters of the station's orbit were 200 × 222 km at an inclination of 51.6 degrees to the equator, with a period of 88.5 minutes. Although the launch had been flawless, the radio-telemetry during the first revolution showed a problem: the cover that had protected the aperture of the main scientific module during the ascent through the atmosphere had failed to jettison. As this module held 90 per cent of the observational apparatus, the station's scientific results would be severely curtailed if the cover could not be released.

---

[1] In accordance with Soviet tradition, the first space station did not bear the number '1'. If the first example of a new type of spacecraft were to be numbered, it would make evident that a series of such vehicles were planned, and the Soviet Union went to great lengths to keep its plans secret.

Right: Three weeks prior to their launch, Yeliseyev and Shatalov attend the 24th Congress of the Soviet Communist Party. Left: Afterwards, in Red Square, with Rukavishnikov.

The Soyuz 10 cosmonauts with the mission's technical managers: Semyonov (left), Bushuyev, Yeliseyev, Chertok, Mishin, Lobanov, Shatalov, Severin and Rukavishnikov.

Since 1966, General Kerim Kerimov from the Strategic Rocket Forces was the chairman of the State Commission for Soyuz missions.

Although the Soviet news outlets featured the launch, other than describing Salyut as "an orbital scientific station" their reports revealed little of its nature and mission. However, the fact that it bore a name at all implied that it was a substantially new type of spacecraft, and the report that three ships had been stationed in the Atlantic to communicate with it was a clue that it might be associated with the manned space programme.[2] TASS said the flight was going as planned, and that an adjustment to its orbit had been made. The routine phrase "functioning normally" was applied to the performance of its systems. The bland statement of its purpose as being "to perfect the elements of the design and on board systems, and to conduct scientific research and experiments in space" simply meant wait and see.

Meanwhile, at Baykonur the final preparations were initiated to launch Soyuz 10. To preclude injuries at this late stage, the crew were prohibited from participating in sporting activities such as basketball, handball, volleyball, tennis and soccer – all of which were popular among the cosmonauts. However, they were allowed to run 10 kilometres per day to maintain their fitness. Anyone coming into close contact with them wore a protective face mask to minimise the risk of passing on any respiratory infections that might jeopardise the planned duration of the mission.

Early on the morning of 20 April the rocket bearing Soyuz 10 was installed on Pad No. 1. Several hours later, the State Commission presented the cosmonauts to space journalists. The prime crew was Colonel Vladimir Shatalov (44), commander; Aleksey Yeliseyev (37), flight engineer; and Nikolay Rukavishnikov (39), research engineer. The backup crew was Colonel Aleksey Leonov (37), commander;

<hr />

[2] The ships were named *Morzhovets*, *Kegostrov* and *Academician Sergey Korolev*.

The DOS-1 station in the Assembly-Test Building at Baykonur. On the left is the transfer compartment, with the smaller (white cylinder in the middle) and larger working compartments (right). On the transfer compartment can be seen two folded solar panels, a rendezvous antenna, and the docking port (far left).

A view of the DOS-1 station in the Assembly-Test Building in which the vehicle has been partially rotated. The 'hole' on the right of the main compartment is the aperture to enable the main scientific equipment to see out. It was sealed by a protective cover for launch.

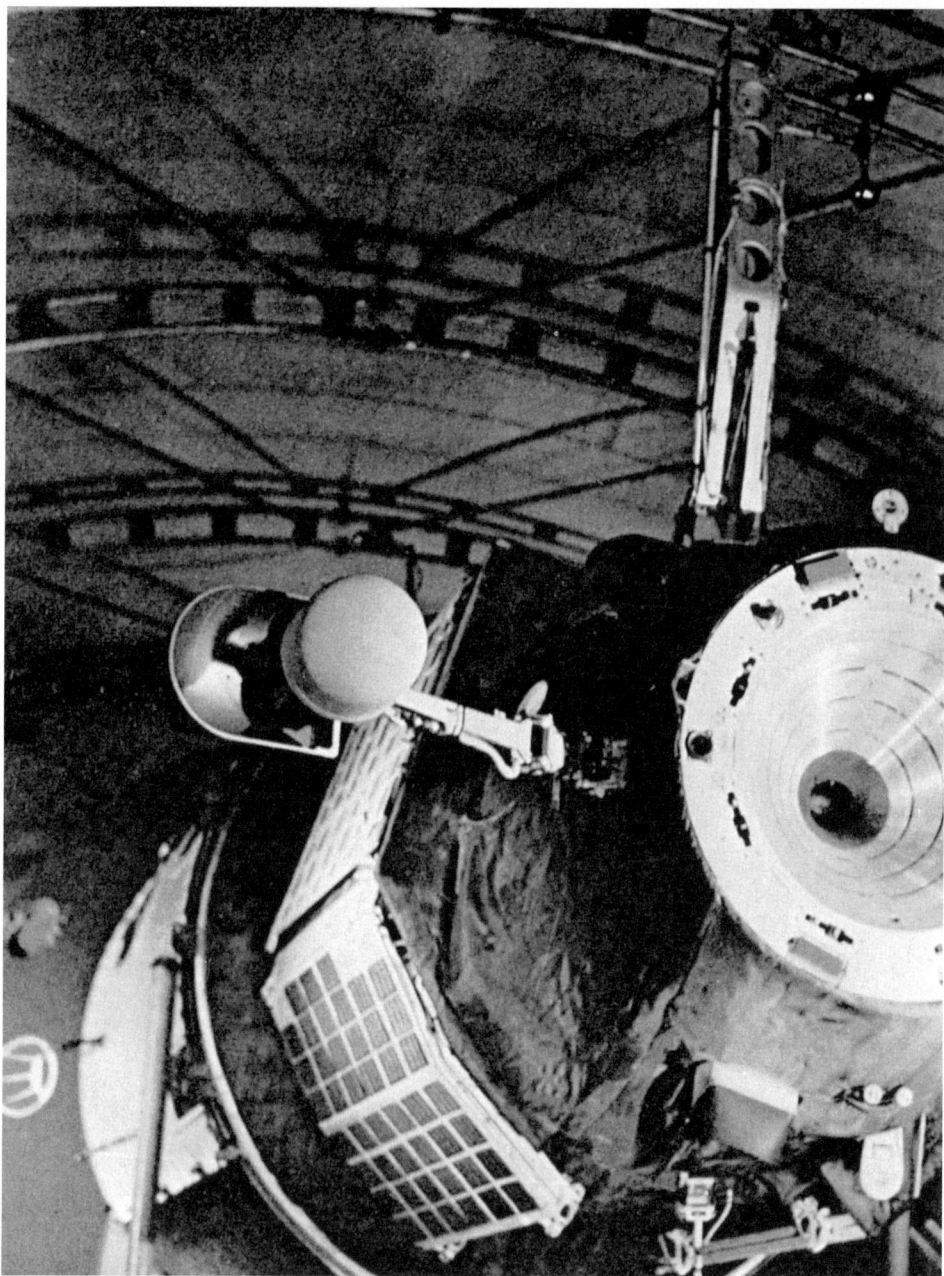

Final testing of the DOS-1 station prior to mating with the Proton launch vehicle. The rendezvous antennas are deployed upward and to the left of the passive docking system. The central hole of the docking cone is to capture the head of the probe on the Soyuz. The TsKBEM's logo is visible on the short conical adapted between the small and large working compartments (far left).

A Proton launch vehicle with the DOS-1 station on the pad at Baykonur. The wide part of the working compartment is exposed, but the narrower part and the transfer compartment are inside a shroud for the ascent through the atmosphere. The white support ring below is jettisoned after orbital insertion. The original name 'Zarya' is visible on the working compartment. Also visible is the white protective cover for the scientific equipment aperture.

Valeriy Kubasov (36), flight engineer; and Lieutenant-Colonel Pyotr Kolodin (41), research engineer.[3] And, of course, there was the support crew: Lieutenant-Colonel Georgiy Dobrovolskiy (43), commander; Vladislav Volkov (36), flight engineer; and Viktor Patsayev (38), research engineer.

Let us take a closer look at each member of the prime crew.

---

[3] Kubasov's mind may have been distracted at this time, because in Moscow his wife was giving birth to their second child: son Dmitriy.

Engineers test systems in the propulsion module of the Soyuz 10 spacecraft in the Assembly-Test Building at Baykonur.

Yeliseyev, Shatalov and Rukavishnikov (right) in the Assembly-Testing Building at Baykonur. In the background is the engine cluster of the Soyuz launch vehicle.

EXPERT IN SPACE RENDEZVOUS

Vladimir Aleksandrovich Shatalov, nicknamed Volodya, was a familiar face for the journalists. When they saw him at the Congress of Communist Party in Moscow a few weeks earlier they had asked him when he was going into space again, and he had replied that he was ready to fly and would launch the next day if permitted. The journalists had laughed, not suspecting that an important event in cosmonautics was imminent. If they had known that a space station was being prepared, they certainly could not have imagined that the tall colonel with the blue eyes had been appointed to command its first crew just two months previously!

Shatalov was born on 8 December 1927 in Petropavlovsk in northern Kazakhstan. When he was two years of age his family moved to Leningrad (now St Petersburg), and there during the Second World War he served in the same brigade as his father Aleksandar, who held the Gold Star of Hero of the Soviet Union, and participated in the legendary defence of that city. In 1945 Volodya completed a special school in Voronezh for future military pilots, and 4 years later graduated from the famous Kachinsko Higher Air Force School. In September 1949 he was flying as a pilot-instructor. He married Muza Andreyevna Yonova, an agricultural engineer, and in 1952 she gave birth to their first child: son Igor. On attaining the best scores at the prestigious Red Banner Air Force Academy, Shatalov became a pilot-engineer in 1956. After a period serving as a deputy squadron commander he gained his own squadron. In 1958 their second child, daughter Yelena, was born.

In May 1960 Major Shatalov was appointed as a deputy to an air force regimental commander, and in February 1961 he became the senior instructor-inspector in the department of military readiness at the 48th Air Army in the Odessa military district. It was while there that he saw for the first time the new Su-7 aircraft. In those days the Su-7 was restricted to the very best pilots. It was an extraordinary plane, capable of flying faster and higher. In addition, it had much better 'air combat' performance. Since becoming a pilot, Shatalov had dreamed of flying in space. When he sat in the cockpit of the Su-7 for the first time, he thought: "I believe that a spacecraft will be similar to this one. Perhaps in the next ten years someone will make the first flight beyond the atmosphere. ... Someone, but not me. I will be too old." On 12 April he took the aircraft up. "As I flew the Su-7, I thought of myself as the pilot of that spacecraft. Although I did not know what it should look like, while at an altitude of 19 km I felt I was in space." On landing, he heard the news that changed his life for ever: Yuriy Gagarin had just landed after orbiting the Earth in a spacecraft named Vostok. "What about me? Am I too late? Gagarin is seven years younger than me! My way to space must be closed. ... Not to worry; they can conquer space without me."

In fact, almost all of the first 20 cosmonauts selected from the ranks of the fighter pilots were aged 25 to 35 and had very little flying experience; typically only about 300 hours in the air. But early in 1962 the Commander of the Air Army, General Kutakhov, asked Shatalov to select his best five pilots for consideration as potential cosmonauts. On reading the selection criteria, Shatalov realised that he was himself eligible to become a candidate. "The selection criteria apply to me," he pointed out.

"So, what?" Kutakhov replied.

"Comrade General, I could become a cosmonaut too! May I put my name on the list?"

"Don't be silly, Shatalov!" Kutakhov dismissed. "What do you need that for? The chance of becoming a cosmonaut is one in a thousand. You would just waste your time. After four or five years, you'll be excluded due to your health. However, here you have a position, rank, authority, a bright future. What more would you want?"

Nevertheless, Shatalov persisted, and was included on the list for the preliminary medical examination. He and Anatoliy Filipchenko, whom Shatalov had nominated, were selected to fly to Moscow where, with the candidates from other squadrons, they spent two weeks being subjected to various medical checks and examinations, with the number of candidates under consideration being diminished day by day. In April 1962 Shatalov successfully passed the final medical examination, then had an

Colonel Vladimir Shatalov, the Soyuz 10 commander.

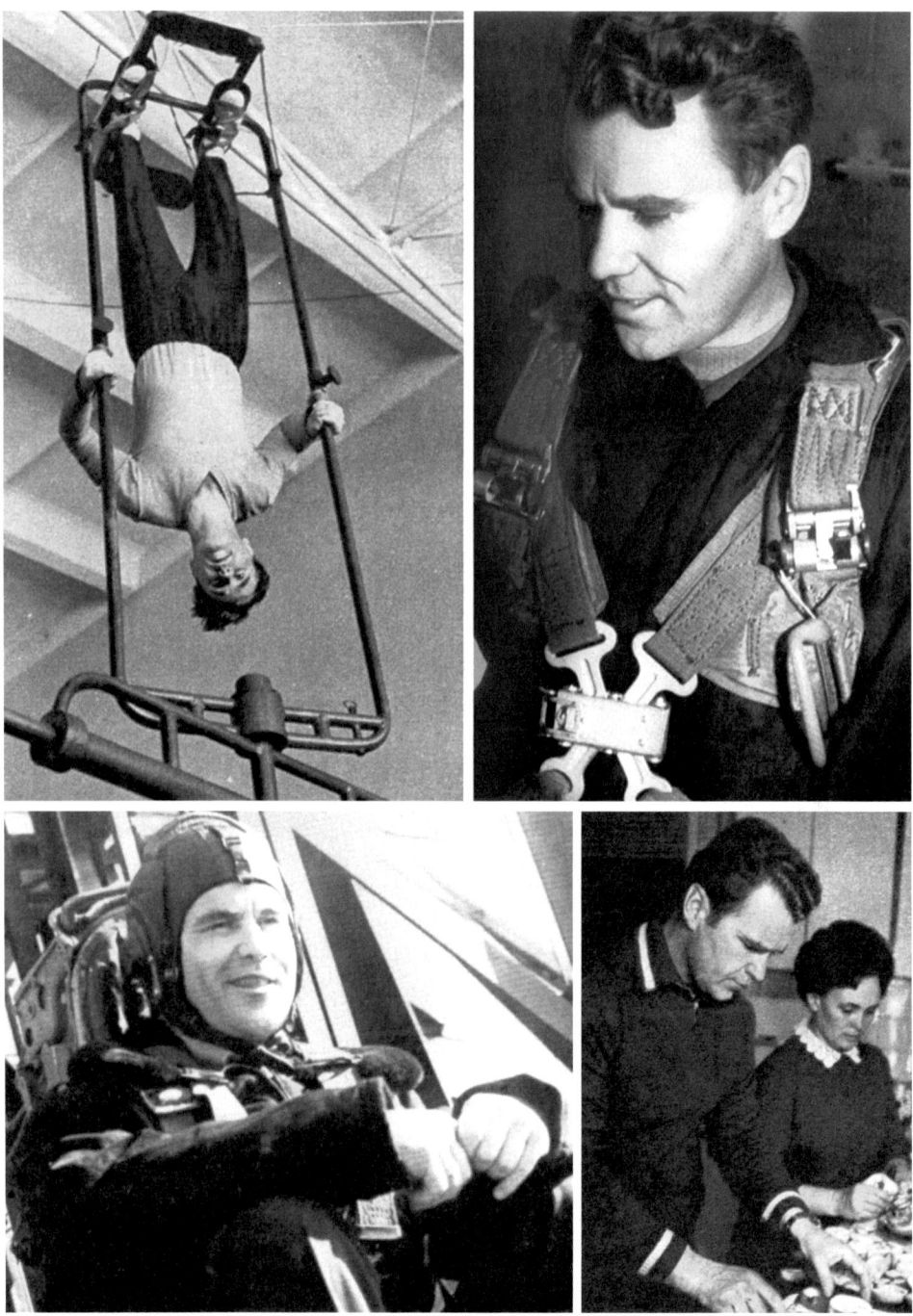

Shatalov in training, and with his wife Muza.

interview with the State Commission that was responsible for cosmonaut selection, whose members included Kamanin and Gagarin. Shatalov returned to his squadron without knowing the outcome.

While Shatalov waited for news, in August 1962 two new cosmonauts – Andriyan Nikolayev and Pavel Popovich – flew a two-spacecraft mission. General Kutakhov, eager to retain Shatalov in the Air Force's management loop, nominated him as a regiment commander. Normally this would have been excellent news, but Shatalov believed that this rank would work against his becoming a cosmonaut. What could he do? Accept the nomination? Or ask about his cosmonaut candidacy? He wrote to the Commander in Chief of the Air Force, Marshal Vershinin, and to his deputy for cosmonaut training, General Kamanin. It is not clear whether this played any role in the issue, but soon thereafter Shatalov received an invitation to travel to Moscow.

By the end of 1962 Shatalov passed the mandate commission, and on the evening of 11 January 1963 he arrived at the TsPK with 14 others as the Air Force's second cosmonaut group. Compared to the 1960 group, the new cosmonauts were generally older, better educated, higher in rank and more experienced. All had an engineering qualification, some were test pilots, and several were non-pilots from the Strategic Rocket Forces who had served at Baykonur – the pilots referred to them as 'rocket men'. When he became a cosmonaut, Shatalov was a Lieutenant-Colonel with more than 2,500 hours of flying experience, which was ten times greater than most of the first group. He had flown 17 types of aircraft, including the newest models such as the Yak-18, MiG-21, Su-7B, Il-14 and Tu-104, and he was also qualified to fly the Mi-4 helicopter.

In those days all the Air Force cosmonauts lived in a 3-storey building, the tallest at Zvyozdniy. On the lower floors were classrooms, a mess room and a recreation room with billiards. There, and in several small buildings scattered among the pine and beech trees, the new cosmonauts took their first steps on the road to space.

The general training at the TsPK was completed in January 1966. During this time, Shatalov had classes in the theory of space flight, studied the systems of the Vostok and Voskhod capsules, made 10 flights in aircraft that simulated weightlessness and performed over 100 parachute jumps. He also served as a communications operator for the Voskhod flights in October 1964 and March 1965. From May to December 1965 he trained as commander of the third (backup) crew, with Yuriy Artyukhin, for a Voskhod flight which was set for November 1965 but cancelled. From January to May 1966 he trained as the copilot of the second (backup) crew for the Voskhod 3 mission which was to last 16–20 days. His commander was the famous test pilot Georgiy Beregovoy, who was a late addition to the 1963 group. On the prime crew were Boris Volynov and Georgiy Shonin, both of whom were members of the 1960 group. The spacecraft passed the pre-launch tests, but there were repeated delays in mating it with the rocket, and in May the launch was cancelled – together with the remainder of that programme. If it had been launched, Voskhod 3 would have been the first manned space flight since the death of Korolev. The new Chief Designer at the TsKBEM, Mishin, with the support of the Kremlin, had decided that to continue with Voskhod would represent a diversion of resources away from the development of the much more capable Soyuz spacecraft.

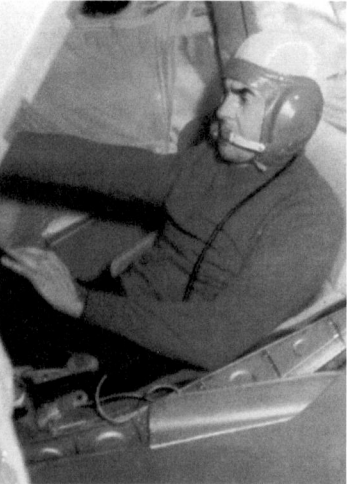

Beregovoy failed to dock his Soyuz 3 with the unmanned target in October 1968.

Shatalov uses cutout models to demonstrate how two Soyuz spacecraft should align for docking.

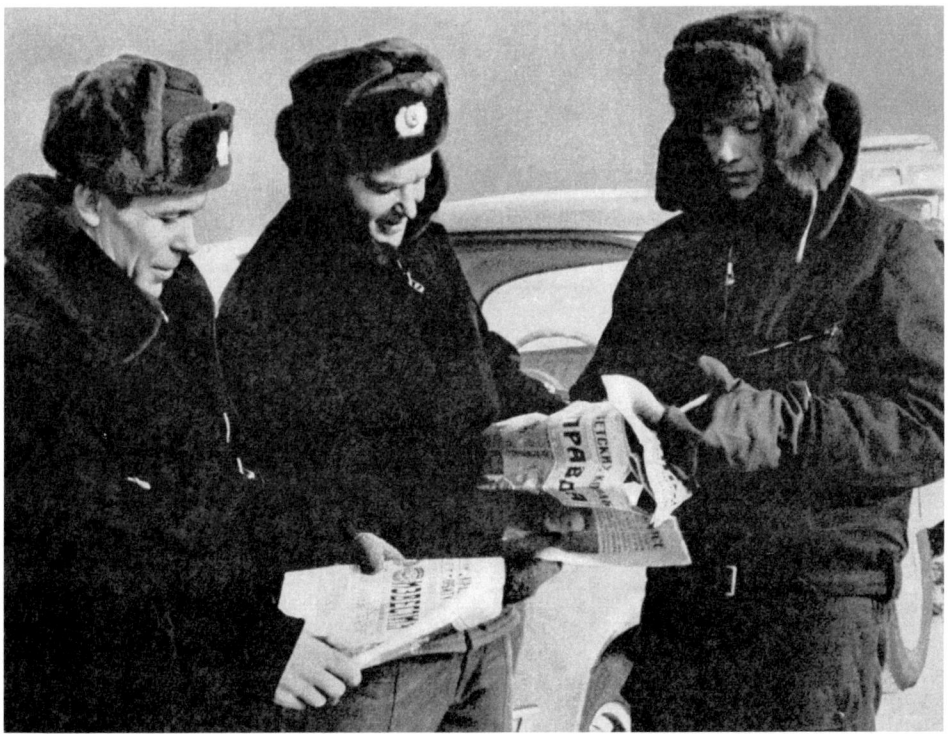

Khrunov (left), Shatalov and Yeliseyev catch up on the news after landing in the Soyuz 4 spacecraft.

Shatalov was immediately transferred to Soyuz, and from January 1967 to January 1968 trained as commander of the third (backup) crew, with Volkov and Kolodin, for the role of a passive spacecraft on a docking mission. In the meantime, he served as a communicator for the Soyuz 1 flight that claimed Vladimir Komarov's life. Having trained for the passive role in a docking, Shatalov was reassigned to train as commander of the third (backup) crew of the active spacecraft, and he completed this training in July 1968.

As the programme prepared to resume manned flights, from August to October Shatalov prepared for a mission that called for docking with the unmanned Soyuz 2. "Volynov, Beregovoy and I were training for this flight," he recalls. "In the final exam, I had the highest score at '5', Volynov had '4' and Beregovoy's score was very low. They decided to set a second examination. Volynov and I both scored '5', but Beregovoy's score was low again. After we flew to Baykonur, Beregovoy went through additional training and sat another examination; he managed to scrape a '4', and was nominated to fly Soyuz 3. Behind this decision was Beregovoy's authority as a test pilot and his long acquaintance with Kamanin. Despite his poor scores, it was thought that it would be good for morale to nominate a famous test pilot who had flown combat missions against the Nazis. Other candidates – including in the early stage Feoktistov – were excluded. Beregovoy tested the systems of the modified spacecraft, but unfortunately he failed to achieve a docking."

Shatalov continued to train to fly the active vehicle in the rendezvous and docking of two manned spacecraft, and on 14 January 1969 was launched as commander of Soyuz 4. On 15 January Soyuz 5 lifted off with Volynov, Yeliseyev and Khrunov, and the following day Shatalov made the first docking between two manned space vehicles, and then Yeliseyev and Khrunov made an external transfer and returned to Earth with Shatalov on 17 January.

The testing of rendezvous and docking, spacewalking and the external transfer of cosmonauts from one vehicle to another was an important contribution of the Soviet lunar landing programme, as that plan called for the cosmonaut who would land on the Moon to spacewalk between the command ship and lander in lunar orbit both prior to and following the landing. As an experienced instructor test pilot and one of only a few cosmonauts able to fly a helicopter, Shatalov spent a short time after the Soyuz 4/5 mission on the N1-L3 programme. He went to the test pilot school for the Mi-8 helicopter, to use this to simulate a lunar landing. Due to the limitations of the N1 rocket, the lunar module would have very little fuel for manoeuvring to select a landing point. In fact, the cosmonaut would not see the ground until the final 30–40 seconds of the descent. Immediately after the spent propulsive stage was jettisoned, he would have to rotate the lander for a view of the surface and set down as soon as possible. The helicopter was used to rehearse this final phase of the descent. Special covers were installed on the cockpit window, and after the instructor had climbed to an altitude of 70 metres he would open the covers to give Shatalov a view of the ground, and then Shatalov took over and attempted to land in a relatively small area in the available time. Shatalov was not actually assigned to an L3 crew; his role was to assess this training.

On his next space mission in October 1969, Shatalov was not only commander of Soyuz 8 but also in charge of the three spacecraft participating in the 'group flight'. This mission came as a surprise, as in August he and Yeliseyev had entered training as the single backup crew for the Soyuz 6 and 8 two-man crews and, with Kolodin, as the backup crew for the Soyuz 7 three-man crew. The prime crew for Soyuz 8 was Nikolayev and Sevastyanov but they failed several simulations, and therefore on 18 September, following the final examinations, the flight was reassigned to the backup crew. The three spacecraft were launched on successive days, Soyuz 8 on 13 October. Its primary objective was to rendezvous and dock with Soyuz 7. During these operations, the cosmonauts on Soyuz 6 were to fly close by and take pictures of the other two vehicles. Unfortunately, the Igla system on Soyuz 8 failed. Shatalov attempted to continue the approach manually, but without success. On landing on 18 October Shatalov admitted that he was happy at his safe return, but wished his flight could have been longer. Although the main task had not been accomplished, the Kremlin celebrated all seven cosmonauts as heroes.

In 1969 Shatalov's former patron, General Kutakhov, had become Commander in Chief of the Soviet Air Force. He wished to assign Shatalov to succeed Kamanin, and Shatalov duly received the nomination from the Central Committee of the Communist Party to become General Director of Cosmonaut Training in the Soviet

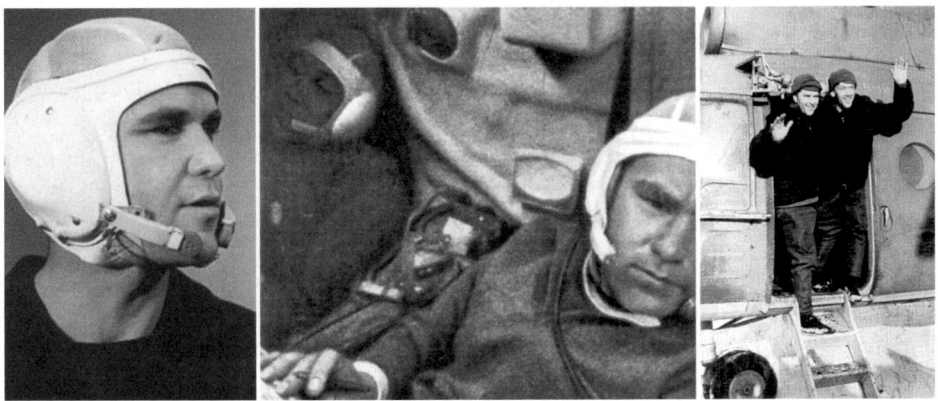

Nine months after his first flight, Shatalov (left) was named to command Soyuz 8. The main objective of docking with Soyuz 7 was not achieved, and after 5 days in space with Yeliseyev (centre), they returned to the Earth (right).

Air Force High Command. However, by late 1969 Zvyozdniy was rife with rumours of the DOS space station programme, and as the only cosmonaut to have achieved a docking in orbit Shatalov's thoughts turned towards a third flight, this time to a space station. He wrote to Kutakhov: "What are one or two space flights for a cosmonaut? I think a cosmonaut should fly at least ten times in space. This is already a profession. I believe I would be more useful as a cosmonaut. Things are still in the early stages, so let me fly. Above all, now that the space station is developing, permit me to work on it." Kutakhov agreed, but asked Shatalov to promise that after completing his third flight he would replace Kamanin. To Shatalov's delight, in 1970 he was nominated to command the crew to fly the first space station mission. He went to Baykonur for the launch of Soyuz 9 in June 1970, and when Nikolayev and Sevastyanov were presented to the journalists the fact that Shatalov was seated beside Kamanin was an intimation of Kamanin's imminent retirement.

When during the first reshuffle Shatalov was reassigned from the first to the third crew, he accepted that he would not have the distinction of visiting the world's first space station, but the second. But then Shonin was dismissed and at the last minute Shatalov got his wish.

"INTERESTING THINGS ATTRACT ME"

Just as Shatalov stood out in the military cosmonauts, so did Yeliseyev among the TsKBEM's civilians. He was the first member of the group of cosmonaut-engineers selected in May 1966 to be assigned to a prime crew: after 3 months of training he was nominated as flight engineer of the crew which, if Soyuz 1 had not encountered difficulties, would have been launched on board Soyuz 2 to perform a joint mission involving a docking and an external transfer. When this task was finally achieved in January 1969, Yeliseyev became the third Soviet cosmonaut to make a spacewalk.[4] On being assigned to a backup crew for the 'group flight' later that year, Yeliseyev had only one month's notice that he would fly on Soyuz 8. The next year, he started to train for his third mission.

Aleksey was born on 13 July 1934 in the small town of Zhizdra – near Kaluga, which is famous for Konstantin Tsiolkovskiy. His father, Stanislav Kuraytis was of Estonian origin, and worked in the laboratory of a shoe factory. Before Aleksey was one year old, his father was accused of being an enemy of the Soviet Union and was sentenced to five years in prison; but stayed incarcerated for almost a quarter of his life.[5] After her husband's imprisonment Aleksey's mother, Valentina Ivanovna, remarried. She worked as a chemist for the Soviet Academy of Sciences, and later gained a PhD degree. When the Second World War began they were in Moscow, and young Aleksey joined his childhood friends in collecting scattered bomb parts from German aircraft. When his mother was transferred to a factory to Siberia, Aleksey was sent to stay with his grandparents in Kazakhstan, where he gained his first schooling. He moved back to Moscow in 1943 when his mother returned to give birth to her second son. In 1950 Aleksey adopted the surname of his mother – Yeliseyev. One year later he finished high school in Moscow. The next step was the Bauman Higher Technical School (MVTU), which was one of the two leading universities in the city for aeronautical engineering and also, as events would show, for educating the future designers of space vehicles. He particularly favoured physics, mathematics, engineering and, of course, chemistry, but was also keen on sports, twice winning the national fencing championship, and he enjoyed the poetry of Vladimir Mayakovskiy.

Following his graduation in 1957, Yeliseyev was assigned to one of the top-secret scientific research institutes to work on theoretical aspects of the aerodynamics of cruise missiles and rockets. The launch of Sputnik shocked him, because he had no idea that there was an institution in the country devoted to the development of a real rocket capable of placing a satellite into orbit. He wanted to join that effort. In the early days, Boris Raushenbakh's team was the leader in rocket dynamics. Some of its engineers were developing the guidance system for an automatic probe intended to travel on a trajectory behind the Moon in order to photograph the region that was never directly visible from Earth. When Yeliseyev heard of such work, he felt he had to join Raushenbakh immediately. However, as he was only in a training post, this was not possible. To acquire the necessary status, he applied to the Moscow Institute for Physics and Technology (MFTI), and in late 1959 was allowed to join Raushenbakh's team – which had been incorporated into OKB-1 and was working for Korolev. In the meantime, Yeliseyev had married Valentina Pavlovna Shpalikova, who worked as a mechanical and hydraulic engineer, and in March 1960 their single child was born: daughter Yelena.

---

[4] The first two Soviet spacewalkers were Aleksey Leonov, on the Voskhod 2 mission in March 1966, and Yevgeniy Khrunov, who made the transfer from Soyuz 5 to Soyuz 4 several minutes ahead of Yeliseyev.

[5] After the death of Stalin in 1953, Stanislav Kuraytis was rehabilitated and granted a PhD degree, but he died soon thereafter.

Aleksey Yeliseyev, the Soyuz 10 fight engineer.

At OKB-1, Yeliseyev was made aware for the first time of a super-secret project to develop a vehicle to enable a man to fly in space: project 3KA, which was later to become famous by the name of Vostok. He worked on the design of the systems to enable the pilot to control the spacecraft. When in April 1961 Yuriy Gagarin became the first man to orbit the Earth, Yeliseyev asked Konstantin Bushuyev, one of the most influential managers at OKB-1, to assist him in becoming a cosmonaut. Although surprised by the young engineer's request, Bushuyev agreed, and in late 1962, soon after his graduation from the MFTI, Yeliseyev went to the Central Air Force Scientific Research Hospital (TsVNIAG) – which the cosmonauts had named the 'Palace of the Lords'. Another OKB-1 engineer, Vitaliy Sevastyanov, was also present. In fact, both men had applied without the knowledge of their boss, Korolev. Yeliseyev suffered a minor difficulty when the pressure in the altitude chamber was reduced to that at an altitude of 5 km, but when the test was repeated he performed satisfactorily. Actually, this first and most complex phase of the medical testing was so rigorous that only 1 in 20 of the candidates passed. Those who did so had a good chance of being selected to become cosmonauts. Yeliseyev advanced to the second phase of medical testing. However, one day Korolev visited the hospital to discuss changes in the medical criteria for civilian cosmonaut candidates. He had believed from the beginning that the best people to fly in space were the engineers who had designed the spacecraft. He had a very difficult meeting with the Air Force military doctors. He explained that the civilian engineers were very busy, and therefore did not have as much time to spend on sporting activities as did the military pilots. He requested that the criteria for the civilians be relaxed. On the other hand, the Air Force doctors, well aware that General Kamanin had never liked the idea of sending civilians into space, insisted that only people in perfect physical conditions should fly. Finally, one of doctors mentioned a civilian engineer by the name of Yeliseyev whom they were in the process of testing and was just as fit as the best military men. Korolev ordered the doctors to remove Yeliseyev from further consideration, and on returning to Kaliningrad demanded an explanation of how one of his engineers could have applied to become a cosmonaut without official approval. Raushenbakh advised Yeliseyev that he should avoid contact with Korolev until the metaphorical dust had settled.

After his unsuccessful attempt to become a cosmonaut, Yeliseyev continued with his task of developing systems for controlling spacecraft. In addition, he took part in calculations of the fuel consumption for two Vostok spacecraft that were launched on a 'double flight' in June 1963, one of which carried Valentina Teryeshkova, who became the first woman to fly in space. During the entire mission, Yeliseyev was on duty at OKB-1, prepared to answer any technical questions from the control team at Baykonur. Undeterred by his dismissal from the medical screening for a cosmonaut, in 1963 he passed a special training course at the TsPK for the Air Force's second cosmonaut group. His first visit to Baykonur was in October 1964, to participate in the preparations to launch the three-man Voskhod. At the same time, his colleagues were planning the first spacewalk. As an expert in spacecraft dynamics, Yeliseyev was included in a group of engineers who studied how a cosmonaut could move in open space, and how the vehicle could be controlled while a cosmonaut was outside. Tests were performed by simulating weightlessness in a Tu-104 aircraft. Following Leonov's pioneering spacewalk in March 1965, Yeliseyev was assigned to work on the development of the control systems of the Soyuz spacecraft, which was not only to operate in Earth orbit but also, in a modified form, undertake the L1 circumlunar mission. On returning at high speed from a lunar trajectory, the L1 was to perform a manoeuvre involving penetrating the atmosphere twice, the first time at about 11.2 km/s, corresponding to its original 'escape' speed, and then, after some braking, it was to re-emerge from the atmosphere on a ballistic arc that would lead to a second entry at a speed more comparable with that of returning from a low orbit. Yeliseyev investigated the deceleration loading imposed by such a manoeuvre. This led to one of the most unusual experiments of his career: testing the ability of a cosmonaut to manually control a vehicle under high deceleration loads. Some of the subjects of these tests were able to survive 26 g for 70 seconds!

As Yeliseyev recalls of his own participation in these trials: "You experience the increase in load as an increase in weight. You feel as if your body was poured with steel. A force is pressing you into your seat stronger and stronger. At 4 g, you can't move your legs, just your arms. At 8 g, your vision goes and you can't see beyond your nose. Then as the loads increase, you can see only a strange granular pattern. At 12 to 14 g, you can't see anything. Breathing is almost impossible. The forces that strain your muscles to avoid blood spouting out reach their maximum levels. You can't do anything; you just hope that you have enough strength left to endure it. The maximum load that I was exposed to was 18 g. Each time that I passed 14 g, I thought it was the end. It is interesting that my sense of hearing remained unaltered. I could hear even during the highest gravity forces. From this I concluded that we should use sound signals to control our lunar spacecraft during re-entry!"

In July 1965 a group of OKB-1's young engineers passed the preliminary medical screening for consideration as potential cosmonauts. Having concluded that the Air Force was inappropriately assessing civilian engineers, Korolev had decided not to send OKB-1's candidates to the 'Palace of the Lords' for the medical examinations, and instead employed the Ministry of Health's Institute of Biomedical Problems. In general the screening was similar to that of the Air Force, but it was shorter and included a psychiatric interview. Yeliseyev was very intrigued by this interview: "I remember that the majority of my colleagues were irritated by the psychiatrist. The interview lasted 3 to 4 hours. He asked us about our grandparents: did we remember our grandmothers and grandfathers and the most common topic of their arguments? He wrote in his notes much more than we actually spoke! Now a question about our other relatives. And again he scribbled. He did not permit smoking although he was himself smoking. We were all very careful during this interview. If we didn't fully concentrate he might infer from our replies that we had a psychological anomaly – and then it would be impossible to disprove his conclusion." But Yeliseyev satisfied the psychiatrist, and on 23 May 1966 was accepted as one of the eight members of the TsKBEM's first group of cosmonaut-engineers.[6]

From May to August 1966 the group undertook basic training at the TsKBEM, as well as parachute jumps, tests in an isolation chamber and simulated weightlessness in a Tu-104 aircraft. In late August, they trained for the possibility of landing on water. This latter training was performed on the Black Sea, and also involved a group of military cosmonauts from the TsPK.

When the cosmonaut-engineers were sent to the TsPK to start training for Soyuz missions, Kamanin intervened, saying that he would not permit them to train there unless they satisfied the standard medical screening! One week later, the 'Palace of the Lords' accepted only Kubasov, Volkov, Yeliseyev and Grechko. On 6 September they started to train as candidates to serve as flight engineer on a mission involving an external transfer between two docked spacecraft. In October, Grechko broke his leg in parachute training and was replaced by Oleg Makarov.

In selecting the crews for the spectacular joint mission that was to introduce the

[6] The eight TsKBEM engineers selected for the first group of civilian cosmonauts were Sergey Anyokhin (commander), Vladimir Bugrov, Vladislav Volkov, Georgiy Grechko, Gennadiy Dolgopolov, Valeriy Kubasov, Aleksey Yeliseyev and Oleg Makarov.

Soyuz spacecraft, it was natural, given his experience in planning Leonov's historic spacewalk, that Yeliseyev should be assigned as one of two cosmonauts who were to undertake the external transfer from Soyuz 2 to Soyuz 1. After his appointment, Yeliseyev received an unusual request: the psychiatrist wished to speak to his wife and mother! The reason, he was told, was that the psychiatrist wished to know how Yeliseyev would react if a serious problem were to occur during the spacewalk. For example, would he become hysterical and respond only to the voice of a woman he regarded as being special to him. Yeliseyev was against the idea. His mother had no idea that he was a cosmonaut. And he did not wish to involve Valentina Pavlovna because their marriage was coming to an end.[7]

Colonel Komarov was to command Soyuz 1. Having flown on the 1964 Voskhod mission, this would make him the first cosmonaut to have a second flight in space. Soyuz 2 would set off 24 hours later with Valeriy Bykovskiy, Yevgeniy Khrunov and Yeliseyev. Komarov would rendezvous and dock. Khrunov and Yeliseyev were to don spacesuits and transfer to Soyuz 1. After undocking, Soyuz 1 would return to Earth at the end of its third day in orbit, with Bykovskiy doing so 24 hours later. It was very ambitious inaugural mission for the new spacecraft. The Soviet Union had not sent anyone into space since March 1965. Meanwhile, the ten American Gemini missions had made the first orbital rendezvous (involving two manned vehicles) and also the first docking (using an unmanned target), had conducted several one-man spacewalks, and had set the endurance record at 14 days, all of which put them in a good position to prepare for a lunar landing mission. However, if all went well the Soyuz 1/2 mission would enable the Soviet Union to catch up in a highly dramatic manner.

Unfortunately, Soyuz 1 ran into difficulties immediately upon attaining orbit. The State Commission ordered the launch of Soyuz 2 cancelled, and everyone focused their efforts on enabling Komarov to return to Earth at the first available opportunity. Although the spacecraft circled the Earth every 90 minutes or so, the rotation of the planet meant that the best opportunities to make a landing in the recovery zone were either after one revolution or after 24 hours. The state of the spacecraft precluded an immediate return. Komarov gained sufficient control of the vehicle to attempt to return the next day. However, just as everyone thought that the worst was over, the parachute system failed, the descent module smashed into the ground at great speed, and he became the first man to lose his life on a space flight. In fact, the outcome could have been worse – when the parachute system of Soyuz 2 was tested, it was found to have the same flaw. If Soyuz 1 had operated correctly in orbit and Soyuz 2 had been launched, then all four cosmonauts would have been lost!

While the design of the Soyuz spacecraft was being modified to rectify the faults, training continued. In December 1967 Yeliseyev was granted an M.Sc. (Technical). In February 1968 Bykovskiy left the docking group to train for the L1 circumlunar mission, and was succeeded by Volynov. Although Yeliseyev was not a member of

---

[7] Later in 1966, Yeliseyev divorced his first wife and married Larisa Ivanovna Komarova, who was an engineer at the TsKBEM.

In April 1967 the first manned Soyuz mission ended in the death of Vladimir Komarov.

Komarov's ashes are interred in the wall of the Kremlin.

any lunar crew, he participated in training involving aircraft flying over Somalia at night, the objective being to enable the cosmonauts to familiarise themselves with the constellations of the southern sky that were to be used in navigating the first of the two atmospheric entries of a returning circumlunar mission. In late May 1968 it was decided to revise the ambitious 'inaugural mission' plan by launching the first of the two Soyuz spacecraft unmanned. Only if this functioned satisfactorily would the second spacecraft be launched with a single cosmonaut to make the rendezvous and docking. Unfortunately, although Beregovoy in Soyuz 3 rendezvoused with the unmanned Soyuz 2, he was unable to dock. However, because both spacecraft had performed satisfactorily, it was decided to proceed with the 'full' mission involving an external transfer.

When Yeliseyev was launched into space in January 1969 on Soyuz 5, more than 27 months had elapsed since he received the flight engineer assignment. He recalls of his 37-minute spacewalk: "Travel along the external surface of the ships proved to be the easiest and most pleasant part of the transfer. It barely required any effort. The 'landscape' gave the sensation of limitless space and freedom. It was similar to the experience prior to a jump from an aircraft, but in this case there was no wind – and there was no concern about the operation of the parachute! I paused in order to memorise what I could see. Below was the ship that we had left. To the left, shaped like the top of a bell, was our descent module. Volynov was inside, alone. Beyond was a module housing the instruments, the engines and the solar panels. To the right was the second ship, of the same type as ours. I could see Khrunov, his torso safely inside the orbital module of Soyuz 4, holding onto my cable. Far, far away, was the Earth's horizon, whitish-blue, passing very slowly. Above the horizon everything was black. Probably the light reflecting from the ships and the filter of my space helmet prevented me seeing the stars and planets. I slowly unclenched my fingers to try to float without holding onto the handrail. By having a safety line, releasing the handrail posed no danger. I walked further – I say 'walked', although I do not know how to name this method of movement. Do I crawl? Swim? Fly? My feet played no part. It was simply hand over hand on the rail – as if passing this to someone. But 'spacewalking' has become the accepted term. When I approached the hatch of the orbital module, Khrunov had entered fully and was against the far wall so as not to interfere with my entry. I freely swam through the hatch, we gathered the cable and informed Shatalov that we were going to close the hatch. The transfer was finished. It remained only to fill the orbital module with air, to verify its hermetic seal, and to remove our spacesuits." Shatalov, Yeliseyev and Khrunov landed the next day, and Volynov the morning after.

After a well-earned vacation in Central Asia, Yeliseyev returned to the TsKBEM, where the final preparations for the next big event were well in hand. The plan was to launch three manned vehicles at daily intervals in October 1969. Georgiy Shonin and Valeriy Kubasov would go up first in Soyuz 6. Anatoliy Filipchenko, Vladislav Volkov and Viktor Gorbatko would follow in Soyuz 7. Then Andriyan Nikolayev and Vitaliy Sevastyanov were to dock Soyuz 8 with Soyuz 7. Yeliseyev was put on a common backup crew. One morning Mishin called Yeliseyev into his office and asked him to fly on Soyuz 8. This took Yeliseyev by surprise: "I knew that the crew

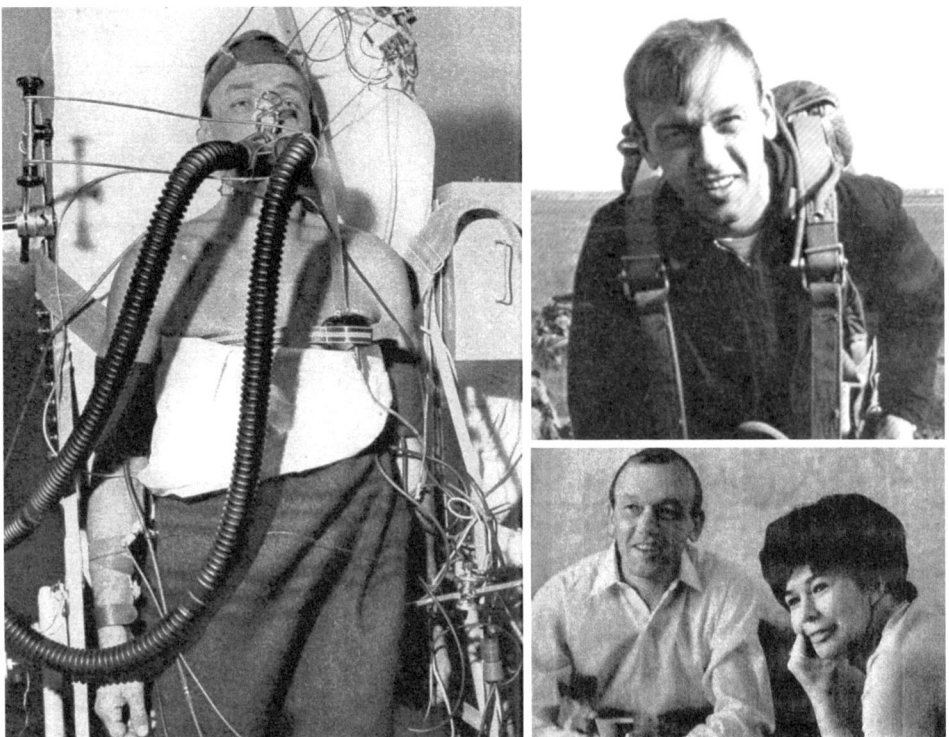

Yeliseyev in medical testing (left), parachute training and with his second wife Larisa.

In late 1966 Yeliseyev began to train for the mission in which two Soyuz spacecraft were to dock and cosmonauts were to make an external transfer from one vehicle to the other. He trained with numerous cosmonauts, including Gagarin (left), Komarov (middle) and Shatalov (right).

The crews of the successful Soyuz 4/5 joint mission in January 1969: Yeliseyev (left),
Volynov, Khrunov and Shatalov.

An artist's depiction of the historic external transfer from Soyuz 5 to Soyuz 4, showing Khrunov (right, in the hatch of Soyuz 4) watching Yeliseyev begin his spacewalk. Note that artist Andrey Sokolov showed the cosmonauts wearing their life support systems as backpacks, whereas in fact they attached the packs to their legs.

A member of the recovery team assists Yeliseyev (waving) from the hatch of the Soyuz 4 descent module, as Shatalov stands on the left. Another member of the recovery team stands alongside the capsule with the warm clothes for Yeliseyev. Khrunov is not visible.

of that ship was ill prepared, but I did not expect such a turn of events. The missions were less than two months off. The three crews had already had their examinations. I looked at Mishin interrogatively, and waited for an explanation."

"I do not want to let that crew fly!" said Mishin. "They work thoroughly badly."

"But, I have not prepared," Yeliseyev pointed out.

"The programme is almost the same as yours from the previous flight, without the transfer. You will succeed."

When Yeliseyev agreed, Mishin telephoned Kamanin: "Nikolay Petrovich, I can't permit the Soyuz 8 crew to fly; they work badly. For my part, I nominate Yeliseyev. Select someone from your stronger cosmonauts."

Kamanin was surprised at Mishin's late intervention, but after a brief objection he agreed that Nikolayev should not fly. The next day Mishin and Kamanin met at the TsPK and, after an unpleasant conversation, agreed to send Shatalov and Yeliseyev. One by one, the three spacecraft were placed into the required orbits, but the Igla rendezvous system on Soyuz 8 malfunctioned and there was no equipment available to control the operation manually. At mission control, ballistic experts improvised a plan to enable Soyuz 8 to perform manoeuvres which would bring it within several hundred

Shatalov and Yeliseyev on the launch tower prior to entering Soyuz 8 (top), and alongside the descent module just after the landing.

metres of its target and then, when the crews could see each other, perhaps they would be able to dock manually. But everyone was aware that the likelihood of success was almost zero. The controllers supplied Soyuz 8 with the manoeuvre data, and where they should look for Soyuz 7. Yeliseyev was to observe through the portholes set at 90-degree intervals around the orbital module, to locate their target visually. Shatalov was to remain at the controls of the descent module and turn his spacecraft as instructed by Yeliseyev until Soyuz 7 entered the field of view of his forward-looking optical periscope. Yeliseyev saw a bright dot travelling against the clouds beneath, but the range was impossible to judge. He recalls: "We wanted so much to dock, and we tried everything that we could. We had to hold visual contact with Soyuz 7 for the entire period of the approach, while attempting to match our speeds. Shatalov attempted to use the orientation engines on a continuous basis, but these small engines were incapable of cancelling the speed difference, and Soyuz 7 flew by and disappeared." There was insufficient fuel to set up another rendezvous attempt. Soyuz 8 landed after 5 days in space, and although the primary objective of docking had not been achieved, the 'group flight' was officially another success in Soviet cosmonautics.

Although everyone was disappointed, there was no time to dwell on this failure, and within 7 months Yeliseyev was back in training in the expectation that his third flight would be a truly historic one.

BETWEEN SPACE AND BIKES

"His ambition is to convert a refrigerator into a vacuum cleaner," joked Shatalov of Nikolay Nikolayevich Rukavishnikov, the Soyuz 10 research engineer who was an expert in electronics and the physics of cosmic rays. Short and skinny, and quiet but with a serious face, Rukavishnikov was a natural technician who loved to repair old apparatus and to devise new things, even once attempting to improve the design of a helicopter.

Nikolay was born on 18 September 1932 in the town of Tomsk in western Siberia, to a family which, before the Soviet era, owned a brickyard and a bike company. He was raised without his father Nikolay, of whom nothing is unknown. His mother Galina and stepfather Mikhail Mikheyev were railway designers and travelled widely. His early interests at school were geography, mathematics and physics, and after his stepfather introduced him to radio equipment he became a radio-amateur. Because he was always on the move, he grew used to changing houses, schools and friends. He entered high school in 1947 in the small town of Angrem in Uzbekistan, where he gained first grade. Then in 1950, in one year, he passed three grades in the town of Kehtaice near the Mongolian border, where his parents were working on a new railway. Upon finishing his schooling at high school No. 248 in Moscow in 1951, he immediately went to the Moscow Institute of Engineering and Physics (MIFI). In May 1957 he graduated with a physics diploma from the faculty for electronic calculators, having specialised in dielectrics and semiconductors. In July he went to work at the Central Scientific Research Institute TsNII-58 in Podlipkah, a village near Moscow which hosted several top-secret research institutes and organisations, including OKB-1, and which later became Kaliningrad. His early work was on the development of one of the first Soviet computers, named 'Ural', and he participated in the testing of automatic control and protection systems for nuclear reactors.

In September 1959 Rukavishnikov transferred to OKB-1, where he worked as an engineer in the department which made automatic controls for interplanetary probes, and between October 1960 and January 1967 he worked on systems for a variety of spacecraft. One task was to develop apparatus to automatically process information which the crew of the L1 circumlunar spacecraft would require if they were to take manual control. In addition, he was involved in testing guidance systems. Later, he led a team which developed experiments in terrestrial studies and solar physics for satellites. Meanwhile, he had married Nina Vasilevna, a mechanic at OKB-1, and in 1965 she gave a birth to their only child: son Vladimir.

Rukavishnikov's first move towards becoming a cosmonaut was when he passed the medical screening in May-June 1964 as one of 14 candidates that Korolev was considering for a Voskhod flight. Konstantin Feoktistov was also a member of this group, and it was he who was launched 4 months later. In May 1966 the TsKBEM selected its first group of cosmonaut-engineers, but after four failed the Air Force's medical screening it was decided to add to the diminished group, and in November 1966 Sevastyanov and Rukavishnikov were selected for medical tests. They joined the group in January and February 1967, respectively. As Rukavishnikov recalls of this time: "Of course, I had to catch up on all the training that other cosmonauts had already passed. This included thousands of hours of intensive training, centrifuge, altitude chamber, simulated weightlessness flights and parachute training." His first parachute jump was scary, because his hood covered his eyes and he was unable to see where he was going to land. Later, however, he was able to joke about it with his colleagues.

Yeliseyev later said that he had not expected Rukavishnikov's selection: "To be honest, when I saw him for the first time I expected that the doctors would dismiss him early on in the medical screening. But I was wrong. It appeared Nikolay was in excellent health." Rukavishnikov was acknowledged to be devoted to his work. As Yeliseyev told a journalist, the new cosmonaut would stay at OKB-1 day and night until his task was done. Rukavishnikov was notable among the civilian cosmonauts for his unusual passions. The first one was bicycling. In the 1950s he had fallen in love with cycles and motorbikes and would ride at any opportunity, day or night. His second passion was travelling. On summer vacations he would leave the group, and disappear into the hills and mountains to explore nature in solitude. And finally, he was an expert in servicing television apparatus, and even made a set for himself!

Rukavishnikov was assigned to the L1 project, in which a two-man variant of the Soyuz would fly on a circumlunar trajectory. The unimaginative name selected for this project was Zond ('Probe'). The commanders were to be Air Force cosmonauts. When three crews were formed, Bykovskiy and Rukavishnikov were chosen for the second. Unmanned missions were flown to test the spacecraft's systems and perfect the two-stage penetration of the atmosphere, but the success of America's Apollo 8 in December 1968, which orbited the Moon ten times, greatly reduced the value of the simpler circumlunar loop and the L1 project was cancelled. The L1 crews also trained for the N1-L3 lunar landing, but with the development of the N1 launcher suffering problems, after the Americans landed on the Moon in July 1969 most of the lunar group were reassigned. In March 1970, Rukavishnikov joined the Contact project as flight engineer for a Soyuz mission commanded by Lev Vorobyev. This

Nikolay Rukavishnikov, the Soyuz 10 research engineer. His official portrait (top left), during theoretical lessons (top right), and celebrating his nomination to the 'first crew' for the DOS-1 station – in the company of fellow DOS cosmonauts Volkov (left), Leonov (obscured by Volkov), Kubasov, Shatalov, Kolodin and Dobrovolskiy.

was to test in Earth orbit the rendezvous and docking techniques for the N1-L3, in order that these would be available if it eventually proved possible to mount a lunar mission. However, two months later he was assigned to the first DOS-1 crew as cosmonaut-researcher.

On the eve of the Soyuz 10 launch, Soviet cosmonaut number 23 had an excellent reason to be happy, because if everything went according to plan then he, the rookie on the crew, would be the first man to pass through the hatch and enter the world's first space station!

Specific references

1. Shatalov, V., *The Hard Roads to Space*. Molodaya Gvardiya, Moscow, 1978, pp. 139–177 (in Russian).
2. Yeliseyev, A.S., *Life – A Drop in the Sea*. ID Aviatsiya and kosmonavtika, Moscow, 1998, pp. 8–34 (in Russian).
3. Lebedyev, L., Lukyanov, B., and Romanov, A., *Sons of the Blue Planet 1961–1981*. Politizdat, Moscow, 1981, pp. 178–188 (in Russian).

4

The drama of the Granites

INTO SPACE

After the press conference at mid-day on 20 April 1971, the Soyuz 10 cosmonauts and their backups went to Pad No. 1 to inspect the 50-metre-tall rocket, enclosed by its service structure. Also present were hundreds of engineers, technicians and the military who managed Baykonur launch operations. The tradition of this gathering had been established a decade earlier, when Gagarin had prepared to ride a similar rocket from the same pad to become the first man to orbit the Earth.

All nine cosmonauts stood in line: the prime crew at one end, then the first and second backups. Behind them were senior people from the TsKBEM, the Air Force, the Strategic Rocket Forces, the Ministry for General Machine Building and the Academy of Sciences. After being given flowers, one by one the cosmonauts were introduced to the launch team. As the spacecraft commander, Shatalov gave a brief speech to thank the launch staff and all of the institutions involved in preparing the mission.

The launch was scheduled for 5.20 a.m. local time (3.20 a.m. Moscow Time) on 22 April. Despite the overnight heavy rain, it was decided to proceed as planned. On arriving at the pad, the cosmonauts rode the elevator up the service structure, entered their craft and strapped into their couches. One by one, the service masts were swung away from the vehicle and people left the pad. The final preparations were conducted from the nearby command bunker, with the cosmonauts participating by radio. But with only a minute remaining before the rocket engines were due to ignite, the umbilical that had supplied electrical power failed to retract from the third stage, and Mishin, who was the technical director for the launch, halted the operation.

As Shatalov recalls: "We were awaiting the command: 'The key is on the Start switch'. But instead from the command bunker we heard: 'Prepare for evacuation! The launch is delayed for a day!' This was nothing new for me. I'd heard the same command in preparing to launch on Soyuz 4. Then I was so disappointed, but this time I accepted it readily. I looked at Rukavishnikov – it was to be his first launch –

and saw how much he was suffering. Probably he was thinking everything was over, so I encouraged him: 'Cheer up, everything will be all right! Tomorrow you will be launched for sure!' Rukavishnikov did not respond. Aleksey joked with me: 'There are always problems with *you*. Everyone else goes on the first attempt, except you! It is clearly the number thirteen!' " In the 10-year history of the Soviet programme, the only other time that a launch had been abandoned after the crew had entered the spacecraft was the first attempt to launch Shatalov in January 1969 – evidently he was jinxed by virtue of being the 13th cosmonaut.

The bunker ordered the cosmonauts to remain seated and await the arrival of the evacuation team. The cosmonauts understood the reason. If there were to be a false signal to the launch escape system, this emergency rocket would instantly draw the orbital and descent modules up away from the remainder of the vehicle, and if this were to happen the three men would need to be safely in their couches. Fortunately, there was no false command. As soon as the vehicle was 'safe' and the service structure reinstated, the evacuation team opened the hatch and helped the three men out. For the two hours that they had spent in the cabin they had been at a pleasant $+28°C$, but outside it was still raining, there was a fierce wind, and the temperature was freezing, so they were given warm clothes for the bus back to the Cosmonaut Hotel.

An inspection by the technicians established that the umbilical tower to the third stage had failed to disengage because rain had accumulated in the connector and frozen it into place. The State Commission decided to retain the rocket loaded with propellant, and to reschedule the launch for the following day. The next night the temperature dropped to $-25°C$, and when the crew returned to the pad just after midnight they wore black leather coats over their lightweight cotton flight suits for protection against the weather. After making a brief report to General Kerimov, the cosmonauts once again entered their spacecraft.

The umbilical again refused to retract, but Mishin, knowing the reason, allowed the operation to proceed, and Soyuz 10 successfully lifted off at 2.54 a.m. Moscow Time on 23 April. As it did so, Yeliseyev called poetically: "The sky is cloudless, clear and starry, and the dawn is breaking. We're ready to go up."

Soyuz 10 entered a slightly higher orbit than planned, its altitude ranging between 210 and 248 km, its plane inclined at 51.6 degrees to the equator and with a period of 89 minutes. At orbital insertion, it was 3,456 km ahead of the Salyut station. The first three revolutions of the Earth were without problems. The plan was to perform an automatic orbital manoeuvre on revolution 4, but this was not possible owing to an error in the programming logic for the command – evidently a problem involving the gyroscopic system. The mission controllers on Earth scheduled the manoeuvre for the next revolution, but the parameters could not be specified until the rate at which the initial orbit was decaying had been determined, which could not be done until the spacecraft was once again in range of the Soviet tracking radars. Once the necessary computations had been performed, the data was read up to the spacecraft, but this left insufficient time for the crew to key in the data and the opportunity for the action was missed. In addition, it seems that the ionic sensors that formed part of the spacecraft's orientation system malfunctioned as a result of contamination of

A rail transporter delivers Soyuz 10 to the pad.

The rocket erected on the pad, but with the split service structure yet to be raised.

The Soyuz 10 crew give a press conference at Baykonur.

The three crews assigned to DOS-1 are introduced to the launch team at the pad: Shatalov, Rukavishnikov, Yeliseyev, Leonov, Kubasov, Kolodin, Dobrovolskiy, Volkov and Patsayev (far right).

Shatalov (centre) reports his crew's readiness for the Soyuz 10 mission prior to going to the pad for launch.

their surfaces – a common problem for spacecraft during their first hours in space, but easily rectified simply by allowing the harsh sunlight to vaporise the thin film of contaminant. Finally, Shatalov made the manoeuvre manually at 1.34 p.m., with an impulse lasting 17 seconds.

With the rendezvous initiated, the cosmonauts were free to open the internal hatch and enter the orbital module. Yeliseyev recalls: "Together with Volodya, I floated into the more spacious orbital module. We advised Nikolay to remain in his seat in the descent module for a certain time, and to move his head as little as possible. He was in space for the first time, and our desire was to help him to avoid a vestibular disturbance. Nikolay felt completely normal, but he didn't immediately master the visual situation. I remember that when I wanted an item from the descent module, I swam in through the hatch head forward, my legs in the orbital module, and I asked Nikolay to give me the item. On hearing my voice he immediately turned his head towards me, and I saw consternation on his face. Then he said: 'To the hell! Could you at least arise in a human manner!' And we both laughed."

In accordance with established tradition, Soviet television did not show the launch until nearly 8 hours afterwards. The 45-minute black-and-white broadcast included a recorded interview with Shatalov, who said the mission would mark a new stage in the exploration of space and contribute to the establishment of space stations and long-duration flights. It then provided a view of Shatalov in the spacecraft, wearing his dark flight suit and a white communications helmet. The radio call-sign for this mission was 'Granit' ('Granite'), and each man had his own numeral. The people heard mission control talking to Rukavishnikov: "Granite 3, please check your radio apparatus before attempting to speak to us again, as you are coming through garbled. Please, Granite 3, don't speak so fast!"

Minutes before a Soyuz launch.

The official Soviet news agency TASS reported that the cosmonauts had started to conduct a programme of "joint experiments" with the orbital station launched four days previously. But there was no indication of the objectives of the mission, or its intended duration. No information had been released about the size of the station, or its apparatus. As far as anyone was aware, everything was going to plan. However, not only had the cover of the station's scientific module failed to release, there were now other problems. On the second day there were indications that two fans inside the main compartment had failed, and since then others had failed. In fact, by now only two of the eight fans were available to ventilate the air in the station. Despite the absence of official information, knowledge of the capability of the Proton rocket enabled the station's mass to be estimated at 15 tonnes, and the *Daily Mail* included an artist's impression based on information from 'Iron Curtain' sources that showed a two-storey cylinder some 3.5 metres in diameter and 7 metres tall with a volume ten times larger than the cramped Soyuz spacecraft – which was correct.

The Soviets did not actually announce that Soyuz 10 would dock; observers in the West knew that officials would not disclose an intention until it had been achieved. As a result of recent newspaper articles by Academicians Keldysh and Petrov, there

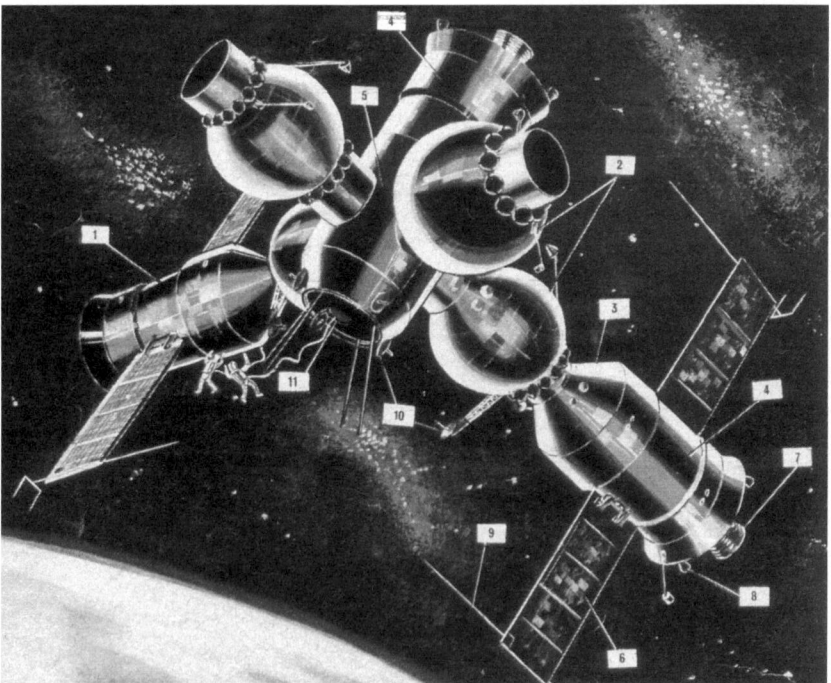

In the absence of official information, Western analysts speculated that Salyut was a 'hub' on which departing Soyuz spacecraft would leave their orbital modules loaded with specialised apparatus.

was a belief that the Soviet Union was following a bold plan, with Salyut as just the first step. In any case, the fact that Soyuz 10 was commanded by Shatalov was a strong hint that a docking was planned. Interestingly, some people thought that the Soyuz 4/5 mission had been a rehearsal, and Yeliseyev and Rukavishnikov would make an external transfer to the station! Some Western newspapers suggested even more implausible theories, including that there was a large centrifuge on the station to simulate Earth's gravity. And there was speculation that Salyut was a 'hub' with four docking ports, and each Soyuz to visit would leave its orbital module in place in order to expand the station's facilities.

FLIGHT CONTROL

The development of the Soviet space tracking network began in the early days of rocketry to facilitate the tracking of intercontinental ballistic missiles in test flights from Baykonur. The system was then expanded and increased in scope to deal with orbital flights. The relatively brief Vostok and Voskhod missions were managed at Baykonur by Sergey Korolev, as the technical director for space missions, with the support of the so-called Operation Group of the Strategic Rocket Forces. The first Flight Control Centre (TsUP) was at Scientific Research Institute No. 4 (NII-4) in Bolshevo, near Moscow. For the Voskhod missions it was relocated to the control centre of the Ministry of Defence's General Staff, which had better communications. Colonel Amos Bolshoy headed the Operation Group of the TsUP in Moscow for all manned space missions until 1966, providing continuous contact with seven ground stations known as Ground-Test Polygons (NIP) which formed a chain that stretched across the Soviet Union. They were at Bear's Lake near Moscow, Kolpashevo, Yeniseysk and Ulan Ude in Siberia, Sarishagan in the south, Petropavlovsk in the Far East and Ussuriysk on the Kamchatka peninsula. At each site, military and civilian engineers analysed the parameters of the spacecraft's orbit derived from radar tracking, and the conditions of its systems from telemetry received during communications sessions lasting at most 12 minutes. The Operation Group relayed the data to the TsUP and provided continuous contact with Korolev at Baykonur. The NIP sites were part of the Command-Measurement Complex (KIK) operated by the Strategic Rocket Forces.

Due to the complexity of the Soyuz programme and the ambitious plans for lunar missions, the flight control system underwent a major revision in the mid-1960s. The TsUP was moved to NIP-16 near Yevpatoriya on the west coast of the Crimea, which had been responsible for controlling automated interplanetary probes. Known as TsUP-E ('E' for Evpatoriya in Russian), it was much more capable than the old TsUP, and it controlled all Soviet manned space missions between 1966 and 1975 – when a new facility was build in Kaliningrad.[1] Some 500 people worked around the clock in three shifts. NIP-16 was the USSR's largest command-measurement site. It was in radio communication with the other sites, and could receive from or transmit

[1] After 1975 TsUP-E controlled only manned military missions to the Almaz stations.

The main room of the Flight Control Centre in Yevpatoriya.

to spacecraft. It had many very distinctive antennas, some of which were very small, similar to domestic television antennas, while others were extremely large. Some of its antennas looked as if they had been constructed in a hurry, others had a beautiful design even although in some cases their construction had taken only a few months – for example the enormous antenna complex that was built to communicate with the first probes dispatched to the planet Venus.

The TsUP-E was established in a small two-storey building. On the first floor was the communications centre, which had apparatus to register the telemetry from the spacecraft in the form of graphs on long rolls of paper. On the second floor was the control room housing the flight controllers, experts on all flight procedures and the civilian experts on the systems of the spacecraft. They jointly compiled a flight plan to be radioed to the crew specifying what must be done on each orbit. Alongside the control room were representatives of the TsPK, with one of the active cosmonauts serving as the communication operator who spoke to the crew in space, and also the military specialists for the technical segment of NIP-16 and, by radio, its sister sites.

The core of the mission management team was the Chief Operative and Control Group (GOGU). The military part of GOGU was responsible for the operation of all ground stations, including the necessary technical support. In 1966 Major-General Pavel Agadzhanov, a veteran of the tracking network, was appointed as head of the GOGU for Soyuz flights. His Deputy was Colonel Mikhail Pasternak. There was a separate GOGU for the L1 circumlunar missions, with Colonel Nikolay Fadeyev in charge of flight operations. The other members of the GOGU were technical people from the TsKBEM. From 1966 to 1968 the technical director for Soyuz missions was Boris Chertok. In this role he was responsible for all decisions relating to each space mission. Prior to this, he had been responsible for the control of interplanetary probes. In 1969 Yakov Tregub, who had commanded the cosmodrome at Kapustin Yar, took over this role. He was Deputy Chief Designer of Complex No. 7, which managed the testing of systems for spacecraft, the training of cosmonauts and flight control. Another member of the GOGU was Boris Raushenbakh, a department chief and expert in the control and guidance systems of

The antennas of the NIP-16 tracking and communication complex in Yevpatoriya. The insert shows personnel from the TsKBEM (Tregub, Bushuyev, Raushenbakh and Chertok), the TsPK (Kamanin, Nikolayev and Popovich) and the Strategic Rocket Forces (Agadzhanov).

spacecraft. His team planned the actions needed for rendezvous, docking and un-docking. For Soyuz 10, the key men were therefore Agadzhanov, Tregub, Raushenbakh and Chertok, with cosmonaut Pavel Popovich communicating with the crew.

In contrast to the American mission control facility in Houston, Texas, which had rows of controllers at consoles and large computers to process data in real time, the main control room at TsUP-E was remarkably unimpressive. On the front wall there was a large map of the world displaying the position the spacecraft in its orbit, and a large black-and-white screen on which television transmissions were shown. The members of the operative group sat around a long table and analysed data traced on rolls of paper. To the side were several controllers. After commanding the Apollo 8 mission in December 1968 Frank Borman made a goodwill tour of the world, and in the summer of 1969 he became the first American astronaut to visit the Soviet Union. On a visit to Yevpatoriya he was so surprised by the modest facilities of the TsUP-E that he presumed the real control centre was somewhere else, highly secret, and perhaps hidden underground!

For the early manned space flights, contact was possible only while the spacecraft was over Soviet territory. During 'silent orbits', when a spacecraft was crossing the oceans or over other continents, the crew would either rest or perform experiments that did not require communication with the TsUP. However, in order to achieve a landing in the prime recovery zone on Soviet territory it was necessary to perform a succession of critical operations leading up to re-entry while over the Atlantic Ocean. To provide communications with the spacecraft during these operations, and during the planned manned lunar missions, a number of Scientific Exploration Vessels (NIS) of the Soviet Academy of Sciences were included in the space tracking and control system. Although some ships had been equipped in the early days to receive transmissions from the unmanned Vostoks, four 'modern' tracking ships were laid down in 1967, starting in June with *Kegostrov*, which had a displacement of 6,100 tonnes. It was stationed off the coast of Africa in the Gulf of Guinea. *Morzhovets* and *Nevely*, which were smaller, operated in the South Atlantic. *Borovochi* operated elsewhere. In addition, three smaller ships were capable of receiving radio signals from spacecraft: *Bezhitsa*, *Dolinsk* and *Ristna*.

Later in 1967 the first of the second-generation ships was added. At 17,500 tonnes, *Cosmonaut Vladimir Komarov* was much larger, with a variety of antennas capable of providing all functions of a NIP ground station, including relaying transmissions between a spacecraft and Yevpatoriya – making it a 'universal' communications ship. For manned flights it was stationed in the North Atlantic, near Sable Island, off the coast of Nova Scotia. In January 1969 it was the first to congratulate the Soyuz 4/5 crews on accomplishing their external transfer. In October that year it participated in relaying a transmission from a manned spacecraft (Soyuz 8) through a Molniya satellite to enable, for the first time, the TsUP-E to communicate with a crew while not over Soviet territory.[2]

---

[2] The Molniya (Lightning) satellite was in a highly elliptical orbit with a 12-hour period and the highest point of its orbit over the Soviet Union.

The tracking ship *Academician Sergey Korolev* (top) and its control room (bottom left). In the TsUP-E, members of the GOGU, General Pavel Agadzhanov and Yakov Tregub (glasses) analyse telemetric data.

In December 1970 the network was augmented by *Academician Sergey Korolev*, which was even larger, having a displacement of 21,460 tonnes and a length of 182 metres. It had over 50 antennas, the largest of which was 12 metres in diameter. In March 1971 it relieved *Cosmonaut Vladimir Komarov* in the North Atlantic, which then concluded its seventh voyage by sailing to Odessa for refurbishment.[3]

Each ship had a TsPK cosmonaut-engineer to communicate with a spacecraft. For example, Yuriy Artyukhin was on board *Cosmonaut Vladimir Komarov* and Anatoliy Kuklin was on *Academician Sergey Korolev*. In addition, for the Soyuz 10 mission, there were experts from the TsKBEM familiar with the design of the DOS docking system to provide advice as necessary. A favourable pass lasted 10–12 minutes. As soon as the spacecraft rose above the ship's horizon, the controllers began to decode its transmissions. The decoded data was transmitted through a

[3] In December 1971 *Cosmonaut Yuriy Gagarin* joined the network. At 45,000 tonnes, it was made the flagship of the fleet. All ten tracking ships had their home ports either at Odessa in the Black Sea or at Leningrad in the Baltic.

Molniya satellite to the TsUP, where it was analysed by the GOGU, which then drew up the necessary commands for transmission to the spacecraft when it came within range of the next station.

For the 18-day Soyuz 9 mission in June 1970, medical experts from the Institute for Biomedical Problems were admitted to the main control room of the TsUP-E for the first time. They analysed data from the medical sensors attached to the bodies of Nikolayev and Sevastyanov, and contributed to the organisation of the crew's time, which was a serious issue on a long-duration flight. The most active periods were while the spacecraft was over Soviet territory, in range of the NIP ground stations. The transmission of data was at its highest rate during such passes. In addition, the crew could submit reports on their observations, comment on specific events and ask questions. Once beyond Soviet territory, they resumed working independently of Earth. By breaking the familiar sleep pattern of the cosmonauts, this organisation upset their circadian rhythm. A major challenge was to ensure that the crew of the first space station were able to work effectively throughout their month-long flight.

NINETY MILLIMETRES FROM SALYUT

On the morning of their second day in space, the Soyuz 10 crew performed systems tests in preparation for the final manoeuvre, which was achieved as planned. When their trajectory brought them within 16 km of Salyut the Igla automatic rendezvous system was activated. When the radar had locked onto the station's transponder the Igla began to steer Soyuz 10 towards its target, with the crew as mere spectators.

Just before midnight on 24 April the control room at the TsUP-E was so crowded that late arrivals had to stand. The GOGU members were seated, as was Popovich at the communications system, but squeezed in around the table, some seated but most standing, were the TsKBEM managers, representatives of the other design bureaus involved in the mission, generals, politicians and members of the State Commission. One of the most anxious was Armen Mnatsakanyan, the Chief Designer of the Igla. This had failed when Soyuz 8 had attempted to rendezvous with Soyuz 7 in October 1969. He had been criticised by the Kremlin, but not punished.

The final phase of the rendezvous had been timed to occur over the Soviet Union, in order to have 'live' communications, but the loudest voices in the control room were those of Mishin and General Kerimov, demanding explanations of events from the members of the GOGU, including wishing to know what would be done if the Igla were to fail!

"Approaching; Soyuz is two seconds in front of the Salyut!"

"Why do you give us seconds? Give kilometres!"

"Granite reports radio lock-on achieved. Igla works!"

General Agadzhanov, the head of the GOGU team, lost concentration and shouted into the microphone: "We understood you – the distance is 10 kilometres. Do not interfere!" In fact, only the first part was intended for the cosmonauts; his directive not to interfere was directed at Mishin and Kerimov, whose interminable calls for explanations were interfering with the work of the controllers, but

Agadzhanov still had the microphone keyed when he spoke these words. The cosmonauts, having no idea of the state of the control room, expressed surprise: "We only reported on the progress of our approach, according to the indicators on the command panel."

One of controllers complained, saying that it would be a miracle if he survived the morning without suffering a heart attack.

General Kerimov, ignoring Agadzhanov's direction, again demanded information. Struggling to remain calm, Agadzhanov offered an apology to the crew: "Igla works, understood! This is to Granite. Distance 11 kilometres. The rest was to our guests!"

On hearing of the increased range, Mishin exclaimed: "How! Firstly 10, now 11? Who is guilty?"

Ignoring Mishin's question, Agazdhanov spoke a series of sentences, some to the crew and others to inform the people in the control room: "The DOS engine started! Granite reports about the work of its engine. The programme for the 81st orbit has been executed. The DOS engine worked for 60 seconds. I'm No. 12: Granite, on the 82nd orbit we await from you the most important reports about the operation of the Igla and the automatic approach."

"Why do you speak so much?" demanded Mishin angrily.

Somebody attempted to calm Mishin by explaining that Agadzhanov was at the same time communicating with the cosmonauts and serving as commentator for the audience.

"Engine works 20 seconds; 25 seconds; 30 seconds; 35 seconds; 40 seconds; 45 seconds."

"Why don't they turn it off themselves?"

"The approach speed is 8 metres per second; steady radio lock-on."

"We see a bright point. Distance 15 kilometres, speed 24."

"Please! Silence in the room!"

"Who will explain to me why they were at 11 kilometres and now the distance is 15? Chertok, Mnatsakanyan, Raushenbakh – why do you sit and do nothing?"

"Igla is working," Mnatsakanyan told Mishin.

"This is a mad house! Only Igla does not go mad," said Raushenbakh quietly.

Fortunately, the chaos in the control room was not matched in space. Soyuz 10 continued its automatic approach without any glitches.

"Distance 11, speed 26.5," reported the crew.

"Distance 8, speed 27.5; distance 6, speed 27. DPO light. Starting to turn."

At this point Mishin exclaimed: "It can't approach at that speed! Why do you do nothing? Tell the crew what to do!"

Knowing that the rate of closure was according to plan, Raushenbakh explained to Mishin: "It isn't necessary to intervene, it will brake now."

The spacecraft had turned and started its braking sequence. The crew continued to report the closure parameters.

"Distance 4, speed 11. We can see the target against the background of the Earth – its flashing navigation lights. Distance 2.5, speed 8."

The medical telemetry showed that the heart rates of Shatalov and Yeliseyev were 100 beats per minute; Rukavishnikov, less active, was only 90 beats per minute.

At 1,600 metres from Salyut the speed was 8 metres per second. At 1,200 metres it had slowed to 4 metres per second. At a distance of about 1,000 metres, the crew could see the station in the optical periscope.

With the approach going smoothly, the mood in the control room improved.

"Distance 800, speed 4."

A few seconds later: "I see the target well and distinctly."

At this point the spacecraft passed out of range of the last station in the chain that stretched across Soviet territory, leaving the people in the control room in a state of apoplexy during the 30-minute wait for the next communications opportunity.

Mishin demanded an explanation from Raushenbakh for why the docking had not occurred while over Soviet territory. Instead of answering, Raushenbakh noted that Soyuz 10 had consumed 80 kg of fuel in making the approach – almost twice the amount planned! When no one appeared to appreciate the implication, Raushenbakh pointed out that if Shatalov failed to dock at the first attempt, the fact that 45 kg of fuel would be required for the descent meant that there would be insufficient to set up a second approach, and the crew would have to prepare for an immediate return to Earth.

Meanwhile, the Igla continued to steer Soyuz 10 towards its target. At 500 metres the approach speed was just 2 metres per second. Never before had any spacecraft approached such a large vehicle in space.

Shatalov recalls: "All the dynamic operations of the ship were conducted without any problems. The only issue appeared at the time that the Igla took control of the approach: the ship would oscillate from side to side periodically, requiring the firing of the correction engines. At a distance of 150 metres I took manual control. It was simpler than on the Soyuz 4 mission. The station grew bigger and bigger – in space, it appeared to be much larger than it had on the ground! When we were very close, Aleksey and Nikolay carefully inspected its docking mechanism, antennas and solar panels."

The final approach was at about 30 cm per second. When the probe on the front of the Soyuz came into contact with the conical drogue of Salyut, the cosmonauts saw the Mechanical Connection indication on their instrument panel. The docking process was automatic, and the crew had only to monitor their instruments as the spacecraft slowly advanced in order to drive the head of its probe all the way into the drogue. There were some vehicle motions, and a scraping noise as the probe slide across the drogue. The probe engaged the mechanism at the apex of the drogue, and began to retract to draw together the two annular collars in order to establish a hermetic seal. The cosmonauts awaited the signal that would indicate that the retraction process was complete. Instead, a warning signal came on to indicate that the mechanism had stalled. How could this be? What had happened?

When Soyuz 10 flew into the next communication zone, Shatalov heard an eager call from Earth, and reported: "I am Granite, I hear you well! At 4 hours 47 minutes we made a manual approach. Contact and mechanical capture passed. The retraction began. But in the 9th minute the SSVP stopped. Retraction not completed. Docking not achieved. We don't understand why. Look at the telemetry. Let us know what to do."

  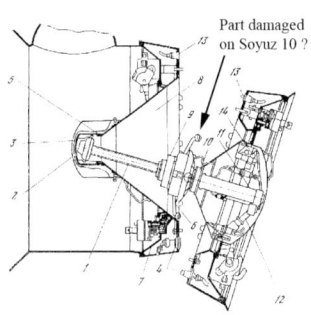

The active docking probe of the Soyuz (left) and the passive docking cone of the Salyut (centre). The diagram shows the lever on the probe of Soyuz 10 believed to have been damaged. At the top of the pin (1) of the probe is the head (2), which is inserted into the nest (3) of the cone (8). On the sides of both docking mechanisms are the connectors for electrical (13) and hydraulic (4) links between two vehicles. The shock absorber (12) is on the base of the probe. (Diagram courtesy Sven Grahn)

Everyone in the control room turned in silent expectation to the people who had designed the docking system. Pale faced, Vsevolod Zhivoglotov, a member of that team, explained that the active probe had touched the cone of the drogue according to plan. The length of the probe was 390 mm in its fully extended state. It started to retract, but when the length was down to 90 mm the mechanism was automatically commanded to halt. To the amazement of all concerned he explained *eight* things that could have gone wrong, including the possibility that one of the lateral levers of the probe had broken off – and he said that a pronounced swinging action just after capture strongly suggested that this had occurred.

Mishin exploded: "Why swinging? What are the dynamics? Raushenbakh! Why were there fluctuations?"

Cosmonaut Popovich, who had continued to talk with the crew, told Chertok that Yeliseyev had just reported that during the retraction process the orientation engines had been firing, causing a strong motion of the ship. For Chertok this was sufficient to indicate what had happened: "It is most probable that the mechanical breakdown occurred because of the large transverse oscillations – we didn't turn off the control system!" As the probe penetrated the drogue, the spacecraft had been deflected and the control system had tried to eliminate the angular deflections. However, the ship was no longer free to manoeuvre, and instead of rotating about its centre of mass, as the control system expected, it swung on the end of the probe and this broke part of the mechanism. In conclusion: "To continue the docking attempt will be futile. We must make a decision about the undocking."

As Shatalov recalled of these dramatic moments: "Just after the capture, the ship swung to the right by 30 degrees, and then to the left. The period of oscillation was seven seconds. We were concerned that we might lose the docking mechanism. We didn't know why this was occurring during the retraction operation. We approached the station with almost no difference between the axes of the ship and the station, so such motions ought not to have happened." The continuous firing of the orientation engines consumed a lot of fuel. "Before docking, the pressure in the tanks was 220 atmospheres, and it was only 140 when the operation automatically terminated. It is unbelievable how much fuel was consumed during this period."

Soyuz 10 was connected to Salyut only by small latches gripping the head of the probe. The disappointed crew were told to do nothing until the next communication session. Meanwhile, the engineers at the TsUP assessed the situation. The next time that the orbital complex appeared over Soviet territory Rukavishnikov was asked to enter the orbital module and check the electrical contacts of the docking mechanism to ensure that the retraction had not been halted by a faulty signal – since if that was the case the docking probe might not have been damaged at all, and the retraction should be able to be resumed. Rukavishnikov was fully familiar with the system. He removed a cover to access the electronics of the docking system, and confirmed that all of the connectors were as they should be. That was the last hope.

'MOM' DOESN'T RELEASE 'DAD'

To dock with the Salyut station was a four-stage automated process over which the cosmonauts had no control. The first stage was the initial mechanical contact, when the head of the active spacecraft's probe touched the interior of the conical drogue. This activated a sensor in the shock absorber on the probe. Then stabilisation thrusters were to slowly force the ship forward to drive the head of the probe into the hole at the apex of the cone, which the engineers referred to as the 'nest'. When the head of the probe penetrated the nest, this initiated the capture stage, and latches in the nest engaged the probe in order to prevent it slipping out. The Apollo spacecraft had a similar system, and American astronauts refer to this as a 'soft docking'. The third stage involved retracting the probe to draw the two annular collars together, to engage latches which would form a rigid bond and establish electrical and hydraulic connections located around the external rim – a status that astronauts refer to as a 'hard docking'. Then the probe would release its head, which would remain in the nest while the 'beheaded' probe withdrew into the housing on the nose of the orbital module. Once air had been introduced to the hermetic tunnel and the seals verified, the cosmonauts could swing back the hatch, complete with the docking assembly, to enter the tunnel and then swing the drogue into the station.

In the case of Soyuz 10, the problem struck between the second and third stages – during the retraction, 9 minutes after the first contact. The only physical connection was the head of the probe in the nest. However, owing to an oversight in planning, the control system of the Soyuz spacecraft was still operating and when this noticed an early deviation in attitude it fired the thrusters in an effort to eliminate the 'error'. If the spacecraft had been free, these impulses would have conformed to the logic of the control system; but it was not free – its probe was confined by the drogue. Upon finding that the spacecraft did not conform to its logic, the control system started to fire the thrusters on a continuous basis in an effort to assert its authority, and this subjected the probe to dynamic forces sufficiently strong to break one of the four levers surrounding its base. The probe was designed for a maximum force of 80 kg, but survived a load of 160–200 kg before failing.

The first error in the design of the docking process was to leave the spacecraft's control system active after the initial capture, because the conditions required by its logic no longer applied. The second error was to make the docking sequence fully automated once it had been initiated by the mechanical contact. Yeliseyev, who had participated in the development of the control system, had realised that the control system was jeopardising the docking process, but had no way to intervene – he was a frustrated spectator.

As Soyuz 10 was a 7K-T spacecraft designed to operate as a space station ferry, it carried air, water and food for just 3 days of autonomous operations. There was no option but to return to Earth as soon as possible.

The task was to separate from the station in a manner that would not damage the drogue. In designing the undocking process it had been assumed that the docking would have been finished and that commands could be directed through the circuits in the collars – which was impossible in this case. What would normally occur was that after the crew had left the station they would seal the Soyuz hatch and then command the latches to release the head of the probe from the nest so that the spacecraft could fire its thrusters to withdraw. However, in this unpredicted situation it was possible that the mechanism would fail to release. Indeed, the first attempt failed, and when Shatalov fired the thrusters his spacecraft simply swung around on its damaged probe.

In the control room General Andrey Karas, the Commander of Space Assets in the Strategic Rocket Forces, said bitterly: "Well, congratulations. You've developed a docking system in which 'mom' doesn't release 'dad'!"

There were two emergency options: one to cut loose the docking mechanism from the nose of the orbital module, and the other to release the orbital module itself. In both cases the only access point to Salyut would be left fouled.

Afanasyev of the Ministry of General Machine Building issued a directive: "This 'amputation' is not suitable. What do you want? To lose the first orbital station? Search for a method by which to deceive your super-clever scheme."

Salyut was saved by Zhivoglotov, the engineer who had appalled the control room by outlining eight possible reasons for the docking failure. After Zhivoglotov had outlined his plan, instructions were read up to Rukavishnikov who, during the 84th revolution, once again entered the orbital module and reconnected a number of the cables to deceive the mechanism into thinking that the release command came from Salyut. The command was issued on the next revolution by the cosmonauts using their command panel – and the latches released the head of the probe! At 10.17 a.m., after 5 hours and 30 minutes of drama, and during the 5th revolution spent in a soft-docked configuration, Soyuz 10 withdrew from the station. The news prompted loud applause in the TsUP. Although Soyuz 10 had not achieved its main objective of boarding Salyut, everyone hoped that the station was undamaged and therefore would be available to a future mission.

For almost half an hour Soyuz 10 flew in formation with Salyut, with Shatalov manoeuvring while his colleagues inspected and photographed the docking system.

Few of these black-and-white pictures were published, and those that were released were of a poor quality. Nor was the television from the spacecraft during this period released. On Saturday, 24 April, Moscow TV declared that the docking had taken place and showed a 30-second clip which was said to be from an automatic camera on Salyut as Soyuz 10 withdrew. The Earth was in the background. The only part of the station that was visible was just in front of the camera, and was brilliantly white. The docking was portrayed as having been successful, with the link-up being only a test in an ongoing programme – there was no suggestion that the cosmonauts were to have entered the station.

THE NIGHT RETURN

Shatalov and Yeliseyev spent their second night in space snoozing, but their rookie colleague, Rukavishnikov, remained awake, watching the Earth and taking pictures. In fact, he had a criticism of the spacecraft: "At a temperature of 20 degrees it is impossible to sleep in the flight suits. It is very cold. During the first night we slept only two or three hours. Instead of sleeping, we sat and shivered! It is necessary to carry sleeping bags." He was disappointed by the failure of their mission. Instead of setting a new record of 30 days in space, the flight would last just 48 hours! How long would he have to wait to receive another opportunity to fly?

On the original plan, the landing after a 30-day flight would have been in daylight – it was this timing which had required the launch to occur at night. To return after two days meant landing in darkness, which was something that the authorities had always avoided. After examining the options, it was decided to make the descent at the first opportunity on 25 April, aiming to return to a site 80–100 km northwest of Karaganda, a town on the Kazakh steppe. Normally, a Soyuz would automatically orientate itself to perform the de-orbit manoeuvre, but on this occasion Shatalov was told to do this manually – although since it would be dark outside he would have to fly 'on instruments'. In case of a problem that prevented the planned manoeuvre, the TsUP investigated the possibility of making it in daylight and landing in Australia, South America or Africa.

Shatalov aligned Soyuz 10 as specified. In normal circumstances, the cosmonauts would be able to make visual checks to verify the orientation, but outside was pitch black – there was not even moonlight to show the position of the Earth. They would be completely at the mercy of the automated systems. At 01.59 a.m. on 25 April the main engine was ignited to start the lengthy de-orbit burn. As the descent sequence was automated, the crew were passengers. After the engine shut down, pyrotechnic charges were fired to jettison both the propulsion module and the orbital module, and Rukavishnikov said that he had seen the flashes. The crew could only hope that the descent module was aligned with its heat shield facing in the direction of travel. As the module penetrated the upper reaches of the atmosphere, it was enveloped in a shockwave of glowing plasma. It was like being inside a neon tube whose colours changed. This awesome sight had been denied to their predecessors who returned in daylight!

Even the veteran Shatalov was astonished:

> As the ablative coating of the ship burns off we can see a real fire around us. To an outside viewer our descent module would have looked like a meteor. The g-forces are increasing. Our breathing is difficult. Around us something is crunching, and the module is shaking. Through the windows we can see a dance of orange and red sparks. The impression is much more dramatic than during a daylight descent. Finally the plasma fades, and a few minutes later the three parachutes deploy: first the pilot chute, then the drogue and finally the main. It was again darkness outside the windows. At an altitude of 5,000 metres we saw the first detail of the surface. Aleksey and Nikolay, who had windows on opposite sides of the cabin, both reported seeing a lake below. We would prefer not to land in the water. When Aleksey again looked out, he shouted "Land!" – just like the lookout of Columbus's sailing ship. Next we heard the soft-landing rockets fire, there was a shock and then – nothing. As there was no motion, we knew that we had landed on soil. Excellent! We shook hands and congratulated ourselves on having made a successful return. Just after we reported by radio that we were down and packed the flight log, we heard knocking on the wall – the recovery team had arrived. Despite the conditions, they had done their job perfectly. They had spotted us during our parachute descent, and as soon as we landed their helicopters had set down alongside.

The landing occurred at 2.40 a.m. on Sunday, 25 April, about 120 km northwest of Karaganda. When it was realised that the descent module might splash into a lake some of the recovery team had donned aqualungs in preparation to jump from the helicopter into the water to attend to the capsule. But then a gust of wind carried it on shore, and it landed 42 metres from the water's edge. Often a capsule would land on its side, but this time it settled in the preferred upright position – as indeed it had for Shatalov and Yeliseyev's previous landings. This first descent in darkness concluded the shortest Soviet space flight for six years.

Shatalov knew before this flight that Soyuz 10 would be his last space mission, as he had promised to accept an appointment to replace Kamanin. In addition, when Soyuz 10 landed Yeliseyev decided not to seek another opportunity to fly in space:

> We landed on the shore of a small lake. The helicopter was already circling, awaiting us. The recovery group included three very restrained and taciturn fellows wearing scuba-diving suits. We felt that these were courageous and disciplined people on whom we could rely. ... As I stood beside the descent module I thought: What next? Should I make one more flight to end this run of failures? ... No.

Several minutes after the landing, the TsUP received a call from one of the rescue helicopters reporting that the cosmonauts were in good health. Finally, the people at the control centre were able to relax. Despite the failure of the main task, everyone was delighted at the completion of this short but tricky flight. However, the Kremlin

Rukavishnikov (left), Shatalov and Yeliseyev appear disgruntled at the ceremony after their night-time landing.

The Soyuz 10 cosmonauts meet their wives at the airport.

Despite their inability to dock with Salyut, the Soyuz 10 crew received medals on a visit to the Kremlin. Standing alongside the cosmonauts are Premier Kosygin and General Secretary Brezhnyev. Ustinov (in profile) is behind Brezhnyev. The mood appears to be rather sombre.

was dissatisfied. On Soyuz 8 the Igla rendezvous system had failed. Although it had worked on this occasion, and Shatalov had steered his ship in to make contact with the station, a fault had interrupted the docking process. This was not good enough! But it was not the fault of the crew, and on their return to Moscow Rukavishnikov received a Gold Star as a Hero of the Soviet Union. His veteran colleagues already had two such awards for their previous missions.

TASS announced the landing without saying why Soyuz 10 had returned so soon. Officially, the crew had fulfilled their assignment. The mission was "a stage in the general programme of work" associated with Salyut. As TASS explained afterwards: "The programme of scientific-technical studies has been fulfilled." That is: "Studies directed at checking the efficiency of perfected systems for the mutual search, long distance approach, berthing, docking and separation of the ship and the station were carried out." For years, therefore, Soyuz 10 was classified as a successful test flight whose objectives had simply been to test the new docking system and to assess how the two vehicles behaved in a joined configuration. The cosmonauts were forbidden to state otherwise. At a press conference broadcast by Moscow Radio on 26 April, Shatalov said the flight was "not extensive in duration, but tense and magnificent in its tasks". He repeated what he had said prior to launch, that the flight represented a stage in a programme to develop orbital research stations. He said: "perfecting new systems for sighting, approaching and docking with an unmanned station were the mission's most important tasks", and "all these tasks were carried out completely". Even when Shatalov wrote his autobiography, *The Hard Roads to Space*, which was published seven years later in the typical Soviet style, he said nothing to imply that his third and final space mission had been anything less than a complete success.

At the press conference Yeliseyev was asked to describe Salyut: "The station is indescribably beautiful. A most impressive piece of equipment with a huge quantity of instruments, all kinds of antennas, a docking system, and 'CCCP' written on its side in large letters.[4] The station was gleaming white, and equipped with a flashing beacon to aid us in our approach." Shatalov added: "Salyut is so heavy that on Earth powerful cranes had difficulty in turning it."

Apart from the crew of Soyuz 10, few people were permitted to talk to journalists about the mission. One such person was Konstantin Feoktistov, one of the station's designers, who stuck to the official line that the objective of the mission had been to test the docking system: "The docking of a relatively small transport spaceship with a large orbiting laboratory proved to be more difficult than docking vehicles of the same size." He said that a new type of docking unit was tested – which was true. In the course of the manoeuvres, Soyuz 10 changed its orbit on three occasions and the station did so four times. Rukavishnikov had conducted "a series of important tests and technical experiments" during the docking – which was certainly true, although Feoktistov did not explain what these "tests" involved and why they were necessary. And he repeated the line that it was never intended that the cosmonauts should enter the station.

Some Western observers speculated that Soyuz 10 had landed after just two days because Rukavishnikov had developed 'space sickness'. The story was that a severe case of vertigo had prevented him from going into the voluminous station. Veteran cosmonaut and space physician Dr. Boris Yegorov was quoted as saying that during one communication session Rukavishnikov told ground control he had experienced "unusual and rather unpleasant feelings" as a result of the increased blood flow to his head – which was undoubtedly true, because this is a consequence of entering a weightless state. Yegorov was also quoted as saying that this crew had to cope with "a considerable emotional load" – which was also true, given the problems that they faced, although the fact that there were problems was a secret. When a *Guardian* correspondent asked Rukavishnikov how he felt in space, he replied: "A lot better than I'd expected in advance! On the first day I felt good, ate and worked normally. The next day I ceased to notice weightlessness. For me, working in weightlessness was pleasurable and joyful – for example, it was possible to catch an object in the air." Shatalov confirmed that Rukavishnikov's status had been good throughout the flight: "I think he felt even better than Yeliseyev and I." And this was confirmed by the in-flight biomedical telemetry: at the vital moments Rukavishnikov's heart rate was lower than for his more experienced colleagues. So much for the story that he had fallen ill and caused the mission to be cut short!

---

[4] In cyrillic the Union of Soviet Socialist Republics (USSR) is Союз Советских Социалистических Республик (СССР). It is sometimes written as *Soyuz Sovetskikh Sotsialisticheskikh Respublik* (SSSR).

"SHOW ME THAT DESIGNER"

A commission led by Boris Chertok decided that the first space station had almost been lost as a result of the error of leaving the Soyuz control system active during the automated docking process. By the end of April, the commission – which included the docking system designers Lev Vilnitsky, Viktor Kuzmin, Vladimir Siromyatnikov and Vsevolod Zhivoglotov – made the following recommendations:

- The speed of contact should be no greater than 0.2 metres per second.
- After capture, the docking probe should not start to retract until the Soyuz was stable.
- The crew must have the ability to control the docking probe.
- Add a panel to the spacecraft to enable the docking process to be controlled manually by the crew.
- Install special levers around the pin of the probe to evenly distribute the potential loads caused by oscillations of the Soyuz.
- Reinforce the levers on the probe to accommodate dynamic forces of 160 kg – twice the previous maximum.

The technical documentation was prepared within 24 hours. The modifications to the docking mechanism would take seven days, and testing was scheduled for early May.

However, Ustinov, Secretary of the Central Committee of the Soviet Communist Party responsible for defence and space, decided that he wished to see the docking system that had almost derailed the ambitious space plan. He and Ivan Serbin of the Industries Department paid a visit to Department No. 439 of the TsKBEM, which had devised the system. The designers laid out their system diagrams, and Mishin gave a simplified account of how the mechanism was intended to operate. When Mishin explained what had prevented Soyuz 10 from docking, Serbin demanded: "And who made it? Show me that designer!"

Vilnitsky, a former military officer and the head of the group which developed the system, stepped forward and said that the docking mechanism was a new system, much more complex than that used for Soyuz 4/5, because it had active and passive elements, was mobile, and in addition to forming electrical and hydraulic links was required to form a hermetic seal to enable the cosmonauts to pass through. He then explained that the design parameters were based on conditions experienced during the previously successful dockings between pairs of Soyuz ships – on two occasions unmanned, and once manned. The case of Soyuz 10 had been different, because the station had three times the mass of the ferry, and the dynamics were more extreme. But he was confident that with the new data from Soyuz 10 it would be possible to rectify the deficiency.

Serbin sarcastically interrupted: "Shall we tell TASS to prepare an announcement that comrade Vilnitsky made a mistake? In a week's time he will correct everything and the next crew will transfer into Salyut through the hatchway?"

"It would be an honour for me to make a TASS announcement," replied Vilnitsky calmly. "The next docking will be normal, I give you my word."

"You give us your word, but if you exceed the deadline you'll still be hoping for complete impunity!"

At this, Ustinov intervened: "The minister will decide whom to punish, and how. Now, show us through which hatchway will the cosmonauts have to climb from the ship into the station?"

With the tension eliminated, Ustinov examined with great interest the elements of the docking system and the internal tunnel. He made an observation about its small diameter – just 0.8 metres – and then asked about the new docking system that was under development for the joint Soviet-American space mission which was planned for 1975.

The testing of the modified docking system in May took a week, and successfully simulated a range of contact speeds and docking angles. Everything was now ready for the next mission to Salyut.

Specific references

1. Chertok, B.Y., *Rockets and People – The Moon Race, Book 4.* Mashinostrenie, Moscow, 2002, pp. 275–288 (in Russian).
2. Shatalov, V., *The Hard Roads to Space.* Molodaya Gvardiya, Moscow, 1978, pp. 217–224 (in Russian).
3. Yeliseyev, A.S., *Life – A Drop in the Sea.* ID Aviatsiya and kosmonavtika, Moscow, 1998, pp. 73–74 (in Russian).
4. Afanasyev, I.B., Baturin, Y.M. and Belozerskiy, A.G., *The World Manned Cosmonautics.* RTSoft, Moscow, 2005, pp. 227–228 (in Russian).
5. Kamanin, N.P., *Hidden Space, Book 4.* Novosti kosmonavtiki, 2001, pp. 295–301 (in Russian).

5

Mutiny at the cosmodrome

OPTIONS

While the engineers at the TsKBEM were modifying the docking mechanism of the Soyuz to eliminate the problem which had prevented Soyuz 10 from linking up with Salyut, on 2 May 1971 Vasiliy Mishin proposed to General Kamanin a revision to the programme. Owing to concern that Salyut's drogue might have been damaged, he proposed that the next mission should carry in its orbital module two spacesuits, identical to those used for the external transfer during the Soyuz 4/5 mission. Once the rendezvous had been accomplished, the spacecraft would 'park' close alongside Salyut and one of the cosmonauts would don his suit and exit the orbital module in order to inspect the station's docking mechanism. He would then cross the gap and, by gripping onto a series of handles on the surface of the station, make his way along to the area of the science module and open the cover that had failed to release immediately after the station reached orbit. As part of this scheme, Mishin proposed that only two cosmonauts should be assigned to the next mission, rather than three. Although he did not mention names, he probably had in mind Leonov and Kubasov, the commander and flight engineer of the second DOS crew. Both were admirably suited to the assignment since Leonov was the first man ever to make a spacewalk and Kubasov, having been Yeliseyev's backup for Soyuz 5, had undertaken training for such activity.

But this was simply unrealistic. First, the TsPK could not prepare cosmonauts for so complex a spacewalk in a time as short as one month. Second, Gay Severin from the OKB Zvezda that had designed the EVA suits and airlock facilities did not have two spacesuits available. Indeed, the inclusion of the exterior hatch on the transfer compartment of DOS-1 was not to enable spacewalks to be undertaken, for none were planned, but was forward planning for the stations that would follow. In late 1970 Kamanin had argued with Mishin to carry at least one EVA suit on board the station, but there had been insufficient time to install the ancillary apparatus and, as a result, Mishin had gone so far as to delete the tanks that would have carried the air to replenish the compartment after a spacewalk. On 3 May, at the meeting with the cosmonauts and trainers at the TsPK, Kamanin directed that Leonov, Kubasov and Kolodin should train according to the initial plan. Although there would be time for Dobrovolskiy, Volkov and Patsayev to train for external work, this was ruled out as the limitations of the 7K-T variant of the Soyuz meant that to accommodate a pair of spacesuits its crew would have to be reduced to two cosmonauts.[1]

On 7 May Mishin suggested to the Council of Chief Designers that regardless of the inability of Soyuz 10 to dock, it should still be possible for two crews to occupy DOS-1. It was decided that testing the modified docking system must be finished by 18 May and that the launch of Soyuz 11 should be scheduled for 4 June. The crew would be Aleksey Leonov (37), commander; Valeriy Kubasov (36), flight engineer; and Pyotr Kolodin (41), research engineer. Their assignment was to spend between 30 and 45 days on board Salyut. Then Soyuz 12 would be launched on 18 July with Georgiy Dobrovolskiy (43), commander; Vladislav Volkov (36), flight engineer; and Viktor Patsayev (38), research engineer. The duration of their mission would be determined by the resources remaining available to the station and the outcome of the first mission.

At the meeting of the Military-Industrial Commission (VPK) on 11 May, Mishin explained what had been learned from the failure of Soyuz 10 to dock with Salyut, and how the docking system had been modified for Soyuz 11. With the support of Kerimov he proposed postponing the launch of Soyuz 11 to 14 June and advancing

The 'first crew' for Soyuz 11: commander Colonel Aleksey Leonov (left), flight engineer Valeriy Kubasov and research engineer Lt-Colonel Pyotr Kolodin. (Courtesy www.spacefacts.de)

---

[1] Soyuz 5 was able to be launched with three cosmonauts and two EVA suits because it was a 7K-OK, as opposed to a 7K-T, and because by serving as the passive vehicle in the Soyuz 4/5 mission it had carried less propellant and no active docking system – it had the lighter passive unit. Furthermore, for half of its three days in space Soyuz 5 had only one man on board. The fact that the 7K-T that would have flown as Soyuz 12 would have been required to carry the extra air, water, food and apparatus needed to sustain the planned 30–45-day visit to DOS-1 would have made it difficult to accommodate in the orbital module two spacesuits and the ancillary air tanks.

Kubasov (standing, left), Leonov and Kolodin at the TsPK in Zvyozdniy. General Nikolay Kuznyetsov, the commander of the Cosmonaut Training Centre, stands on the right. (From the private collection of Rex Hall)

Kolodin (left), Leonov and Kubasov in front of the Soyuz simulator. (From the private collection of Rex Hall)

Soyuz 12 to 15 July, with each flight lasting 30 days. But Kamanin refused. If, as he had been advised, the station's resources would last no longer than the end of July or start of August, this would put the final crew at risk. He suggested that the main objectives of the Soyuz 11 mission should be to successfully dock and gain entry to the station; the duration of the mission was a secondary issue that should be decided by how events progressed. The majority of the commission, including Smirnov, its chairman, agreed that the key issue was that the cosmonauts should enter the station. In addition, Smirnov said: "There is no pressure on you regarding the date of launch, and the 30-day duration is not essential. Nevertheless, we must ensure the safety of the cosmonauts. Conduct the necessary calculations, checks and tests. If you have full confidence that the flight will have satisfactory results, report this to the Central Committee. You know that comrades Brezhnyev and Kosygin will consent to this mission only after you have assured its success." The next day the ballistics experts said that 6 June was the best launch date in terms of illumination conditions during the docking – if something were to prevent the docking, the spacecraft would be able to make a daylight landing. The maximum duration that would permit a landing at dawn was 25 days. In view of Kamanin's reservations, Mishin accepted 6 June as the launch date.

When Kamanin was asked by his boss, General Kutakhov, about the risk of the Soyuz 11 crew being lost, he replied: "We wouldn't lose the crew, but I don't have a firm conviction of a successful docking, cosmonaut transfer into the station and its activation." Kamanin outlined the potential sources of difficulty, including the poor visibility from the Soyuz, a failure of the automated systems and the strength of the docking mechanism. But he rejected Kutakhov's suggestion that a letter be sent to the Central Committee to say that the Air Force had reservations as to the likelihood of the forthcoming flight succeeding. Kamanin said: "I will do everything possible to avoid losing the crew, and to make possible the accomplishment of their task, but the Chief Designer and the Strategic Rocket Forces must be held responsible for the reliability of the technology."

On 14 May, at the traditional pre-flight meeting with Ustinov at the Kremlin, the main message to the TsKBEM was similar to that from Smirnov: "Launch Soyuz 11 only if you are certain that the preparations are satisfactory. We are not rushing you. The State Commission will set the final date." With these words, Ustinov carefully washed his hands of any responsibility for the potential failure of the mission.

SHADING ON THE LUNG

After a brief rest, hunting near the town of Vladimir and fishing on the Bear Lakes to the west of Zvyozdniy, on 21 May both crews flew to Baykonur with Kamanin and Beregovoy. In the Assembly-Test Building (MIK) the engineers had installed a docking command panel to Soyuz 11, and the cosmonauts rehearsed using it – they now had control of all docking operations until the final stage. In the meantime, one of the Igla rendezvous system units failed during tests. It was replaced, but the TsKBEM managers were concerned about the system's reliability. Then both crews returned to Moscow for their final training at the TsPK. Although the crews had flown to Baykonur in separate aircraft, Beregovoy decided to extend his visit to the cosmodrome and they had to break with precedent by returning in a single aircraft. At the TsPK Gubaryev, Sevastyanov and Voronov (the third crew) were already a month into intensive training. However, Sevastyanov had to break off in order to go to the Air and Space Exhibition in Paris, leaving Gubaryev and Voronov to train alone. If all went to plan, after backing up Soyuz 12 this crew would be the first to visit DOS-2 in the new year.

Although the launch date for Soyuz 11 was only a few weeks off, much remained to be done. Not only had Leonov's crew to fly somewhat earlier than expected, they had also to train to use the revised docking system. The spacecraft was loaded with an additional 10 kg of fuel to allow extended docking manoeuvres, and as a further precaution its resources during autonomous flight were increased from three to four days.

The two crews flew to Baykonur in separate planes on 28 May, accompanied by a large number of experts from Moscow and members of the State Commission. The crew of Soyuz 10 were also present to assist with the final preparations. Two days later, on 30 May, the cosmonauts celebrated Leonov's 37th birthday, and on 1 June they marked Dobrovolskiy's 43rd birthday – no one could know that it would be his last. Later that day, Mishin arrived from Moscow after an unpleasant meeting with the N1-L3 lunar programme expert commission headed by Academician Keldysh. Mishin's dilemma was that he desperately wished to push on with the development of the N1 rocket and start manned lunar missions, but was obliged to spend much of his time on the DOS programme – for which he was the technical manager. After addressing the well-known limitations of the N1, Keldysh had told Mishin that a lunar landing in 1973 was unrealistic and that the lunar project should be reviewed in detail with the members of his commission to devise a new plan for presentation to the Kremlin. On arriving at Baykonur, Mishin did not bother to explain this bad news to his deputies. In fact, everyone was pleased to find him brisk and fresh after having recently spent three days in hospital.

On 2 June the crews discussed with Chertok, Feoktistov and other representatives from the TsKBEM the docking procedures and potential failures of the automatic systems. They also discussed issues relating to the time that the station had spent in space – the possibility of toxic agents having accumulated in its atmosphere, food spoilage, water contamination and erosion of the seal of the hatch between the two spacecraft. After both crews had spent approximately half an hour in the descent module rehearsing, Soyuz 11 was installed on its rocket ready for transport to the pad. Meanwhile, Salyut continued to orbit the Earth, awaiting its first visitors. That evening the cosmonauts exercised and played chess to relax. Kubasov, Kolodin and Volkov liked tennis; Patsayev soccer; Leonov did not mind and would play anyone at anything. After a movie they retired to bed.

Although there was a general feeling that all of the procedures had been assessed and the cosmonauts and the spacecraft were ready, there were still some concerns in relation to the rendezvous technique. After Soyuz 10 Yeliseyev was appointed as deputy to Yakov Tregub, responsible for flight control. As an expert on the control

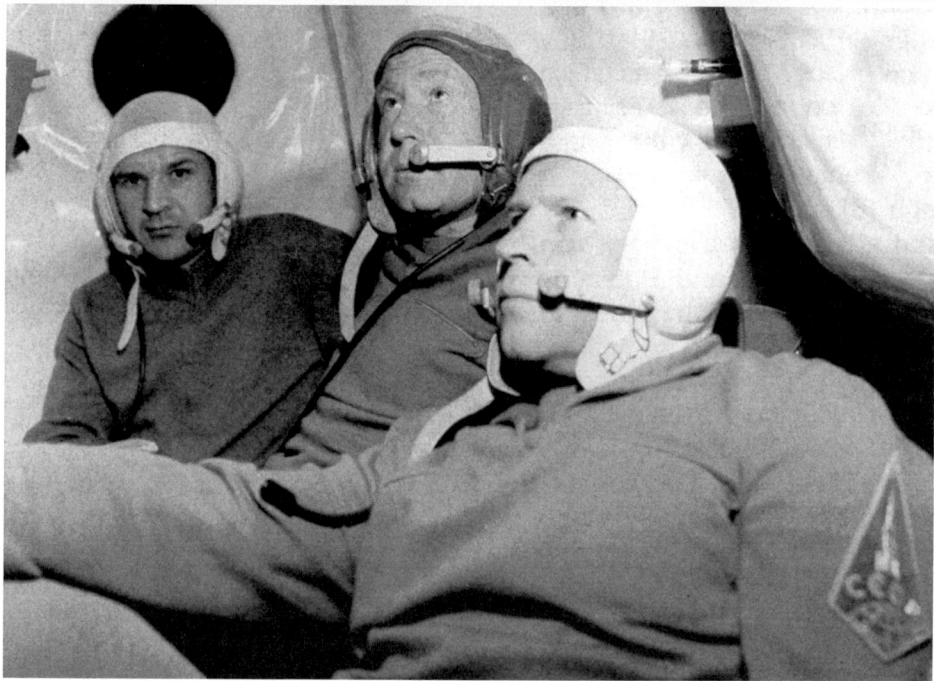

Kolodin (foreground) was the single rookie on the 'first crew' for Soyuz 11. Leonov (in the middle) was the first man to make a spacewalk. Kubasov was a veteran of the Soyuz 6 mission. (Top picture from the private collection of Rex Hall. Bottom picture first published in Spaceflight magazine by the BIS).

There are not many photos showing the 'first' and 'second' crews for Soyuz 11 in joint training. In this case Kolodin, Leonov, Dobrovolskiy and Patsayev (partially obscured on the right) are being shown equipment for the Salyut space station.

system he had demanded of Chertok and his own former boss, Raushenbakh, who had remained in Moscow, precise figures to enable the cosmonauts to monitor the operation of the Igla in different rendezvous scenarios. This information took the form of graphs showing the permitted variance of the rate of approach as a function of the range to the station. Whenever their speed 'touched' a limiting line on the graph, the control system should automatically fire the thrusters either to accelerate or decelerate in order to remain within the 'corridor'.

After lunch on 3 June Kerimov informed the State Commission of a Politburo meeting at which Brezhnyev and Kosygin had asked for another check to ensure that Soyuz 11 would be able to dock and that the crew would be able to enter the station. Afanasyev, Keldysh, Bushuyev and Smirnov had told the Politburo that Leonov's crew would fly the mission. Kosygin asked if they were well prepared, and Smirnov

Soon after arriving at Baykonur, Leonov recommends a chess move to Volkov as Dobrovolskiy looks on.

replied affirmatively. Noting that France had announced its intention to conduct an atmospheric nuclear test in the Pacific, Brezhnyev asked whether this would pose a risk to the cosmonauts, and Bushuyev said it would not.

On hearing this, Severin, known for his jokes, suggested to the Commission: "We should ask the cosmonauts to report how a nuclear explosion looks when seen from space."

"Why?" someone asked.

"To enable them to decide for themselves whether it is sensible to return to Earth once the nuclear war begins!"

The Commission decided to install Soyuz 11 on the pad at 6.00 a.m. the following morning, 4 June, and schedule the launch for 7.55 a.m. on 6 June.

While the State Commission discussed the forthcoming launch and laughed about the French nuclear test, the cosmonauts were having a routine medical examination. The mood changed suddenly when an X-ray scan showed an unusual dark spot on Kubasov's right lung which had not been present on a scan in February. Could it be tuberculosis? When an additional scan confirmed that he did indeed have something on his lung, the physicians announced that he would not be able to fly the mission. Kubasov was one of the first civilians to have passed the Air Force's medical screening for cosmonaut selection; he was one of the strongest cosmonauts; he was fit and healthy – only the previous evening he had run 5 km and then played tennis. Although Kubasov insisted that he was feeling perfectly alright and was ready to fly, the physicians ruled that he was unfit to fly.

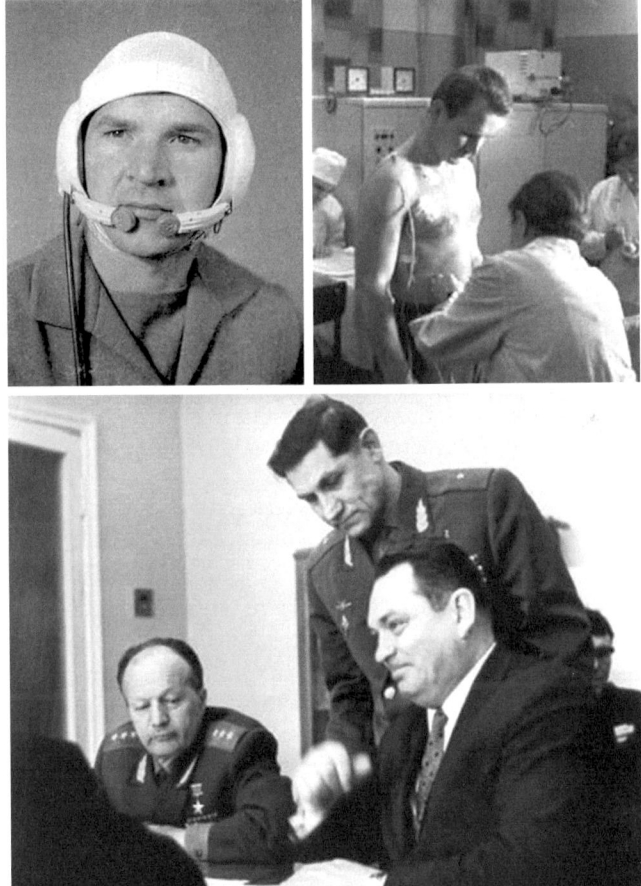

Kubasov in portrait and undergoing medical screening for the Soyuz 11 mission. Below: After Kubasov was grounded by the medics, Kamanin and Mishin (in the foreground) and Kuznyetsov (standing) argue about who should fly the mission. (From the book *Hidden Space*, courtesy www.astronaut.ru)

This was unprecedented. In 1969 the original Soyuz 8 crew had been replaced as a result of poor scores in the training examinations, but that was almost two months prior to the mission. In this case the cosmonauts were already at Baykonur with just three days to the launch date. Who should fly? Representatives of the Air Force, the Ministry of General Machine Building and the Ministry of Health had all signed a document which specified that in the event of a cosmonaut on a prime crew being medically disqualified prior to travelling to Baykonur he should be replaced by his backup. However, there should be no individual replacements once the crews were at the cosmodrome – the plan was to replace the entire crew with its backup, which meant that Leonov's crew would have to be replaced by Dobrovolskiy's crew. That was the rule ... but the situation was difficult. When Dobrovolskiy's crew was first assigned, this had been in the expectation that it would fly to DOS-2 in 1972. As a result of the inability of Soyuz 10 to dock with Salyut, and the desire to make two visits to DOS-1, Dobrovolskiy's mission had been advanced by one year. Now they faced setting off with only a few day's notice and being the first to attempt the new docking procedure. In contrast to Soyuz 10, which included two veterans, one of whom (Shatalov) was the only cosmonaut to have previously made a docking, only one member of Dobrovolskiy's crew (Volkov) had flown in space.

Kamanin called a meeting of the senior Air Force representatives present at the cosmodrome – cosmonauts Shatalov, Leonov, Kolodin and Dobrovolskiy, General Kuznyetsov, who ran the TsPK, General Goreglyad, who was Kamanin's long-time aide, and the medical staff. They analysed the new situation and, after weighing the factors, decided that the best solution was to reject the rule and instead to substitute Volkov for Kubasov in Leonov's crew. When Kamanin suggested this to Kerimov and Mishin, they agreed. But a short time later Mishin rang Kamanin to say that he had changed his mind – he had discussed the matter with the Kremlin, which was of the opinion that they must follow the rule and assign the mission to Dobrovolskiy's crew.

Interestingly, only a few people at the cosmodrome were aware of what was afoot. In particular, Chertok, who after Mishin was the most senior TsKBEM man present, found out only late in the afternoon when he was stopped outside the dinning room by Severin, who complained about having to replace the couches, flight suits and medical belts – which would not be easy to do now that the spacecraft was installed on the third stage of the launch vehicle and within its aerodynamic shroud. Chertok was dumbfounded. He and Severin went into the dinning room to talk to Shabarov, who was responsible for testing manned spacecraft; he had heard nothing. Severin was astonished: "Is it possible that your boss didn't consider it necessary to consult you about such a fundamental issue? To replace a crew at just two days notice. This is something that has never been done before – not here, nor in America.[2] Will we once again perform an experiment 'for the first time in the world'?" At this point, Mishin called and asked all the managers of the TsKBEM and representatives of the Institute of Biomedical Problems and the Ministry of Health to meet at 11.00 p.m. in

---

[2] In fact, two days before Apollo 13 was due to launch in April 1970 NASA had exchanged a member of the prime crew with his backup, owing to a medical concern.

the MIK. This was to be the civilian equivalent of Kamanin's consultation with the Air Force representatives.

Dr. Yevgeniy Vorobyev, a physician, explained that an X-ray scan had revealed a shading on Kubasov's right lung about the size of a chicken's egg. He also pointed out that the Air Force was responsible for ongoing monitoring of the cosmonauts' health, and that Kamanin and his medical staff were responsible for answering any queries about to the late discovery of this ailment. A senior member of the Ministry of Health then pointed out the failure of the TsPK to discover that cosmonaut Pavel Belyayev had developed a bleeding ulcer, with the result that he died in hospital in January 1970. At midnight, the TsKBEM managers agreed to put Soyuz 11 on the pad the next morning and to replace the crew facilities once it was in place. Severin said that although it would not be straightforward to do this work through the hatch of the orbital module, doing so should take no longer than five hours. Although the TsKBEM managers took it for granted that the crew would be swapped, the formal nomination of the crew was the responsibility of the State Commission.

At 7.00 a.m. on 4 June, shortly after Soyuz 11 had been installed on the pad, the State Commission gathered in the MIK for a meeting which would be remembered forever. General Kerimov, the chairman, reported that Kubasov was not going to fly; the basis of this decision was the medical report that declared him to be unfit. This took some people completely by surprise, because the previous day the Kremlin had confirmed the crew. Dr. Vorobyev explained the situation: "During the X-ray scan, physicians noted a shading on Kubasov's lung. They took layered roentgenography and calculated that the infiltration is located at a depth of 9 cm. It is deemed to be serious and active." He added that although an examination of Kubasov's blood was generally satisfactory, there was an increase in eosinophils, which are the white blood cells of the immune system.

Kerimov asked Kamanin for his thoughts about the crew. Taking into account the complexity of the planned mission, Kamanin said that Volkov should fly instead of Kubasov. "Leonov has already been in space. He has even spacewalked. Volkov has flown on Soyuz and he will be able to manage the mission objectives." It was a simple case of replacing one experienced cosmonaut with another.

However, Mishin thought differently: "We object! I consulted with our comrades. We have the document signed by the Air Force that in a case like this we have to change the entire crew. The backup crew passed their training with good scores. A new and unharmonised crew would be worse than the backup one. We categorically insist on the replacement of the entire crew." Mishin was supported by Chertok and Shabarov, and even by General Ponomaryev, who was the Deputy Commander-in-Chief of the Air Force, and by General Kuznyetsov, the head of the TsPK who was aware that his elderly and unpopular boss was soon to retire. Those members of the Commission whose role was to ensure that the spacecraft was launched on time and was able to accomplish the planned mission, and so were not particularly interested in who flew, abstained from the debate.

The State Commission decided to replace the entire crew, and told Kamanin to inform the cosmonauts of this. Kamanin did not object. He knew the rules. Perhaps in different circumstances he would have challenged Mishin, as he had often before.

But he was tired of disputes with the Chief Designer and also of misunderstandings with General Kutakhov, his new boss. The mission of Soyuz 11 would conclude his decade in charge of cosmonaut training. Although Yevgeniy Bashkin, a training instructor from the TsPK, pointed out that his team had worked primarily with the prime crew at the expense of the backup crew, this was not intended as support for Kamanin's case. Shabarov asked for permission to leave the Commission early with Severin and Feoktistov, as they had a lot of work to do on the spacecraft. However, because its business had been decided, Kerimov concluded the session. After a brief breakfast Severin and his team went to replace the apparatus in the spacecraft, and Feoktistov's group made the relevant calculations to allow for the change in overall weight of the crew.

In the afternoon, several top-level medical experts flew in from Moscow. After a detailed analysis of the documentation of Kubasov's ailment, and taking additional scans, they confirmed the symptoms of tuberculosis.

MISHIN, VOLKOV AND LEONOV

Let us return to Mishin and the decision to swap the entire crew. When speaking of this issue at the State Commission he repeatedly used "we" rather than "I". Who else was involved in taking this decision? It is clear from Chertok's memoirs that Mishin did not consult either Chertok or Shabarov, his most senior deputies present, as they heard the news from Severin, who was from a different design bureau! The discussion between Severin, Chertok and Shabarov occurred late in the afternoon of 3 June, several hours after the medical examination. The events during those hours are still unclear, but based on the memories of some of the people present, as well as upon later events, it is possible to construct a reasonable scenario of activities by the Air Force people under Kamanin and by the TsKBEM staff headed by Mishin, and this indicates that the decision was made very quickly. If Mishin did not consult his two principal available deputies, what about Moscow?

At 9.00 a.m. on 4 June, immediately following the State Commission's meeting, Bushuyev telephoned Chertok from Moscow. As we have seen, Bushuyev had gone to the Politburo with Afanasyev, Keldysh and Smirnov the previous day to report to Brezhnyev on the preparations for Soyuz 11. Bushuyev gave Chertok a summary of the meeting, and told him that Afanasyev would arrive at Baykonur that afternoon for another test of the modified docking system. But Bushuyev, who was Mishin's second deputy and therefore the third man in the TsKBEM structure, had no idea of the crew change. On hearing of it from Chertok he became agitated: "How dare you decide to do it without consulting us in Moscow! We have reported to the Politburo that Leonov's crew will fly. We confirmed how well they were prepared. And you – because of Kubasov – have replaced them all! Look at the situation in which you have placed Afanasyev, Smirnov and Ustinov! Now they must urgently report again. Afanasyev will be with you in three hours and he won't thank you for it either." It is therefore clear that Mishin did not consult Bushuyev, his most senior deputy having a responsibility for manned spacecraft.

In fact, there was only one man in Moscow whom Mishin was obliged to consult: his old patron, Minister Afanasyev, who in turn would have sought the blessing of Ustinov. Although this *must* have occurred, Bushuyev was clearly unaware of it. It is difficult to prove the case, however, as the leaders of the Soviet space programme made many decisions orally. If there are any documents about this dramatic change, they remain secret in the Kremlin's archive.

Mishin based his objection to Kamanin's suggestion on two elements:

- the document signed by the Air Force stating that once the crews were at the cosmodrome they would not be replaced on an individual basis; and
- his suspicion that if Volkov were to be substituted for Kubasov at this late stage then the crew would not be as harmonious as it would have been with Kubasov, making it inferior to a crew comprising Dobrovolskiy, Volkov and Patsayev, who, even though they were less experienced, had been in training *as a crew* for some time.

On the other side, Kamanin thought that a crew consisting of Leonov, Volkov and Kolodin, with two veteran cosmonauts, would be more capable of completing such a complex mission successfully.

But perhaps Mishin and Kamanin were each driven by a simpler motivation. After much debate, it had been agreed that the first and third crews would have one TsPK cosmonaut (in command) and two TsKBEM cosmonaut-engineers; and the second and fourth crews would have two military cosmonauts and one civilian. But the first crew had not been able to dock with the station, and Mishin and Kamanin may each have sought to interpret this agreement in his own favour: Mishin wishing to fly his two engineers and Kamanin wishing to have two military cosmonauts. Applying the rule of exchanging the entire crew would favour Mishin. Discarding the rule and replacing Kubasov by Volkov would favour Kamanin. Volkov would fly regardless of how the dispute was resolved. The basic issue was which community would have two of its cosmonauts on the crew – the TsPK or the TsKBEM.

At noon on 3 June, immediately after the medical report which grounded Kubasov, Kamanin and his Air Force people decided to reject the rule and instead substitute Volkov for Kubasov in Leonov's crew. Initially, Mishin accepted the plan, but soon telephoned Kamanin and told him that after a conversation with Moscow (actually Afanasyev and probably Ustinov) they must exchange the crew. Officially, the State Commission was responsible for considering the views of Kamanin and Mishin and formally nominating the crew. But with the exception of Mishin, and at a later stage Kerimov, no members of the Commission had been involved in this decision; Kamanin was excluded of course. Mishin did not mention an official document – he simply said to Kamanin that Moscow supported the crew exchange. In fact, to achieve his goal Mishin had used the document between the Air Force and the TsKBEM which specified that once the crews were at the cosmodrome they would not be replaced on an individual basis. When Soyuz 10 failed to dock, it appeared that Mishin had missed the chance to have two civilian cosmonauts on the first crew to board Salyut; but now, thanks to Kubasov's ailment, if he could get the crew exchanged, he had a second opportunity to send two of his cosmonauts.

However, someone was missing in this chain of events: Volkov – the man who may well have played the most crucial role. The Air Force people certainly did not consult the civilian, and Mishin initially accepted Kamanin's plan without seeking the opinion of Volkov. But as Mishin thought about it more deeply, it is reasonable that he would have talked the matter over with Volkov, and possibly also Patsayev, prior to making his call to Moscow.

After Mishin called Kamanin to say that Moscow had consented to the crew being swapped, Kamanin informed the Air Force staff. Leonov exploded. He could not accept this. As a member of the original cosmonaut group, and the first man ever to spacewalk, he urged the Air Force to demand that he fly with Kolodin and Volkov. He had trained for the mission for almost a year. He knew Salyut thoroughly. As a passionate artist, he had even arranged for the station's cargo to include his painting apparatus. While in space he wanted to paint the Earth, the stars, the Moon, distant nebulas, and his colleagues at work in the station. It was his mission. Naturally, he had the full support of Kolodin, who was eager to make his first flight.

The famous journalist Yaroslav Golovanov, who knew many of the cosmonauts well, recalled the atmosphere at Baykonur as follows: "It is hard to describe what was happening in the Cosmonaut Hotel. Leonov was so furious that he was simply growling. If he could, he would have strangled Kubasov. Poor Valeriy could not understand what was going on. He was feeling perfectly well and, after all, it wasn't his fault. In the evening Kolodin visited me, completely crushed. With a glass in his hand he said: 'Yaroslav, you know, I will never fly in space.' And he was right. ...

"I will never fly to space," complained Kolodin (left) in frustration at the decision to ground Leonov and himself along with Kubasov. On the other hand, Volkov (on the right, with Kolodin) was happy to gain the chance to fly this important mission. (Kolodin's photo – first published in *Spaceflight* magazine by the BIS)

Leonov urged the replacement of Kubasov with Volkov. It looks as if he succeeded in convincing the generals, but then Volkov became obstinate, saying: 'If a change is necessary, then change the entire crew.'"

This definitely shows that Volkov was behind the decision; Mishin was merely its executor.

However, in one of his interviews Kubasov said something else: "They intended to move Volkov from the backup crew to take my place, but Leonov categorically opposed this idea."

Are we to believe Leonov did not wish Volkov to be on his crew? It was true that of the cosmonaut-engineers Volkov was the most critical of his military colleagues owing to their lesser technical qualifications. In training at the TsPK for his first flight, he sometimes behaved as if he were the leader of the crew with two military cosmonauts. In fact, Kamanin once told Filipchenko, the real Soyuz 7 commander, to restrain Volkov in the Soyuz simulator. Of course, Leonov would have known of this. In normal circumstances, Leonov would not have been keen to have Volkov on his crew. But Leonov knew that the only way that he would fly on Soyuz 11 was if he accepted Volkov as his flight engineer. Volkov, however, had a choice. He had a guaranteed ticket to fly. If he flew under Leonov's command he would be the only civilian on board. If the backup crew flew, then not only would he fly with the men with whom he had trained, but because Dobrovolskiy and Patsayev were rookies he would enjoy the status of a veteran. So for Volkov the choice was simple. And there is another unusual aspect to Kubasov's claim. He was close to Volkov: both were from Moscow; they were the same age; they graduated from the Moscow Aviation Institute; they worked together for years at OKB-1; they successfully passed all the cosmonaut examinations and medical tests and were chosen for the TsKBEM's first group of cosmonaut-engineers. As much as he may have sought to protect Volkov, by making his claim Kubasov actually raised an old and never documented story of a complex relationship between Leonov and Volkov: allegedly, when the crews for the DOS missions were first nominated Leonov belittled Volkov, pointing out that although a veteran he was only on the *third* crew, and hence had no chance of flying to the first space station.

On 4 June the State Commission confirmed what Volkov and Mishin desired: the replacement of the entire crew. When they heard of this from Kamanin, Leonov and Kolodin continued to complain. Having two cosmonauts, both military officers, one a space veteran and the other a rookie, dispute the decision of a State Commission was a remarkable moment for the centralised and totalitarian Soviet system – both unprecedented and incomprehensible. Kamanin, who was always on the side of *his* cosmonauts, acceded to the pressure imposed by Mishin, who was able to rely upon the rule signed by the Air Force stating that once the crews were at Baykonur there would be no individual cosmonaut substitutions. Having lost the support of the Air Force and his closest colleagues at the TsPK, Kamanin did not wish to pursue the matter further. But Leonov and Kamanin did. Lacking the support of their generals, they went directly to the only man who could have the decision changed: Mishin. In the 2004 book *Two Sides of the Moon*, which Leonov co-authored, he summarised the conversation with Mishin ahead of the final meeting of the State Commission on the evening of 4 June. Leonov says that Mishin warned him: "Don't forget that you shared a room with Kubasov. Perhaps you drank from the same glass. We can't take the risk of you becoming ill while in space." In hindsight, Leonov acknowledged Mishin was correct. But at the time he could not accept the decision. He and Mishin exchanged some rather unpleasant words. Just before the State Commission convened, Mishin advised Chertok of his difficult conversation with Leonov and Kolodin – during which Kolodin said that he had known all along that he would not fly: "To them, I am the 'white crow' – they're all pilots and I'm a missile man."

That was true: among the 15 members of the 1963 group of Air Force cosmonauts, Kolodin was one of four who were not pilots. He had served at both the Baykonur and Plesetsk cosmodromes in the Strategic Rocket Forces. As a 'missile man' at the TsPK, he did not think he had much chance of ever being assigned to a prime crew in competition with the Air Force officers, some of whom had test pilot experience. Fellow 'missile man' Eduard Buynovskiy has said that when the cosmonauts of the second group arrived in Zvyozdniy they were immediately separated into pilots and non-pilots. In addition, Kolodin was notable for the curiosity of having lost half of his left thumb in an accident! According to Leonov, Kolodin had a particularly hard time. In 1964–65 Kolodin was Leonov's second backup in preparations for the first spacewalk. He was appointed as a general backup for the 'group flight' of October 1969 along with Shatalov and Yeliseyev, but when the two-man crew of Soyuz 8 was replaced Kolodin was not needed. Now, when he was on the threshold of space, it was decided that he should be stood down! Kolodin reportedly tried to convince Mishin to substitute him for Patsayev on Dobrovolskiy's crew. Of course, Mishin refused, and Kolodin, almost with tears in his eyes, warned ominously, "History will not forgive you for what you have done."

It is interesting that in his published diary Kamanin did not write in detail of his conversations with Leonov and Kolodin. He said simply that Leonov's entire crew reacted incorrectly and in an inappropriate manner. According to Kamanin, their behaviour was totally unacceptable and did them no honour. However, they were not the alone in this. As Kamanin put it: "They are guilty for that, as are many Big Chiefs who added fuel to the flame."

JOURNALISTS AND THE NEW CREW

The final meeting of the State Commission started at 6.00 p.m. on 4 June. In the past these sessions had been fairly ceremonial in nature because all the details had already been resolved and the purpose was to confirm readiness for the launch. The most interesting part of the session was always the presentation of the cosmonauts. However, this occasion set a precedent. Dobrovolskiy, Volkov, Patsayev, Leonov, Kubasov and Kolodin were seated in a line behind a long table. The men who had been assigned to fly looked serious, almost anxious. According to Leonov they even appeared to have been frightened by the sudden change in their schedule. Established in mid-February, Dobrovolskiy's crew knew that they should have had more time to train – another month at least, to prepare themselves for the *next* visit to Salyut. Leonov's crew held their heads low, clenched their fingers, and appeared nervous. Behind them sat the Soyuz 10 crew. The tension in the air was oppressive.

Kamanin introduced the prime crew. As he announced the names, the cosmonauts stood up. Dobrovolskiy briefly said that his crew was ready to conduct the assigned tasks.

At the session, Leonov was very disappointed: "At the previous session I thanked the Commission for their trust. Now I can only express my regret about what has happened."

How about Kubasov? He felt especially guilty because his ailment meant neither of his crewmembers would be allowed to fly.

After just 20 minutes the meeting was concluded.

The famous journalist Mihail Rebrov was present: "I recall the intense silence in the room of the State Commission during the announcement of the decision. Then an explosion of protest! Leonov and Kolodin defended their right to fly the mission, saying that they knew the station better, that they had trained for longer, and that the promotion of Volkov from the backup crew would not have complicated their task. However, the State Commission had made its decision: the backup crew would fly. On the faces of the two crews you could feel the tension, envy . . . Everything had happened unexpectedly and painfully. Kolodin suffered more than the others. The anger was apparent on his face."

Reportedly, Kubasov approached Chertok and apologised. "I believed I had only caught a cold – that it would pass in a week and nothing would be visible on the X-ray scan." No one could console him. The great irony is that the diagnosis of the physicians proved to be spurious. A more detailed medical examination in Moscow showed him to be healthy! It was decided that he must have an allergy to the spray applied to the trees at Baykonur. Many years later, however, Kubasov revealed that the pollen from the trees flowering in the late-season spring had initiated his allergy. What was certain was that the dark spot on his lung wasn't the onset of tuberculosis.

Another irony is that the comprehensive medical screening failed to establish that Patsayev had a chronic kidney inflammation.

Thus, the incorrect diagnosis of tuberculosis symptoms on Kubasov's lung led to a healthy cosmonaut being grounded and one with a chronic medical problem being launched into space!

At 7.00 p.m., shortly after the conclusion of the State Commission, Dobrovolskiy, Volkov and Patsayev gave their press conference. Sitting between Kamanin and the Soyuz 10 crew, they were now relaxed and replied to the questions enthusiastically. The journalists knew Volkov as a veteran, but Dobrovolskiy and Patsayev were new. There were so many journalists, with so many questions, that as the room became uncomfortably hot Volkov suggested that they go outside, which they did, and the session was concluded with the crew sitting on a bench with their jackets off and their sleeves rolled up.

The Soviet space journalists knew that they were expected to ask only about the cosmonauts' lives, their backgrounds, their families, and stories about their training;

"You could feel the tension between the crews," observed a reporter at the dramatic meeting of the State Commission when the 'second crew' was named to fly instead of the 'first crew'. Leonov, Kubasov and Kolodin sat dejectedly with Dobrovolskiy, Volkov and Patsayev.

The mood was more relaxed at the press conference following the State Commission meeting: Yeliseyev (left), Volkov, Dobrovolskiy, Patsayev, Kamanin and Shatalov.

"Soyuz 11 was a difficult assignment". Volkov (left), Dobrovolskiy and Patsayev after the press conference. (From the private collection of Rex Hall)

not about the upcoming mission, the major tasks and the planned experiments. Once the mission was underway TASS would publish all that was necessary. The excited cosmonauts spoke willingly, often simultaneously. They jumped from topic to topic. They began by speaking about themselves, in particular about their early years, switched to their training, and then returned to their childhood years. At one point, Patsayev spoke of Korolev. Despite tradition, Dobrovolskiy felt obliged to offer an insight into their mission. As he put it: "Soyuz 10 inaugurated work with the orbital station. Our mission is to complete the next stage of the work begun by Soyuz 10." The official story was that the Soyuz 10 crew had not been meant to enter Salyut. Volkov said the Soyuz 10 mission was "rather successful", and that "Soyuz 11 has a difficult assignment". Of course, the journalists knew that with these words Volkov was saying that complex manoeuvres and a docking operation were to be attempted, and that the crew would board the station. After half an hour, the cosmonauts drew the unusual conference to an end because they had to prepare flight documentation.

Dobrovolskiy (left), Volkov and Patsayev meet the launch team at the pad. (The lower pictures are from the book *Hidden Space*, courtesy www.astronaut.ru)

Cosmonaut Nikolayev (second left) and Mishin (centre) embrace the cosmonauts for the
Soyuz 11 mission in relaxed mood. Nikolayev had handled their training at the TsPK.
Patsayev is on the left, and Dobrovolskiy and Volkov are on the right.

But before they departed Volkov and Patsayev shared a cigarette offered by one of
the journalists.

Interestingly, after the press conference Patsayev went to Leonov's room in the
Cosmonaut Hotel to apologise. He especially respected Leonov, and was not happy
at being nominated for his first mission into space in such circumstances.

On 5 June the final crew for Soyuz 11 were introduced to the launch team at the
traditional preflight meeting. Almost 3,000 people gathered at the base of the rocket.
There were generals, officers, soldiers, technicians, engineers, politicians, designers,
and even some people from the other launch pads. For the first time there were a lot
of women present – Korolev's colleagues say that he had not wanted woman on the
pad, believing that they would be unlucky for the forthcoming mission. But on this
occasion it appeared that everyone at Baykonur wanted to see the cosmonauts who
were to lift off the next day for one of the most important missions in the history of
cosmonautics. They formed a ring in front of the rocket, with the cosmonauts in the
centre, holding flowers.

Visibly excited, Dobrovolskiy said: "While on my way here, I prepared a speech.
Now, seeing your smiles, I'll simply say, dear comrades and friends, thank you very
much for your effort. We will do everything that is necessary to complete our task."

It was traditional for the prime and backup crews to take a brief walk in homage
to their predecessors, but Leonov did not wish to participate, and therefore his crew
remained in place. In fact, Leonov and Kubasov had not even wished to attend the
ceremony. When Kamanin had told them that they must do so, Kubasov replied: "If
I am healthy then I must fly. If I am sick, I should not be there."

When the ceremony was over, the journalists went to see the cosmonauts' room in
the Cosmonaut Hotel. It was not very large, but contained three beds, chairs and a

table in the middle draped with a white tablecloth. There were also three displays of flowers which the cosmonauts had picked nearby to freshen up the room. There was a fridge with mineral water, but despite the weather the physicians had ordered the cosmonauts not to add ice to the water. In the meantime, the prime crew had a final session with Mishin and his engineers. At this meeting an unusual photograph was taken showing the three men in an embrace with Mishin and ex-cosmonaut General Nikolayev, who was now one of the training leaders at the TsPK.

Mission commander Lieutenant-Colonel Georgiy Dobrovolskiy, flight engineer Vladislav Volkov and research engineer Viktor Patsayev found themselves on the threshold of space earlier than they or anyone else had expected.

Specific references
1. Kamanin, N.P., *Hidden Space, Book 4*. Novosti kosmonavtiki, 2001, pp. 313–316 (in Russian).
2. Chertok, B.Y., *Rockets and People – The Moon Race, Book 4*. Mashinostrenie, Moscow, 2002, pp. 305–315 (in Russian).
3. Scott, David and Leonov, Alexei. *Two Sides of the Moon – Our Story of the Cold War Space Race*. Simon & Schuster, 2004, pp. 259–262.
4. Novosti kosmonavtiki, No. 3, 2005 (Interview with Valeriy Kubasov).

6

Dobrovolskiy, Volkov and Patsayev

BETWEEN THE SEA AND SKY

"About space?" began cosmonaut Dobrovolskiy when asked the inevitable question by the famous space journalist Alexandar Romanov, "I must admit, I never dreamt about it." The interview occurred in July 1969 during the tour by Frank Borman, the first American astronaut to visit the Soviet Union. All the TsPK's cosmonauts – veterans and rookies – gathered in Zvyozdniy to meet the man who commanded the Apollo 8 mission which orbited the Moon in December 1968. Dobrovolskiy had been seated on one of the rear benches, and Romanov took the opportunity to talk to him in advance of Borman's arrival. "I'm from Odessa," Dobrovolskiy continued, "where people dream only of the sea and travelling across the ocean. The majority of my friends joined the navy."

"But you became a cosmonaut?" Romanov prompted.

Dobrovolskiy smiled and looked at the sky: "First, there was aviation. I devoted ten years of my life to aviation; my happiest years. I cannot imagine myself without flying. It is an awesome feeling to sit in the cockpit of a plane which is totally in your control. And in front of you – blue heavens! I am still in love with the sky. I'm not saying I 'love' it, rather I'm 'in love' with the sky. But, now I don't fly so much, and without flying, I'm like ..."

"But in space nothing is really blue," Romanov interjected, "it's black."

"In space something else attracts you. I wish so much to look at the Earth from the altitude of space! Gagarin was the first to see her blue aureole. Now we call her the Blue Planet. Listen to how that sounds: the Blue Planet! I don't think the beauty of space could ever extinguish our love for Earth. Do you remember the song with the lyric: 'Anywhere carried away by our rockets, we always return to you, the blue Earth'?"

Although it was brief, this interview painted an accurate picture of Dobrovolskiy. His colleagues, to whom he was Zhora, said he was born to fly. His flying biography includes the phrase: "He flies calmly." This is a very rare description to hear, even when talking of the best pilots. Yes, peace and wellbeing are probably the real words to describe the character and life of Dobrovolskiy, a surname which, fittingly, means "a man of goodwill". Blond, tall, broad-shouldered and tough, he was kind-hearted and had a contagious belly laugh. His accent remained 'broad Odessa', and he had a sense of humour typical of someone from that region. At the Air Force school, his friends nicknamed him 'Odessa', and he was proud of it.

Life was tough on Georgiy Timofeyevich Dobrovolskiy.[1] He was born on 1 June 1928 in Odessa on the coast of the Black Sea, in Ukraine. His family was Russian, and lived in the suburb of Blizhniy Melnitza ('the mills neighbourhood'). His father Timofey Trofimovich left when Zhora was two years old, and he was raised by his mother Mariya Alekseyevna. "She is a marvellous woman," he said of his mother. "She represented an ideal. She faced hardship. Without a husband, she had to work to feed us. Firstly she worked in a shop, then in a cannon factory. No matter how tough her life was, I never saw her complain, be sad, in bad mood or in despair."

As a little boy Zhora often asked about his father, but his mother did not say and only later did he learn from relatives that his father had been a member of Soviet counterintelligence, and one day had left home and never returned.

Zhora liked his hometown, the sea and the sky. Lying on the shore, he spent hours watching the ships pass by and enjoying the beauty of the cloud formations. He ran and played with his friends on streets which, 20 years previously, a student of the technical high school had walked – Sergey Korolev. Four years prior to Zhora's birth, Korolev designed a light plane and dreamed of rockets and space. A quarter of a century later, they met.

"I wasn't born with a silver spoon in my mouth, and no good fairies brought me

Lt-Colonel Georgiy Dobrovolskiy, the Soyuz 11 commander.

[1] Георгий Тмиофеевич Добвольский

"He was fanatically strict with himself." Dobrovolskiy was a serious person during training, but relaxed with friends. With General Beregovoy (right) whom Dobrovolskiy especially respected.

In February 1971 Dobrovolskiy joined Patsayev (left) and Volkov (right) as the commander of the 'third crew' for DOS-1. This is a rare photograph of them in the descent module of the Soyuz simulator.

September 1970 to mid-February 1971 he trained with Sevastyanov and Voronov. When cosmonaut Shonin, commander of the first DOS crew, was dismissed after an indiscretion, Kamanin replaced Shonin with Shatalov and gave Zhora command of the third crew. Zhora was expecting to fly the first mission to DOS-2 along with Volkov and Patsayev. But then fate intervened with the failure of Soyuz 10 to dock, and Zhora's crew found itself backing up the prime crew and in line to fly to DOS-1 in July 1971. When a member of the prime crew was grounded for medical reasons just days before the scheduled launch, Zhora's crew got the opportunity to become the first crew to board Salyut!

Marina Dobrovolskiy was 11 years old when her father left Zvyozdniy to travel to Baykonur: "He had never discussed his business things in my presence. However, I remember him saying the word "soon"; he said this to my mother prior to the flight. When he left on these trips he always had a smile, knowing that he would soon be back home. I was never anxious, but I was always eager for him to return." Asked if she had any premonition of the forthcoming tragedy, she replied negatively.

When a journalist asked Zhora about his feelings on the eve of his first flight into space, he said: "I would say that every space flight is a form of combat, because in a very short time you must give it your maximum, your experience, your knowledge – everything you have accumulated over your entire life. For those people who wish to participate in such combat, space is the right place."

Q: "And where is the sense of life?"

"That is a difficult question. Probably it is in the accomplishment of the highest goals. Without motivation, life is mere existence."

On the night before launch he was asked: "Are you excited Georgiy?"

"Yes, I am. All the time I think only of the launch, the flight and the experiments. We have prepared for every anomalous situation, but to be honest I am as excited as if I were about to approach a terrible enemy. Space does not forgive mistakes."

"THE UNIVERSE WAS ALIVE"

At the start of his autobiography *Stepping into the Sky*, Vladislav Nikolayevich Volkov,[7] known as Vadim to his friends, wrote: "To be honest, I was not preparing to be a cosmonaut. In fact, I never dreamt or fantasised about space. As a boy, I had no idea about Tsiolkovskiy." Although the book was not published until 1972, he saw the page proofs several days before he set off on the Soyuz 11 mission.

Standing 179 cm tall, broad shouldered and a real sportsman, Vadim was the most sympathetic of the civilian cosmonaut-engineers recruited by the TsKBEM in 1966. He played soccer in the Moscow championship for many years, firstly as a member of the Moscow Aviation Institute where he was studying, then in Club Burevesnik. He also played ice hockey and handball, and was skilled in athletics. He was even a boxer for a brief period. Among the cosmonauts he was one of the best

---

[7] Владислав Николаевич Волков

Vladislav Volkov, the Soyuz 11 flight engineer.

tennis and chess players, and an excellent guitarist. In contrast to his colleagues, he also had an intense sense of humour, and readily burst out laughing.

Vadim was born on 23 November 1935 in Moscow. He was short and skinny, but impetuous. When his parents moved, he had to change school and meet new pupils. When on the first day the strongest pupil stole his breakfast, the obviously weaker Vadim immediately fought for it. The other pupils respected him for this. It was one of his main characteristics, coincidently incorporated into his surname – 'the wolf'. He later fought for his thoughts and beliefs, even against much stronger opponents. His father Nikolay was an aeronautical engineer and this mother Olga worked in an aircraft factory, and he inherited from them his love of aircraft and the sky. Their house was near Tushino airport, so from his backyard he could watch a variety of different types of aircraft take off and land, and during parachutist displays the sky above would fill with the coloured parachute canopies. Many pilots and engineers employed at the airport lived in the neighbourhood. From time to time, his uncle, Pyotr Kotov, a combat pilot from the Second World War, would visit his house and Vadim would stare in amazement at the medals and decorations on his chest. Pyotr was his mentor. In such an environment it was natural that Vadim should select the sky as his destiny. But what would be his profession? To be a pilot like his uncle, an aeronautical engineer like his father, or something else? In fact, Vadim wished to be a test pilot, but on his uncle's advice he decided to study aeronautical engineering. After finishing at Moscow high school No. 212 in 1953 he enrolled at the famous Moscow Aviation Institute, one of two leading aerospace faculties, which his father

"As much as he was serious at work, the rest of the time he was serene and always laughing." Left: Volkov with his wife Lyudmila, together with Viktor Patsayev and Viktor's daughter Svetlana. Seeing this photo for the first time, Svetlana said: "It is good to see my father smiling". (From the private collection of Rex Hall) Right: Volkov attending to a bicycle with his son Vladimir.

had attended.[8] There he fell in love with Lyudmila Birykova, who was training to be a food-processing engineer. They were married in early 1957, and in February 1958 had their only child: son Vladimir.

On graduating in early 1959, Vadim became an electro-mechanical engineer for aircraft missiles. But in April 1959 he transferred to Department No. 4 of OKB-1, where he had worked as an apprentice during his studies. Vladimir Syromyatnikov, one of the docking system designers, knew him very well prior to his becoming a cosmonaut: "Vadim was capable, but not a very friendly person; he was astute and jealous. We liked to play soccer and ice hockey. We often used to play together against teams from other departments of our design bureau. I must admit that he was a good player, but he was too selfish. When he had the ball, we didn't expect a pass – Vadim would either score a goal or he would miss, but he wouldn't pass. After his graduation they moved him to the central design bureau, because Korolev wanted to reinforce the basic departments of OKB-1 with new people. But this job did not fit with his temperament and he was moved to the organisation department, where he became a deputy to one of the leading designers. Because he continued to play soccer and ice hockey, I met him often."

When Vadim joined OKB-1 the development of the Vostok spacecraft which was to carry the first man into space was well advanced. Later, he was involved in the design of the control system for the modified form of the rocket intended to launch the Voskhod spacecraft. In addition, he worked on the design of the R-9 ballistic missile. In September 1961 he became deputy to the leading Vostok engineer, and in February 1962 deputy to the leading designer for the Voskhod spacecraft, which was a modification of the Vostok intended to perform advanced missions while the new Soyuz spacecraft was under development. Vadim would

---

[8] Other notable Moscow Aviation Institute students were Mishin, Kubasov and Sevastyanov.

later say: "I am proud to have been involved in the Vostok spacecraft which carried Gagarin on the first manned space flight, and in its modification for Voskhod; there are my tracks."

During those years Vadim often met Konstantin Feoktistov, who was one of the leading spacecraft designers. The first cosmonauts were young military pilots who lacked strong technical backgrounds. Although Korolev had some of his engineers, including Feoktistov, instruct the pilots who would soon became famous heroes, the engineers argued that it would be better if they themselves could be permitted to fly in space to assess the performance of the systems they had designed.

On Vadim's first visit to the Baykonur cosmodrome, two Vostoks were launched on consecutive days in August 1962, carrying Nikolayev and Popovich respectively. "It is a marvellous picture. Here, one can see a grand creation of what started on a paper in the design bureau. And, although the involvement is immeasurably small, one feels immense pride in knowing that one belongs to those who conquer space," Vadim observed.

While working at OKB-1 he did not lose his desire to fly aircraft. The famous test pilot Sergey Anyokhin suggested that he enroll in an aero-club. He did, and received a diploma as a sports pilot, which enabled him to fly solo on Yak light planes.

The selection of the Voskhod crews began in early 1964. The first mission, which was scheduled for October, was to have a crew of three: the commander would be a military cosmonaut, the flight engineer would be an OKB-1 engineer and the third place would go to a medical doctor. In May Korolev met the 14 engineers chosen as candidates for this historic mission, and Vadim was among them. He successfully passed all the medical examinations preparatory to the special training for the flight, but when the shortlist was posted on 11 June his name was not on it. Dissatisfied, he went to Korolev to complain. He pointed out to the famous Chief Designer that he was healthy and fit, could pilot an aircraft and had even made parachute jumps. To finish, he suggested Anyokhin as a character reference. Korolev replied calmly: "You are still young. There is time. It is impossible to send everyone on spaceships. Somebody also has to design them." But such words could not satisfy Vadim. He was so disappointed that in the tramcar he told Syromyatnikov that he was going to finish the soccer referee school and became a professional referee! Of course, it did not happen. However, Vadim was included in the team of engineers responsible for recovering cosmonauts from a spacecraft after it had landed. Interestingly, in March 1965 he participated in the recovery of Belyayev and Leonov, who came down in a snow-laden forest far from the planned site. At that time, Volkov and Leonov could have had no idea that their paths would cross six years later in the manner that they did!

In the meantime, based on the list of Voskhod candidates, 12 engineers passed the medical examinations in July 1965 for further consideration as cosmonauts. Finally, in March 1966, shortly after Korolev's death, Mishin, with the support of Minister Afanasyev, signed the document ordering the recruitment of cosmonaut-engineers. The selection committee was chaired by Mikhail Tikhonravov, and on 23 May 1966 eight men were chosen, and because Anyokhin was the most senior engineer he was nominated as the group leader.[9] They were to train for Soyuz flights in Earth orbit and the L1 and L3 lunar missions. Whereas the military cosmonauts were housed in Zvyozdniy, the civilians were accommodated in one of the TsKBEM motels, not far from the design bureau. They began with parachute jumps, flying with instructors in MiG-15 jets, altitude chamber testing, and simulating weightlessness in the Tu-104 aircraft. In August 1966 they trained for recovery from a Soyuz descent module on the Black Sea. This was the first joint training with their military counterparts. One month later, Kamanin sent the TsKBEM's group for medical screening by the Air Force, which only Kubasov, Volkov, Grechko and Yeliseyev passed. They joined the military cosmonauts at the TsPK in training for the mission scheduled for early 1967 in which two Soyuz spacecraft would dock and two cosmonauts would make an external transfer from one to the other. But in October 1966 Grechko broke his leg parachuting, and so Makarov – one of the four civilians who had failed the Air Force screening but had continued to train at the TsKBEM – took the Air Force test again, was passed, and in November joined Kubasov, Volkov and Yeliseyev at the TsPK.[10] Having had some of its men rejected by the Air Force, in November the TsKBEM selected two more, who joined in early 1967.[11]

In November 1966 Kamanin and Mishin agreed eight men to form two prime and two backup crews for the spectacular introduction of the Soyuz ship. Yeliseyev was to be the flight engineer on the prime crew for Soyuz 2, with Kubasov as his backup. However, when Soyuz 1 was launched with Vladimir Komarov on board, it ran into trouble and the launch of Soyuz 2 with a three-man crew, which had been planned for the following day, was cancelled; and when Komarov brought his ship back the parachute failed to open and he was killed.

As regards Vadim, from September 1966 to December 1968 he trained to be the flight engineer of the passive spacecraft. When a third (backup) crew was formed in January 1967 he trained for this role with Shatalov and Kolodin. In August 1968 Kamanin replaced Shatalov with Anatoliy Kuklin, giving Shatalov the active ship.

After the docking and external transfer was accomplished in January 1969, the managers of the TsKBEM and the Air Force developed a programme of missions for the remainder of that year: in April-May Soyuz 6 was to undertake a 7-day solo flight, and in August-September Soyuz 7 and 8 were to dock and remain joined for three days. When the plan was submitted to Ustinov he wrote on it: "This is too thin, it must be thicker." He meant that something spectacular was required to offset the likelihood that the Americans would make the first manned lunar landing in July. But what could be done? After the cancellation of the manned circumlunar flights, the only option was a mission in Earth orbit. And the docking and external transfer had already been achieved. However, someone at the TsKBEM remembered an idea

---

[9] In fact, Anyokhin was a colonel in the Air Force and a former test pilot. Interestingly, despite losing an eye in 1945 during a test flight, 21 years later he was nominated by the TsKBEM as a civilian cosmonaut and given command of the group of cosmonaut-engineers.

[10] It is impossible to prove, but it is likely that Kamanin ordered the Air Force doctors to pass only half of the cosmonaut-engineers sent to the TsPK by the TsKBEM, in order to minimise the number of civilians available to compete with his military cosmonauts for flights.

[11] These two were Nikolay Rukavishnikov and Vitaliy Sevastyanov.

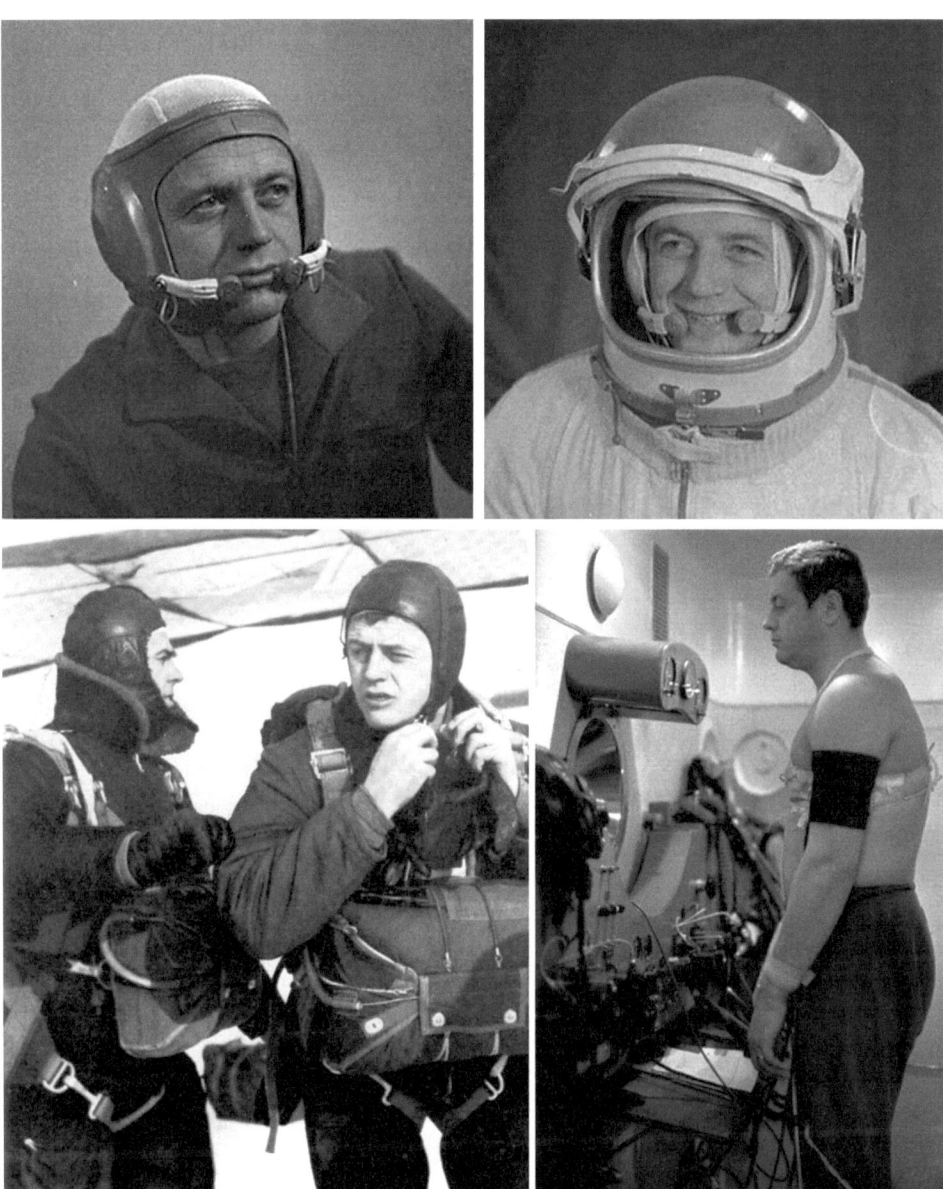

Several rare photos of Vadim Volkov: wearing a helmet-phone (top left) and a spacesuit (top right); preparing for parachute training (lower left) with Beregovoy; and undergoing a medical examination (lower right).

The crews of the 'group flight' of Soyuz 6/7/8 in October 1969: Kubasov (left), Shonin, Volkov, Filipchenko, Gorbatko, Shatalov and Yeliseyev.

Three Soyuz spacecraft undergoing testing at Baykonur. Soyuz 8 with its active docking mechanism is on the left.

once mooted by Korolev. Encouraged by the success of Gherman Titov's 24-hour flight in August 1961, Korolev had proposed launching three Vostoks on successive days. Although this had been judged too demanding, it had been decided to launch Vostoks in pairs – which had been done successfully on two occasions. So why not now attempt a triple flight? One would adopt a position from which it would be able to film the other two performing the docking! By the end of February the TsKBEM and TsPK managers had drawn up such a programme.

Nominated as the flight engineer of Soyuz 7, the passive spacecraft, Vadim joined two military cosmonauts: Anatoliy Filipchenko, commander; and Viktor Gorbatko, research cosmonaut. The training was intense, and occasionally Filipchenko had to restrain his energetic flight engineer, who, as the designer of some of the spacecraft systems, was eager to play a greater role in the simulator. But in his autobiography Filipchenko chose not to mention the difficulties that he had faced with Volkov in training.

Shatalov, who was in overall command of the three-ship flotilla, wrote of Vadim during this time: "As much as he was serious at work, at other times he was serene and always laughing – he could tell jokes and funny stories constantly for hours. He knew how to deal with people of all ages. At weekends, we would spend time in the forest with our families. There, Vadim was continuously surrounded with our kids. I envied him his easy and simple way of communicating with kids. He sang the songs about pirates and thieves, played soccer and climbed trees. He liked to be the centre of attention. He was always smiling."

The three spacecraft were launched one by one, on 11, 12 and 13 October 1969, but owing to a problems with the Igla system on Soyuz 8, an automatic rendezvous with Soyuz 7 was not possible. Although an attempt was made to accomplish this manually, the result was unsuccessful and, with its fuel running low, Soyuz 8 had to give up the chase.

Although Vadim was disappointed not to have docked, he greatly enjoyed his first flight in space, which lasted five days. He had taken with him a lump of Sevastopol soil that his son Vladimir had given him. In addition to the planned experiments, he made TV broadcasts. In fact, he was the first accredited journalist in space, because he had earlier written several articles for the newspaper *Krasnaya Zvezda*,[12] which he signed 'Vladimir Volkov'. While in space, he kept a personal diary. Filipchenko was impressed: "Volkov noted every interesting event during our flight, as well as what happened down on Earth. Even now, knowing how many experiments he had, I observe with pleasure how he succeeded in managing his time and in describing so nicely his impressions and thoughts."

On the second day of the flight, as the other crewmembers rested, Vadim watched Earth from an altitude of more than 200 km. When he experienced something very unusual, he immediately recorded it in his diary: "Orbit 47: There are in the world events that I would describe as 'momentary sparkles', meaning that a man does not immediately understand them. Such a 'sparkle' for me was the Earth's voice. Below, it was night-time. I looked at the onboard globe. Our ship was over South America. I

[12] *Red Star*, the newspaper of the Soviet Army.

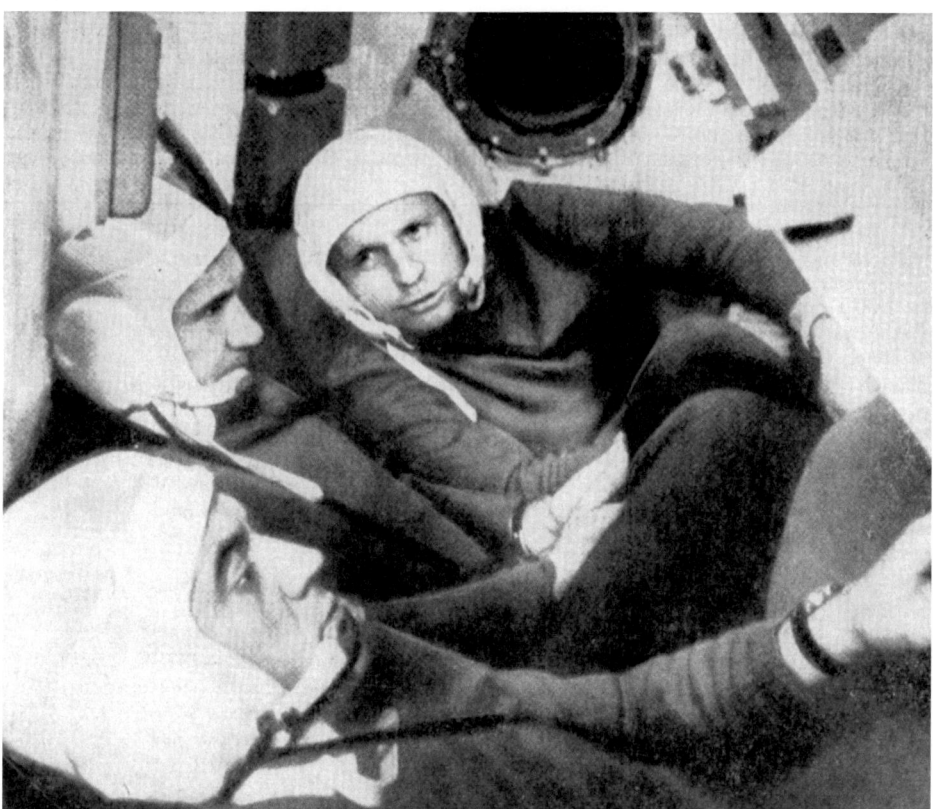

Volkov's first space mission was as the flight engineer of Soyuz 7, with Lt-Colonels Filipchenko (in the middle) and Gorbatko (rear).

controlled the operation of some instruments and, from time to time, I would look away from the panel towards the Earth in the darkness. In the headset, I could hear a characteristic background noise. I had an impression that behind me, above my ear, there was a giant invisible man breathing. Then it was absolute silence. And suddenly, out of the darkness, I hear the barking of a dog! A dog is barking! Is it an illusion? I strained my hearing and searched my memory of all known sounds of the Earth. There is no doubt – it is a dog barking! The sound was barely audible, but full of life. Then it occurred to me that this is the voice of Layka. And then, clearly, I heard a baby cry! Other voices. And again a baby crying. The universe was alive. The Earth was flying past underneath. Somewhere on the Earth, a baby was crying. Somewhere a mother was gently calming her baby. The dog was barking to protect them. It made little sense, but it was possible to feel it; possible only once in a life time. ... Orbit 50: I was watching the sunset. Before the final part of the solar disk disappeared, suddenly several layers of the atmosphere appeared above the horizon. It was red just above the horizon, then orange, then dark blue and finally the black of space. Stars were visible shining through this pattern. Then it all became grey. In the constellation of Scorpio there was a subtle crescent of the ash-coloured Moon. I could clearly see the constellations of the Southern Cross and Centaurus, which are not visible from the northern latitudes of our home. I recall the science fiction books: perhaps one day we will have the chance to fly to the stars?"

In May 1970 Vadim was nominated as the flight engineer of the third DOS crew, with Shatalov and Patsayev. The training for a mission to the space station was very different from his previous experience because, in addition to the Soyuz simulator, the cosmonauts had to familiarise themselves with the much more complex systems of the station.

When in February 1971 the rookie cosmonaut Dobrovolskiy took over command of the third crew, Vadim became its only veteran. The failure of Soyuz 10 to dock with Salyut, and the change of crew on the eve of the Soyuz 11 mission, resulted in Vadim and his crewmates being launched somewhat earlier than they had expected.

In his book, Volkov wrote:

> Your hours in space are not eternal, they will end. Some time, unfortunately, they must end; the hours of your life. But only for the time being. There will be others.

The day before they left for Baykonur the cosmonauts held a big party. As usual, Volkov was the centre of attention. He was smiling and singing. Several years later, Viktor Patsayev's wife, Vera, recalled the days before the flight and that last party, and said that Vadim told her of having had a premonition that he would die in space.

On landing after 5 days in orbit, Volkov (centre) embraces Gorbatko (left) and Filipchenko. In contrast to his colleagues, for the first few minutes he had some difficulty in standing.

In May 1970, Volkov became a member of the 'third crew' for DOS-1.

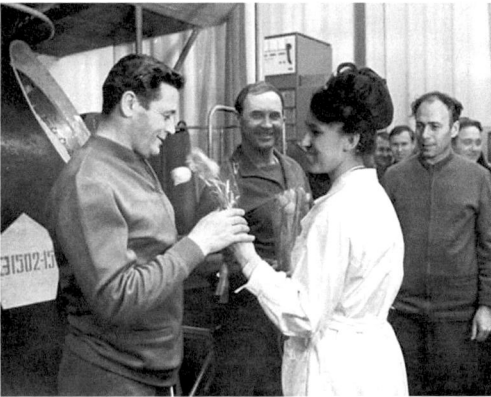

Volkov in training. In an aircraft (left; from the private collection of Rex Hall). On the right Volkov receives flowers as he, Dobrovolskiy and Patsayev conclude their training. (From the book *Hidden Space*, courtesy www.astronaut.ru)

Volkov (left) laughing with Dobrovolskiy.

"I WOULD LIKE SO MUCH TO EXPLORE"

The research engineer of the Soyuz 11 crew, Viktor Ivanovich Patsayev,[13] was tall and skinny, had green eyes, was going bald, and was so quiet that his presence was often overlooked. Cosmonaut Shatalov observed: "Viktor was the total opposite of Vadim. He was also an engineer, also a top expert. But in contrast to Vadim he was reserved, quiet, self-controlled and humble – he didn't talk much. He liked his job. He was an expert in scientific instruments and related apparatus. He passionately wished to fly in space to test and work with different devices, the majority of which had been designed with his participation. He avoided conflicts, and was never in a hurry to tell anyone what he was thinking about. He knew how to listen to all sides. He would prove his views not by mere words, but by logic and indisputable facts." The docking system designer Syromyatnikov said of him: "Viktor was a designer in a neighbouring department, working on the development of the elements for radio-antennae devices. Self-controlled, sometimes a little bit slow, he didn't play games with us – he wasn't interested in sports. I remember wondering how was it possible to select as cosmonauts people who didn't participate in sports. In my naivety, I had thought that it was a job only for real sportsmen."

Viktor was born on 19 June 1933 in Aktyubinsk, a city in northern Kazakhstan not far from the border with Russia. His father, Ivan Panteleyevich, was a director of the local bakery, but at the time of Viktor's birth was doing regular national military service in the army. After that, he was appointed to head one of the departments of the State Security Service. Viktor's mother, Mariya Sergeyevna, described her first child as follows: "Viktor is the replica of his father in appearance and in character – especially when he grew up to match his father's stature. Above all, Viktor liked sincerity and honesty." When Viktor was four years of age the family moved to the small town of Alga, and there, just before the Second World War, they had their second child: daughter Galina. Although Viktor was different from the other kids of his age, he spent his childhood in the same way as everyone in his neighbourhood. Every day during the hot summers he would go with his friends to the Ilek River to swim, fish and collect crabs, and would return home in the late afternoon with a full basket of crabs which, with his grandfather, he would prepare for dinner. In those hard times, fresh crabs were a special pleasure. After the long and severe winter, the steppe would turn green in the spring. Viktor liked to explore with his friends, and one day found a 'kurgan' ('barrow') of a Tatar warrior.[14] Standing on the tomb, he informed his friends of the history the these powerful conquerors from the Far East. Once, he returned from the steppe carrying a young eagle that had broken its wing. He looked after the injured bird for months, and when its wing had healed the eagle flew back to the steppe.

As a self-educated person, Viktor learned to read when he was five. In those times, children began school at the age of eight, but he wanted to go earlier. As his mother recalled: "He went to school when he was only five. In fact, we couldn't separate him

---

[13] Виктор Иванович Пацаев

Viktor Patsayev, the Soyuz 11 research engineer.

from my nephew, who was several years older. Viktor even sat with him on the same bench. The teacher decided not to send him home. In the class photograph, he is present as an equal member of the class. Then when he turned six, Viktor said firmly: "Now I am starting the school seriously!" His father said he should remain at home at least another year, as he was too young. We forgot this conversation, and the next September when the time for school arrived of course we didn't buy books, pencils or copybooks because Viktor was still one year too young for the first class. On that day he disappeared, returning with tears in his eyes. 'What happened to you Vitya?' I asked. 'They didn't allow me to enter school,' he replied. Then I thought we had not understood our son sufficiently. I called my husband. He talked with the school director, who agreed to allow Vitya to start school a year early. We thought Vitya would stay there only a few days and return home, but we were wrong again. Actually, after setting an examination the director wished to enroll Vitya straight in the second class. However, his father was against this. By the end of spring 1940, Vitya finished the first class with excellent scores."

As the stories of the early childhoods of Dobrovolskiy and Volkov describe their characters, this one of Viktor's eagerness to attend school shows his main attribute: his determination to accomplish his goals. His sister grew up under his influence. In 1976 she wrote the book *Courage of Aspiration* about her brother, in which she immortalised him. At the beginning of the book, she wrote: "I remember my brother as a tall boy with large green eyes, often with a book in his hands. ... Once he had started, he would read for hours and nothing could separate him from his book." Of

[14] The tatar warriors were from Mongolia.

his school subjects, he preferred mathematics and the natural sciences, like physics, biology and chemistry. He also liked literature and painting. He was very neat and tidy. His schoolbooks were immaculate. During the school holidays he used to help his grandfather to mow. One time, he brought home a little fox. But after it got into the chicken house he had to release it.

When the war started, his father was called to the front and in October 1941 was killed near Maloyaroslavtsa, about 130 km from Moscow. He was buried in a mass grave. Many years later, Viktor took his children to the place where his father died defending Moscow. The loss of his father deeply affected Viktor, who was less than eight years old. He became reserved and more serious, as though he had matured before his time. In the spring of 1943 he made his first Chkalovets airplane using a design from a pre-war technical magazine. He carved the wooden body of the plane using his grandfather's knife, and made the propeller from a tin can. The first flight was not very successful, but he learned the concept of centre of mass. His second plane left the backyard and continued across the street.

"It flies, it flies!" Viktor cheerfully shouted.

"Who flies? What flies?" demanded his grandfather urgently, hurrying from the house. "You little devil! You terrified me to the death!"

One year after the war, Mariya Patsayeva and her children moved to the village Kos-Istek with Ivan Volkov, her second husband, who had four children. From the start, Viktor got on well with his two stepbrothers, who were of his age. Together, they enrolled in the No. 45 railway high school in Aktyubinsk, and it was there that Viktor finished his seventh and eighth grades. In 1948 they all moved to Nyestorov, on the Baltic in the Kaliningrad region. Many of the buildings were still in ruins and there remained many unexploded weapons lying around. They lived in a damaged building. His stepfather worked in the bank and his mother in the bakery. In 1950 Viktor completed high school as (in his own words) "an average pupil". But he still very much liked physics, astronomy and mathematics. With some school friends he made a small telescope. It may have been his first views of the Moon, the stars and the planets that triggered his desire to fly in space. As he said to one of his friends after using the telescope: "I would like so much to explore. I'll travel around the world, then continue into space. How much must you need to know to do something like that, I wonder!"

On completing high school, Viktor had to decide what to do next. One idea was to enroll in geology. He sent in an application to the Institute of Sverdlovsk, but as he travelled to take the entrance exam he found that he could not buy the ticket for the train and had to spend a night in the Moscow railway station, which caused him to miss the exam. Instead, he decided to try to enroll at the Moscow Geology Institute Ordzhonikidze. He sent several optimistic letters home, but while his score in the exam was good, he did not gain sufficient points to be accepted. Nevertheless, the Institute suggested to the candidates who just fell short that they should attend the Penza Industrial Institute. Viktor was disappointed at his failure, but did not wish to be a further burden on his mother and, although not yet 18 years of age, he decided to live independently and enroll at the Penza Industrial Institute. He had in mind to make a second attempt to sit the exam for the acclaimed Moscow Geology Institute.

During his first year of study, a course was introduced on calculators and analytical machines. It represented a major challenge, but Viktor applied for and was accepted to study this new technology of computers. This was a key point in his career, as he decided to remain in Penza rather than reapply to the Moscow Geology Institute.

Although Viktor had a small scholarship, it was inadequate; even living alone. He studied during the day and unloaded trains at the railway station at night. However, he was one of the best students in his class. His friends said that he did not need to spend much time preparing for an exam; he would simply attend it and gain a pass! He became a member of the Institute's Scientific Society, and each year he would attend the science and technique conference. He presented at least one paper on the design of radio-technical apparatus.

From his earliest childhood, Viktor had developed a love for writing, reading and literature. In particular, he liked the science fiction novels of Tsiolkovskiy and the classics of Lyermontov, London and others. He would visit the library often. When a student, he wrote articles about movies for publication in the local newspaper, and reviewed the literature of ancient China. He had an impressive literature style. It is not too far from the truth to say that Viktor was one of the best journalists, writers and reviewer among the young physicists in the Soviet Union in the early 1950s. In addition, he liked music, history and art. Many years later, when a cosmonaut, he travelled with his colleagues to one of the eastern cities. In touring its galleries and museums, he served as a guide for the cosmonaut group. He was a great planner and organiser. As a student, his only group sport was handball. He preferred individual sports, particularly skiing, chess and biking. He was a sharpshooter, and competed as an archer in the national championships and in the Spartak games. His vacations were taken with his mother and sister. Galina wrote: "He was frequently thoughtful and reserved. I remember, on vacations he would lay on the grass in the backyard for hours with his hands under his head. He used to think intensely about things. He liked to be on his own."

Viktor graduated with distinction on 12 June 1955. The title of his final exam was 'The Design of A Harmonic Functional Analysis Device' and it was 117 pages! He wanted his graduation exam to be something special which would far surpass the required standard, so he prepared it not only as a student but also as a scientist. One week before his 22nd birthday he graduated as a mechanical engineer.[15] He went to work as a design engineer at the Central Aerological Observatory of the national Hydro-Meteorological Service, located in the town of Dolgoprudniy. He designed instruments to be carried on the balloon-borne packages and 'sounding' rockets that were used to gain data to identify the physical characteristics, chemical composition, temperature, humidity, pressure, radiation and magnetism of the upper regions of the atmosphere. The Observatory was the leading institution in the Soviet Union for atmospheric exploration. Initially, Viktor had difficulty adjusting to the job. Being a loner, he avoided speaking with his colleagues about his problems. And because he

---

[15] By this time, Viktor could already speak German. He mastered English several years later, while working at OKB-1.

did not keep a diary the only evidence we have of his thoughts during this time are his letters to his mother and sister and, after his marriage in late 1956, to his wife Vera Kryazheva.

Interestingly, in her book Galina Patsayeva described in detail Viktor's friends and colleagues and every important event in his life, but said nothing about how he met Vera, about their marriage, or even about their children. Vera was a researcher in the Central Scientific Research Institute for Machine Building (TsNIIMash) in the Kaliningrad district of Moscow. It hosted one of the ballistics groups which supported the mission control centre at Yevpatoriya. Their son Dmitriy was born in the autumn of 1957, and daughter Svetlana in February 1962.

In his letters, Viktor described his dissatisfaction with the working environment of the Observatory, and in particular his disappointment at the role of young graduates in the organisation. However, he did not let his dissatisfaction disturb his work. In contrast to other young engineers hired straight from university who needed at least a year to familiarise themselves with the new environment, in less than two months he had been accepted as an equal employee. His first assignment was to develop an apparatus to measure sky brightness, and this was later installed in a meteorological rocket. His managers recognised his skill, working habits and commitment, and in

Viktor Patsayev as a student (left, from the book *Boldness of Aspiration*, courtesy www.astronaut.ru), and tending to a cine-projector with his daughter Svetlana.

January 1956 promoted him to senior engineer in the group investigating the upper atmosphere. He found this much more to his liking. In a letter to his sister he wrote: "The new job has its own characteristics. It also involves annual expeditions. Now, I am much happier. ... I like the fact that I will be able to see the world. Our design group isn't large, there are just six of us. ... I cannot say that we have many great successes, although there have been some." He participated in expeditions that fired rockets from different sites, including deserts. He continued with his scientific work, too, publishing papers on the design and testing of scientific instruments that were highly regarded by experts. Even so, his colleagues observed that although his work was brilliant he was always dissatisfied, always thinking that he should have been able to achieve better. He was working on the design of instruments to analyse the chemistry, temperature and the magnetic field of the upper atmosphere. As he put it: "Actually our small group of young engineers and technicians are initiators of the new ideas."

However, by now something else was interesting Viktor. He met Korolev for the first time in September 1957, when Korolev gave a presentation to the conference dedicated to the 100th anniversary of Tsiolkovskiy's birth. That same day, Viktor read an article in *Pravda* about rocketry signed by 'Prof. K. Sergeyev', which was the pseudonym used by Korolev to hide his true identity from Western spies. A few months later, when Sputnik was already orbiting the Earth, Viktor met Korolev at a seminar. The 24-year old engineer went to Korolev and, after introducing himself, asked if he could transfer to his design bureau. Korolev, who liked young engineers who spoke their mind, asked about his current work. After Viktor had outlined his experience in the design of meteorological rockets, Korolev said that he would find him a position, and in November 1958 Viktor moved to OKB-1 to work as a design engineer. "I was lucky to work with extraordinary people – with the real engineers. They were totally devoted, not wasting even a second of time. They also knew how to inspire others. We all strongly wished to do something new, something unusual." He worked as a designer engineer for elements of the spacecraft, including its life support system, until November 1961, and then in January 1962 became an acting manager of one of the sections of OKB-1. Later (together with Vadim Volkov and other young engineers) he became a member of the recovery team responsible for the evacuation of cosmonauts after landing. At that point, Viktor started to think of becoming a cosmonaut. After discussing it with his wife, he once again went to see Korolev. In contrast to previously, Korolev was reserved. Too many of his young engineers were expressing their interest in leaving their design work in order to join the cosmonaut team. He asked Viktor why he wanted to became a cosmonaut. As usual, Viktor carefully explained his thoughts, saying that he believed that he would be more useful if he personally tested his equipment in the spacecraft. Korolev said 'Good. We'll solve that', and the meeting was over. However, it took a long time to gain permission to form a group of civilian cosmonaut-engineers, and when the first group was announced in May 1966 it contained Vadim Volkov, who was Viktor's friend from the Kolomenskiy flying club, but not Viktor.

Svetlana Patsayeva remembers this time with nostalgia: "We lived until 1967 in a communal flat – three families in a 3-roomed apartment with a shared kitchen and

The Patsayev family in 1966. Wife Vera (left), daughter Svetlana, son Dmitriy, Viktor and Zinaida Nikolayevna, Vera's mother. (Copyright Svetlana Patsayeva).

conveniences. . . . I recall my childhood as a happy time. There were four children in one apartment. We played together all the time. We celebrated our holidays with the neighbours in the flat, and lived very happily and harmoniously. Our parents were young and happy. Previously, they had lived only in the student hostel. Now it was one room in a wooden house, where in winter the strong frost froze the water in the teapot. Nevertheless, we were all happy, merry and friendly! It is not surprising that so many folk were dreamers, inventors. Papa was merry, good and a great inventor. He loved to play with us children, although he had very little time left for the family. He would come home late and continue his work at home. All the tasks in the new apartment he did with mom. He liked sports and trained us. During holidays, we all rode bikes together, and skied in the forest in winter. We had football with friends on the meadow. We would light a fire for a barbecue. In the evening, he would read a book to us and then tell a story. We discussed much with him about mathematics and astronomy. We loved to fantasise together."

In early August 1967 Mishin signed the document to recruit research engineers to participate in the N1-L3 lunar landing programme. On 18 August, after passing the medical examinations at the IBMP, Viktor was accepted into the TsKBEM's second group of civilian cosmonauts with Vladimir Nikitskiy and Valeriy Preobrazhenskiy.

---

[16] The members chosen from the first group were Anyokhin, Bugrov and Dolgopolov – but Dolgopolov left before the lunar training commenced.

Some snaps taken during Patsayev's training. Lower left: In the vestibular testing chair. Lower right: working with a telescope at the Byurakan Observatory. (From the book *Boldness of Aspiration*, courtesy www.astronaut.ru) Top: A picture taken at Yevpatoriya in the summer of 1969 in training for the Contact programme. Patsayev and Dobrovolskiy (second and third from the right), are accompanied by cosmonauts Makarov, Belyayev, Vorobyev (first, second and third from the left in the first row), Rukavishnikov (first on the right) and Klimuk (second row, first from the left). (From the book *Triumph and Tragedies of Soviet Cosmonautics*, courtesy www.astronaut.ru)

At that time, he had been the director of Section No. 324 at the TsKBEM for almost two years. Mishin chose three engineers from the first civilian group and three from the second – including Viktor – for lunar training.[16]

Ten days after joining the cosmonaut group, Viktor completed his examinations at the flying club and became a sporting pilot of the Yak-18 aircraft. In February 1968 he began to fly with an instructor pilot seated behind him, and on some flights he conducted complex manoeuvres that subjected his body to a force of 5 g. Later, he made parachute jumps. With the other research cosmonauts, he trained to simulate weightlessness using a Tu-104 aircraft. In addition, there was a special platform for investigating the dynamics of landing on the Moon. They visited the Zvezda design bureau, and Viktor participated in tests in which he donned the bulky Krechet-94 lunar space suit and rehearsed walking in conditions approximating lunar gravity on a surface expected to be similar to that of the Moon. He was also involved in testing the LK lunar module of the N1-L3 programme. This was to take a single cosmonaut down to the Moon while his colleague remained in the main ship in orbit. Although not members of the lunar cosmonaut group, Viktor and his colleagues contributed to testing the equipment required for the lunar landing mission. In addition, research cosmonauts assisted with the Mi-4 helicopter that was modified to simulate the final phase of a lunar landing.

Viktor's best friends among the civilian cosmonauts were Nikolay Rukavishnikov and Oleg Makarov. According to his son, Dmitriy, Viktor was also a close friend of Vladimir Nikitsky, who was assigned to the lunar training group but had to leave the civilian cosmonauts in May 1968 after being injured in an automobile accident. They all lived in the same neighbourhood, and their families were also friends. His sister Galina recalls: "When he moved to work at the Korolev design bureau, Viktor began to tell me of Tsiolkovskiy, Kibalchich and Tsander. In December 1968, while walking on the frozen Volga River, he said: 'Such people, you can meet only once in a hundred years.' He respected them all so much. Also, he spoke of the flights of Gagarin, Teryeshkova and Beregovoy. My mother told me that Viktor was working with the cosmonauts. But we didn't know he was already a cosmonaut himself!" However, Svetlana Patsayeva thinks that her mother did know: "In our family, we didn't talk about it. Papa was often away on business trips, but I didn't know where he went. Mom certainly knew of his appointment to the group of cosmonauts and of his training trips, but they didn't discuss that with us."

Cosmonaut Yevgeniy Khrunov recalled of Viktor's first days at the TsPK: "When Viktor began training, we saw immediately how well this quiet and unpretentious man could work. His modesty was incredible. I remember, during the first months, I sat next to him in the cafeteria. And apart from "hello" and "see you later", he did not say anything during a period of two weeks. Later, I understood that he remained silent simply because he did not wish to make much noise."

In May 1969 Viktor joined the Contact group, which was training to test in Earth orbit the rendezvous and docking system to be used in lunar orbit by the spacecraft of the N1-L3 programme. It was during this time that Viktor met Dobrovolskiy for the first time. Viktor was assigned as flight engineer for the active spacecraft on the first test. His commander was to be Lev Vorobyev, until Vorobyev was replaced by

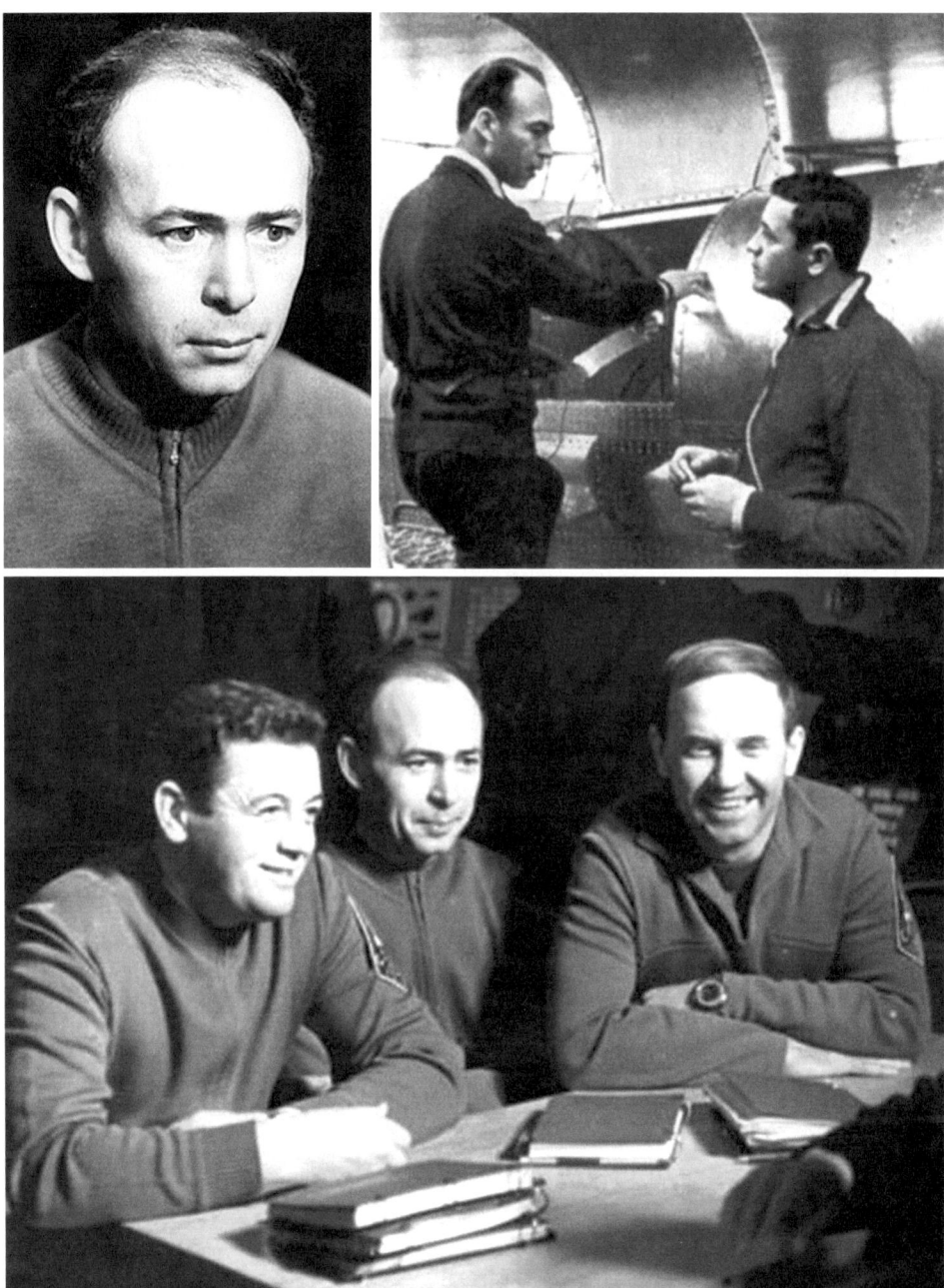

Patsayev in training for the 'third crew' of DOS-1. The photo at top-right shows him with Volkov preparing a centrifuge. (From the book *Boldness of Aspiration*, courtesy www.astronaut.ru)

Rare photos showing Dobrovolskiy, Volkov and Patsayev in the final phase of their training at the TsPK in Zvyozdniy.   Behind are the operator's controls for the Soyuz simulator. Bottom: the cosmonauts with General Nikolayev, who was in charge of their training. (Courtesy Peter Pesavento).

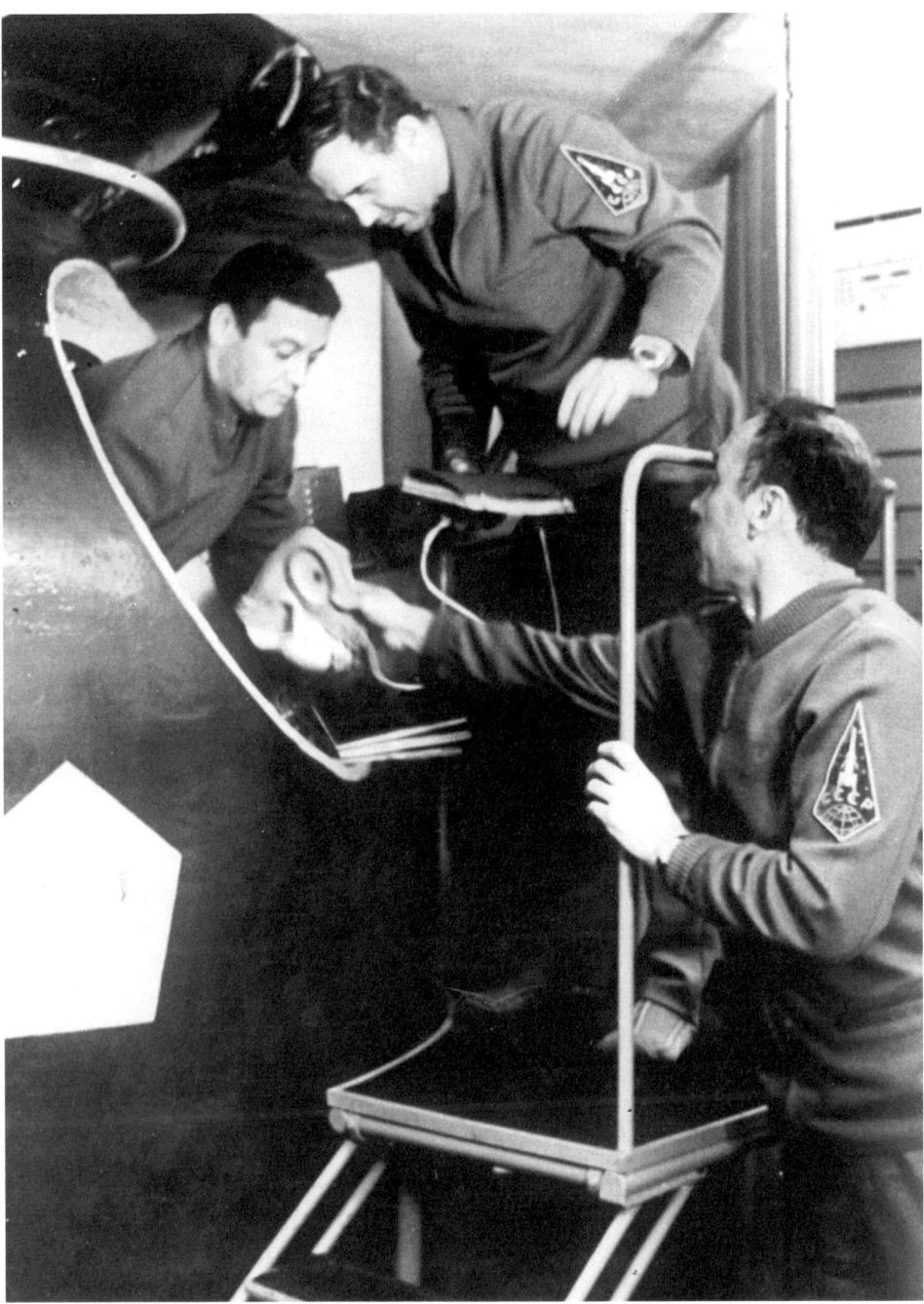

Volkov (in hatch), Dobrovolskiy and Patsayev prepare for their final training in the Soyuz simulator. The 'mission patch' is particularly well presented on Patsayev's arm. (From the private collection of Rex Hall)

"An ordinary journey." Prior to departing for the cosmodrome as the backup crew for Soyuz 11, Patsayev (left), Dobrovolskiy and Volkov visited Lenin's office. (From the private collection of Aleksandar Zheleznyakov)

Volkov (left), Dobrovolskiy and Patsayev at Baykonur, now the prime crew for Soyuz 11.

Pyotr Klimuk. Then in early 1970 Mishin assigned Viktor as research engineer on the third crew for the DOS-1 station. Viktor was delighted. His rank in the crew was a minor issue. Although he preferred individual sports, at work he was definitely a 'team player'. "Your position in the crew – flight engineer, researcher, physician or commander – isn't important. In order to work well together, we have to believe in and respect one another, and we must celebrate the achievements of our crewmates. That is the foundation of a crew." At first, he trained with Shatalov and Volkov. But in February 1971 Dobrovolskiy was made commander.

Cosmonaut Viktor Gorbatko, who flew with Vadim Volkov on Soyuz 7, wrote of Patsayev: "When Viktor began to train for a flight he visited us in Zvyozdniy more often, but he was almost invisible. In the medical room, on the sports field, in the cafeteria, he was always so quiet. Some of the staff at the TsPK did not realise who he was until he actually flew in space! He simply had no desire to be the centre of attention. I think he was the opposite of Vadim. I would watch them in Zvyozdniy. Volkov would wave his hands about to demonstrate something. All the time, Viktor would be quiet, looking down, but then he would say something softly and Vadim would fall silent – the conversation was over."

Viktor spent the Labour Day holiday of 1–2 May 1971 with his family and friends in the countryside. This would prove to be the last time that Svetlana Patsayeva saw her father. "According to a family friend, when they played football father was the goalkeeper. However, he did not stand between the posts, he climbed onto the cross bar and hung upside down. Our friend explained that later he understood that father had done this to assist in preparing himself for weightlessness. According to mom, when he left for Baykonur he had little expectation that he would fly." Her brother Dmitriy adds: "We knew nothing of the flight and the orbital station. All that was secret. Our mother knew only some details. Our father said only that he was leaving for a short trip related to his work. And it was planned as an ordinary journey, just like the previous ones."

But when Kubasov was grounded by an ailment, the mission was assigned to the backup crew. On the eve of his launch on Soyuz 11, Viktor told journalists: "The profession of cosmonaut cannot be anything except attractive. Space exploration is something new and very interesting. I think this flight is a logical continuation of my life."

Specific references

1. Davidov, I.V., *Triumph and Tragedies of Soviet Cosmonautics*. Globus, Moscow, 2000, Chapter "Полет продожается" (Flight Continues) (in Russian).
2. Volkov, Vl., *We Pace into the Sky*. Molodaya Gvardiya, Moscow, 1971, pp. 121–123 (in Serbian).
3. Patsayeva, G.I., *Boldness of Aspiration*. Molodaya Gvardiya, Moscow, 1976, pp. 5–64 (in Russian).
4. Lebedyev, L., Lukyanov, B. and Romanov, A., *Sons of the Blue Planet 1961–1981*. Politizdat, Moscow, 1981, pp. 190–204 (in Russian).
5. Burgess, Colin and Doolan, Kate (with Vis, Bert) *Fallen Astronauts, Heroes who Died Reaching for the Moon*. University of Nebraska, 2003, pp. 181–186.

Interviews with the author:

Marina Dobrovolskaya, 24 May 2007
Svetlana Patsayeva, 1 August 2007
Dmitry Patsayev, 5 September 2007

7

Home in orbit

LIFT-OFF

When Georgiy Dobrovolskiy, Vladislav Volkov and Viktor Patsayev were woken up at 3 a.m. on Sunday, 6 June 1971, it was still dark at the Baykonur cosmodrome. They briefly exercised, shaved, had a light breakfast – their last meal on Earth – and then the final medical checks. An hour later, after brief reports of the status of the rocket and Soyuz 11 spacecraft, the State Commission gave the 'green light' for the launch, and the rocket was fuelled. In contrast to previous missions, this time there were no backups to ride with the crew to the pad. However, they were accompanied on the bus by the Soyuz 10 cosmonauts and some of the officials from the TsKBEM and the TsPK. Just before 5 a.m., with dawn breaking, the bus drew up to the pad, where members of the State Commission, designers, engineers, technicians, military officers, pad workers, TV crews and reporters were waiting.

The cosmonauts wore grey cotton flight suits. Traditionally, military cosmonauts wore officers' caps, but this time all three men wore pilot caps with a badge on the front depicting the Soviet coat of arms. In addition, each man had on his left arm a tall triangular patch with a dark-blue background, a yellow rocket rising from the Earth towards yellow stars, and the letters 'CCCP' below. In contrast to American astronauts, the majority of Soviet cosmonauts did not wear 'mission patches'. The first Soviet patch was designed for Valentina Teryeshkova in 1963, and it was sewn on the blue garment that she wore inside her bright orange pressure suit; it depicted a dove and a small laurel branch. The second patch was created by Aleksey Leonov, a passionate space artist, and was worn on the right arm of his space suit during his historic spacewalk in 1965. Khrunov and Yeliseyev both wore Leonov's patch for their external transfer from Soyuz 5 to Soyuz 4, because they wished to retain it as a symbol of spacewalking. The early cosmonauts had 'CCCP' in large red letters on their white helmets, but their successors in cotton suits did not have a patch, a coat of arms or even a flag. But in mid-1970, training for the DOS missions, Leonov put his old patch on his flight suit. Kolodin did likewise, but Kubasov seems not to have joined in.[1] As their backups, Dobrovolskiy, Volkov and Patsayev also accepted it. In addition, Volkov had a small rectangular white badge on the left side of his chest.

After Dobrovolskiy had made a brief report to General Kerimov, the chairman of the State Commission, the cosmonauts, surrounded by journalists and pad workers, walked to the rocket, which was lit by floodlights. This was another contrast to the NASA way, whose astronauts don their suits in a building 8 km from the pad and, upon emerging, simply wave to friends and reporters on their way to the van which drives them to the pad, which is clear apart from the team whose task is to assist the crew into the spacecraft. At Baykonur, the departing crew walks through the crowd, speaking to individuals, even joking. Dobrovolskiy, Volkov and Patsayev halted in front of the stairs to the elevator, turned and waved. Reading from a piece of paper, Dobrovolskiy began a speech: "I bow my head to all of you for your attentiveness, for your effort." Then as they smiled to the numerous TV and photographic cameras, Volkov whispered to Dobrovolskiy: "It's time to go."

Volkov led the way up the steps, with Patsayev and Dobrovolskiy following. The liquid oxygen boiling off from the rocket blew in small clouds past the cosmonauts. At the top of the steps, Dobrovolskiy turned and called to the crowd: "Don't worry. Everything will be normal; everything will be normal!" The three men paused at the door of the elevator to wave. Even Patsayev smiled. Then they disappeared into the elevator. When they emerged on the platform leading to the hatch in the side of the orbital module of their spacecraft, they posed for one of the photographers, which is another notable detail of this mission because cosmonauts did not generally pose on this platform. The result was one of the most extraordinary photographs taken of this crew. As a final farewell, the three men stood at the railing of the platform and waved their caps at the crowd.

Technicians assisted first Volkov, then Patsayev and finally Dobrovolskiy to enter the spacecraft. Access was through the side hatch of the orbital module, then down through the interior hatch into the descent module, which contained their couches.[2] When Volkov had taken his place, he switched on the cabin lights and ventilators. Once Patsayev was in place, Dobrovolskiy joined them. The technicians bid them farewell and hermetically closed the hatch between the descent and orbital modules, then the external hatch of the orbital module. With the cabin sealed, the silence was striking. Each cosmonaut donned a cap of white net which included earphones and a small microphone on each side. There was still an hour remaining to the time of launch.

As they settled in, Volkov turned to Dobrovolskiy, they smiled at each other, and Volkov pointed towards Patsayev, who was quietly gazing out of the small porthole by his left shoulder. Finally, Patsayev turned his head inwards to his colleagues, and smiled.

With 30 minutes to go, the twin sections of the service structure split the eight levels of the wrap-around walkway and swung down to leave the rocket exposed on

[1] To be precise, if all three members of Leonov's crew wore the patch, then in the case of Kubasov it is not apparent in the photographs available to the author.

[2] The commander's couch was in the centre, the flight engineer to the commander's right and the research cosmonaut to the commander's left. Spanning the cabin in front of them was a panel of instruments, switches and indicator lights.

Volkov (left), Dobrovolskiy and Patsayev take their last steps on Earth as they walk in-step across the pad to their rocket.

Pre-launch speeches, jokes and farewell waves to the crowd gathered on the pad.

After a final wave from the top of the service structure, with the nozzles of the solid motors of their vehicle's launch escape system in the background, the Soyuz 11 crew pose for a photograph moments before boarding their spacecraft.

the pad. Ten minutes later, the topping-off of the liquid oxygen tanks ceased and the kerosene tanks were pressurised.

The control room was in a bunker some 2 km from the pad. A black-and-white TV monitor showed the cabin, but because the camera was located above Volkov's head only Dobrovolskiy and Patsayev were visible. The communication officer was Leonov. Also present was Afanasyev of the Ministry of General Machine Building, who had just arrived from Moscow. The traditional radio call-sign for the TsUP was Zarya.[3] The call-sign for the mission was Yantar.[4]

At 7.40 a.m., Dobrovolskiy reported: "This is Yantar, and we're ready to go up."

With just 40 seconds remaining, the rocket was switched to internal power and its automatic sequencer was activated. Twenty seconds later, the umbilical arm swung away.

---

[3] 'Zarya' means 'Dawn'.

[4] Spacecraft radio call-signs were stones – Soyuz 10 was 'Granit' ('Granite') and Soyuz 11 was 'Yantar' ('Amber'); Dobrovolskiy was 'Yantar 1', Volkov was 'Yantar 2' and Patsayev was 'Yantar 3'.

The cosmonauts could hear the final commands. They tightened their seatbelts. The fuel tower withdrew from the vehicle.

Volkov, the only veteran, who knew the launch commands very well, again joked: "Let us wave farewell to them." They waved to the TV camera. Then: "The key is on the Start button."

A second later, the launch operator repeated this command. Dobrovolskiy turned to Volkov: "It looks like you want to go early."

The final commands:

"Ignition!"

"The main!"

"Start!"

The vehicle had a central core stage and four strap-on boosters, each with a main engine. A turbopump in each of the five segments began to feed fuel and oxidiser. At 7.55 a.m., pyrotechnic charges were simultaneously fired to start the five main engines. The rocket was not actually supported at its base; the core was held by four arms located just above the top of the strap-ons. As soon as the thrust overcame its 310-tonne mass, the rocket began to lift. This released the supporting arms, which immediately swung out like the petals of a flower in order to clear the way for the protruding strap-ons. One way or another, Dobrovolskiy, Volkov and Patsayev were now committed.

The launch was perfect.

Dobrovolskiy reported that there was very little vibration.

After initially rising vertically, the rocket pitched over to a northeasterly heading, and as it ascended through the atmosphere it passed almost directly over the town of Baykonur, some 350 km from the launch site.[5] At 115 seconds, at an altitude of about 45 km, the ring of solid-rockets at the top of the 6-metre-tall escape tower were fired to pull it free of the vehicle. A few seconds later, the four 20-metre-long strap-ons shut down and were jettisoned. The core continued and, with half of its propellant used, it accelerated rapidly. At an altitude of 80 km, above most of the atmosphere, the shroud which had protected the spacecraft from aerodynamic loads was jettisoned. At 288 seconds, at an altitude of 175 km, the 30-metre-long core shut down and was jettisoned. The four-chambered engine of the 8-metre-long third stage started immediately, and by the time that it had built up to its full 35-tonne thrust the framework interstage had been jettisoned too. Soon the third stage began to pitch over further in order to increase its horizontal speed.

At 8.02 a.m., the report from the spacecraft was: "Temperature is 22°C, pressure is 840 millibars, all is well."

At 8.04 a.m., far north of China, the third stage shut down. Dobrovolskiy reported: "Orbital insertion. Commencing separation, stabilisation. Antennae and solar wings deployed. On board everything is in order. Feeling normal."

The parameters of the initial orbital were: altitude 185 × 217 km at 51.6 degrees to the equator and a period of 88.3 minutes. At the time, Salyut's orbit was 212 × 250 km in the same plane with a period of 88.8 minutes.

---

[5] When Sputnik was launched in 1957, this town was the nearest large population centre on the track of the rocket's ascent, so the launch site came to be known as the Baykonur cosmodrome.

Soyuz 11 on the pad, minutes before ignition.

The first Western site to detect signals from the new spacecraft was the Kettering Grammar School in Northamptonshire, England, which picked up a signal within 10 minutes of launch. Geoffrey Perry, the senior science master who led the tracking team, was able to announce the launch of a spacecraft carrying three men more than an hour ahead of the Moscow news report. In fact, because the Kettering team had noticed that after a period of silence lasting almost five weeks Salyut had recently made several manoeuvres and started to transmit signals, they had been awaiting the launch.[6]

THE FIRST ORBITS

Once they had settled down in orbit, Dobrovolskiy began a diary in his notebook, starting with the launch and his impressions of weightlessness:

> The launch went normally. A smooth flight. We felt some swinging and vibration but it wasn't a problem – not too strong. Before separation of the last rocket stage, the loads increased. Then, in an instant, silence! The interior of the cabin became brighter. On board clocks and the globe instrument started after a few seconds. After separation, there was a great deal of dust floating in the cabin. The ventilator worked, but we also collected the dust with the aid of wet tissues. . . .
>
> We have had communications with Earth twice.
>
> At 11.43 we heard the TASS announcement of our launch. On board the ship, everything is all right. After separation from the rocket we all had an unpleasant feeling, but we feel better now. It was just as if someone was trying to pull off our heads. We felt our neck muscles strain. It was as if everything in our body had moved up, and our heads seemed heavier. These feelings were weak while we were in our seats, but still present. In these moments, the forehead and the top of the head seemed to be so heavy. I had the feeling that everything inside my body had moved up.

At the onset of weightlessness, the internal organs that had been held in place by gravity were free to migrate upwards inside the chest and the body fluids that would normally be drawn into the legs tended to accumulate in the upper body, giving the impression of a swollen head.

On the second orbit Dobrovolskiy reported to Earth that the spacecraft's systems were working as they should, and that the crew was feeling well. It was the time to unfasten their belts and enter the orbital module.[7] Volkov, the flight engineer, was the first to leave his couch. After checking instruments on the control panel which indicated the composition, pressure and temperature of the air in the orbital module, he opened

---

[6] The Kettering team's first success was Sputnik 4 in 1960. Its achievements included detecting signals from the Voskhod spacecraft in October 1964 prior to the completion of its initial orbit; identifying the location of the Soviet cosmodrome at Plesetsk in 1966; and the first Western detection of signals from the first Chinese satellite in April 1970.

[7] Although commonly described by Western observers as the 'orbital module', the Russian term for this part of the Soyuz spacecraft, *bitovoy odsek*, is more appropriately translated as 'habitat module'.

the hatch and, like a fish, floated through. The descent module was a bell shape with a 'free volume' of 2.5 cubic metres, but the spheroidal module was some 2.2 metres in diameter and had a volume of just over 4 cubic metres.[8] After they had followed him, the rookies delighted in floating in this weightless 'aquarium' – little children once again. The descent module had two small portholes, but the orbital module had four larger ones located at 90-degree intervals. After the initial novelty of weightlessness wore off, the three men went to the portholes to observe the Earth. It was a beautiful sight. The fact that the planet is a sphere was very obvious. They were flying over the Pacific Ocean, and the Sun was reflecting off the surface of the water. Directly below, the colour of the ocean was deep blue. Towards the horizon, it changed to dark grey. Far away, it was in darkness.

"The sea is always beautiful, even from space, and we can't live without it," said Dobrovolskiy. For him, the first moments in orbit were a flashback to his childhood on the coast of the Black Sea.

Then the cosmonauts returned to the descent module and Dobrovolskiy reoriented the spacecraft to enable the Sun to fully illuminate the solar panels.

On the fourth orbit, at 1.50 p.m., Soyuz 11 successfully made its first manoeuvre to start the rendezvous with Salyut. Then control was transferred from Baykonur to the TsUP in Yevpatoriya.

Because the mission began on a Sunday morning, the families of the crew were at home.

Marina Dobrovolskiy has said that her father was often away from home and she never knew where he went. She recalls when she heard that he had been launched into space: "Of course, I was happy for my father. However, I wanted so much that he should return as soon as possible. We were given a brief note that he'd written to mom, my sister and I, in which he said that we shouldn't worry and that everything would be good."

Svetlana Patsayeva was attending a Young Pioneers' camp, and the news was not entirely unexpected: "I felt that father had some important and very serious work. For me, he was the great authority, and I was really not surprised that he was flying in space."

Viktor's mother Mariya and stepfather Ivan, who at the time lived in the village of Rozhdestvo on the Volga River, were not even aware that Viktor was a cosmonaut! Mariya was in the kitchen and Ivan was fishing when the national radio announced the news:

> In accordance with the programme of near space exploration, at 7.55 a.m. Moscow Time on 6 June 1971 the Soviet Union launched the spacecraft Soyuz 11. At 8.04 a.m., the spacecraft reached the planned orbit. The crew is: commander, Lieutenant-Colonel Dobrovolskiy, Georgiy Timofeyevich; flight engineer, Hero of the Soviet Union, Volkov, Vladislav Nikolayevich; and research engineer Patsayev, Viktor Ivanovich.

---

[8] The 'free volume' of a module was that which was available to the crew after all of the apparatus had been installed.

When she heard the name of her son, Mariya Patsayeva exclaimed so loudly that a neighbour rushed in to see if anything was amiss. When Mariya explained the news about her son, both cried. Others soon arrived and the celebrations began. Someone placed a table against the front wall of the house bearing a notice stating that in this house lived the parents of cosmonaut Viktor Patsayev.

Meanwhile, in space one of the major tasks for the crew on the first flight day was to familiarise themselves with being weightless. They also monitored their vehicle's systems and performed the preliminary preparations for the rendezvous and docking with Salyut. Then it was time to rest – they had awakened early for launch. This phase of the mission was timed to coincide with the period in which Soyuz 11 remained out of contact with the Soviet communication stations for a prolonged time, which was from 3.40 p.m. through to 1.30 a.m. the following day, 7 June.

From Dobrovolskiy's notebook:

> 6 June 1971: Vadim and I slept in the orbital module, in our sleeping bags in a heads-down orientation.[9] Viktor remained in the descent module, stretched across the couches, also in his sleeping bag. We slept from 6.30 p.m. until 12.00 midnight, which was less than usual, but our impression was that we had a good rest. When we returned our heads to the normal position, they again started to swell.
>
> Vadim and I looked in the mirror, then at each other, and smiled – "we have swollen up like bulldogs". We awakened Viktor and made another communication session. Everything is all right on board. Vadim suggested washing our faces with wet tissues. We did so, and returned to our work. At 2.48 a.m., when we flew over the equator, we heard music coming from the direction of Antarctica.

"THE STATION IS HUGE"

After confirming that there were no problems with either the spacecraft or the crew during the first hours of the mission, at 3.00 p.m. on 6 June Kamanin and Shatalov took off in an IL-18 with a dozen Air Force flight control and docking specialists. At 5.00 p.m., the TsKBEM team set off in another IL-18. This group comprised the leading specialists in the spacecraft's systems, namely Mishin, Chertok, Shabarov, Feoktistov and Yeliseyev, accompanied by Minister Afanasyev and some members of the State Commission. After about 4.5 hours in the air, the planes landed at the military airfield near the town of Saki, in Crimea. The passengers were immediately driven to Yevpatoriya. Already at the TsUP were cosmonauts Nikolayev, Gorbatko and Bykovskiy, who had been assigned by Kamanin to talk to the crew. Meanwhile, cosmonauts Leonov, Kubasov, Kolodin and Rukavishnikov flew from Baykonur to Moscow on a third plane.

---

[9] The managers had accepted Rukavishnikov's suggestion that sleeping bags be carried on the Soyuz.

The Chief Operative and Control Group (GOGU) for the Soyuz 11 mission had five members. General Pavel Agadzhanov was in charge. Yakov Tregub, technical supervisor, was responsible for analysing the signals from space and preparing the commands to be transmitted to the two spacecraft. When the specialists from Baykonur arrived at the TsUP, Agadzhanov and Tregub confirmed that everything was normal on both Soyuz 11 and Salyut, and that the crew were resting as planned. Based on the biomedical telemetry and the reports from the cosmonauts, Volkov was in the best condition; his body obviously 'remembered' the weightlessness of his 5-day flight in 1969. Yeliseyev, who had just left the cosmonaut group in order to become Tregub's deputy, took a seat next to Tregub. He was also able to communicate with the crew. The final members were Boris Chertok and Boris Raushenbakh, experts in the spacecraft's guidance, control and electrical systems.

At 6 a.m. on 7 June, as Soyuz 11 made its approach to Salyut, the main control room on the second floor of the TsUP building was packed. Although flight control required only the members of the GOGU and five specialists for data analysis, the command-measurement complex, communications, telemetry and medical support, there were almost 100 people present – many of whom were not directly involved in mission control but had been drawn by the significance of the upcoming event. When the overcrowded room became too stuffy, someone opened the windows and a fresh sea breeze made conditions more tolerable. The communication session was to start at 7.25 a.m., and last for 23 minutes. As the time for contact approached, there was a marked increase in tension.

In the Flight Control Centre at Yevpatoriya, members of GOGU team follow the docking operation in space. In the first row of the left picture are Chertok (glasses) and General Agadzhanov (profile). Beyond Chertok are Tregub (white shirt) and Raushenbakh (black suit). Cosmonaut Gorbatko is in the foreground, with his back to the camera. In the right-hand picture are Minister Afanasyev (left) in the main control room at Yevpatoriya, and Semyonov, the TsKBEM's DOS leader (centre), and Bugayskiy (his counterpart from the TsKBM).

After two manoeuvres, Soyuz 11 was known to be in the ideal orbit to achieve the rendezvous with Salyut. When the range was 7 km, the Igla automatic system was to establish radio contact with the station – a milestone known to the cosmonauts as 'radio capture'.[10]

At 7.26 a.m. Yeliseyev called the crew: "Here is Zarya. Yantar, how do you read us? On line!"

"This is Yantar," came the immediate reply. "Everything is going according to the programme. Radio-capture passed. The automatic approach is progressing. At 7.27 we are distance 4, speed 14." The distance was given in kilometres and the speed in metres per second.

"Understood," replied Yeliseyev. "Everything is normal. Continue reports."

"At 7.31, the SKD fired for 10 seconds. Distance 2.3, speed 8." By SKD he meant the correction engines.

Judging by the radio, it was Volkov making the reports. The stress was evident in his voice.

"Speed is decreasing. I can see a bright point in the VSK. Distance 1,400, speed 4." The VSK was the forward-looking periscope, and Salyut could now be seen in it as a bright point of light. The distance was now being reported in metres. "At 7.37, distance 700, speed 2.5. We have turned. I can see the Earth. Again, there is radio-capture!"

When the radio fell silent, some of the members of the State Commission turned towards the GOGU people in expectation. The NIP-13 ground station at Ussuriysk on the Kamchatka peninsula still had the spacecraft's signal, but it was only static. Yeliseyev called nervously: "Yantar, this is Zarya. I do not hear you."

At first there was continued radio static, but then: "Distance 300, speed 2. I can see the station excellently in the VSK. Roll alignment starts. The docking cone is very clearly visible. Roll alignment ended. Distance 105, speed 0.7. Manual control activated." Now that the Igla had brought the spacecraft almost to a halt 100 metres from the station, Dobrovolskiy had taken control for the final approach. Meanwhile, the station had oriented itself to face its front end towards the newcomer.

Yeliseyev called: "Yantars, when you close in, inspect the docking mechanism." He wanted the crew to look for any damage caused during Soyuz 10's unsuccessful attempt to dock.

"Yes, understood. Distance 50, speed 0.28. The DPO is firing." By DPO he meant the orientation engines. "The cone is clean. It is clearly visible. Distance 20, speed 0.2. The ship is stable. We're going to dock!"

A few seconds later the spacecraft passed out of range of NIP-13 and headed out across the Pacific Ocean. The next communication session would begin at 8.56 a.m. If all went well, the docking would be achieved on the station's 795th orbit, and on the 16th orbit of the Soyuz 11 spacecraft.

Leaving only those responsible for the analysis of telemetric data in the control room, the visitors left the building to attempt to relax after the almost unbearable tension. Just as in the case of Soyuz 10, when Soyuz 11 had flown out of radio range

---

[10] American astronauts would refer to this as 'lock on'.

An accurate painting of Soyuz 11 about to dock with the Salyut space station above the Kamchatka peninsula. Although the scientific instrument aperture on the main compartment is shown facing away from the Earth, in fact the protective cover had failed to release, rendering the instruments unusable. (Painting by Andrey Sokolov and cosmonaut Aleksey Leonov)

it was only a few metres from the station with everything progressing smoothly – but look what had happened on that occasion!

As the time for the next communication session neared, everybody crowded back into the control room to hear from the cosmonauts whether the docking had been successful.

This is how Dobrovolskiy described the moments leading up to and immediately following docking:

> At 7.24, the approach regime began. ... By a distance of 150 metres, the ship had aligned itself with regard to the main axis, placing the station in the centre of the periscope.
>
> At 100 metres, we switched to the manual regime. Speed: 0.9 metres per second. ... After switching, the station began to move to the right in the periscope. ... I began to decrease this lateral speed. ...
>
> I had the feeling that the left controller was insufficient, so I switched to the right one and slightly raised the ship ... and then with the left controller I succeeded in reducing the lateral speed. At 60 metres I reduced the speed to 0.3 metres per second. ... Mechanical contact at 7.49.15. We were stable. The docking occurred at 7.55.30. There were no vibrations or shaking. We almost did not feel the final contact.

Yeliseyev began to call just before the communication session was due: "Yantar, here is Zarya. On line!" Silence. He repeated his call several times.

Suddenly, the operator responsible for receiving TV signals excitedly announced: "There is television! Docking achieved! The picture is outstanding!"

Yeliseyev continued his calls: "Yantars, I'm calling you for the fifth time! Why do you remain silent?"

"Zarya, we report. There were no oscillations during the docking. The programme is complete! We will check the hermetic seal and equalise the pressure according to the programme. We have opened the hatch between the descent and orbital modules and moved into the orbital module. Everything is normal."

The control room was instantaneously abuzz and someone started to applaud, but Agadzhanov told them not to celebrate until the cosmonauts had entered the station. There were still many things to check. The hermetic seal of the docking mechanism had to be verified, the tunnel pressurised, and the hatches opened. Finally, there was the question of the station's atmosphere – had the problems with the ventilator fans during Salyut's first few days in space allowed the air to become toxic.

On the next orbit Volkov established communication before Yeliseyev could call: "Zarya, everything is normal. We are still in the ship. All pressures are within the limits specified by the table. We do not have any remarks. Permission to open the hatch?"

Yeliseyev looked at Tregub who nodded his head: "Open the hatch!"

"Zarya! At 10.32.30 we sent the command to open the hatch. The signal 'Closed' remained. If it doesn't open, we'll use the crowbar."

"Yantars, all goes excellently. Well done! Don't be disturbed. Work calmly."

"Zarya! The opening regime is executed. But the indicator didn't light. Evidently, it did not reach the terminal. However, Yantar 3 has opened it and is about to pass through!"

At 10.45 a.m. on 7 June, 26 hours 50 minutes into the flight of Soyuz 11, Viktor Patsayev entered the world's first space station.

"Yantars, attention!" called Yeliseyev. "The First will talk with you." Brezhnyev, the First Secretary of the Communist Party of the Soviet Union, was on a telephone line to Yevpatoriya. Some people in the control room were surprised that he wished to congratulate the crew so early, with only one man in the station.

The cosmonauts were also surprised: "Zarya, wait! Yantar 3 is in Salyut. Don't start until – Zarya, Yantar 3 has returned! There is a strong smell in Salyut! He will put on a mask and go in again!"

Realising that this was an inopportune moment for Brezhnyev to make his speech, Minister Afanasyev called the Kremlin and deferred the relay with the station to the next orbit.

Mishin was nervous: "All conversations and commands to space must be through me!"

Dobrovolskiy called: "When we opened the hatch, we peered through. The station is huge – there seems to be no end to it! After our compact spaces!"

"Yantars, activate the air regenerators. Communication is ending. We'll pick you up on the next orbit. We are all as happy as you are. Congratulations!"

The 25-tonne orbital complex comprising Salyut and the Soyuz 11 spacecraft left the communication zone. The orbital parameters were 212 × 249 km. The TV which had been recorded from space by Yevpatoriya was sent to the Kremlin, but was not yet released to the national television network.

In the meantime, Mishin asked the doctors to investigate whether the strong smell which had been reported posed a risk to the cosmonauts' health, but the doctors had no idea of the source of the smell and therefore were unable to offer any advice.

Before the opening call of the ensuing communication session could be made, the black-and-white screen of the control room came to life and showed Patsayev and Volkov inside the station. When the cosmonauts heard the sound of the controllers celebrating, they looked towards the camera and waved.

"They heard our ovations!" observed someone in the control room.

"Yantars, here is Zarya! The State Commission and Operative Group congratulate you most sincerely. You are the very first crew on a DOS. We suggest that you take a meal, get some rest, and tomorrow morning we will start the programme."

The only problem so far was the smell, and Patsayev had activated a system that would cleanse the air. Soon after launch on 19 April, six of the eight ventilator fans had failed and during the time that the station had been unmanned the air had grown stale with the smell of the burned insolation on two of the fans. Initially, Mishin had blamed Leonov's painting tools, but Dobrovolskiy said that the brushes and paints were safe in their box. Patsayev found small tracers used by technicians to identify the air flow during pre-launch preparations. After restoring all eight fans to service,

An unusual depiction of Soyuz-Salyut in an undocked state showing cosmonauts in both vehicles (top). The large conical housing for the main scientific instruments has been edited out. A view from an automatic TV camera as Soyuz 11 approached Salyut (bottom left). A TV view of Patsayev (left) and Volkov just after they entered the station (bottom right).

Patsayev and Volkov rejoined Dobrovolskiy, to sleep in their own spacecraft while the regenerators cleansed the air in the station.

While the station was flying outside of the communication zone, the control room was empty. In the evening, the State Commission met and decided that if everything went according to the programme the crew would return to Earth on 30 June – the maximum duration allowing a daylight landing. If successful, this would exceed by five days the record set by Soyuz 9. At the same time at Baykonur, the final preparations for the third launch of the N1 lunar rocket were in progress, and this

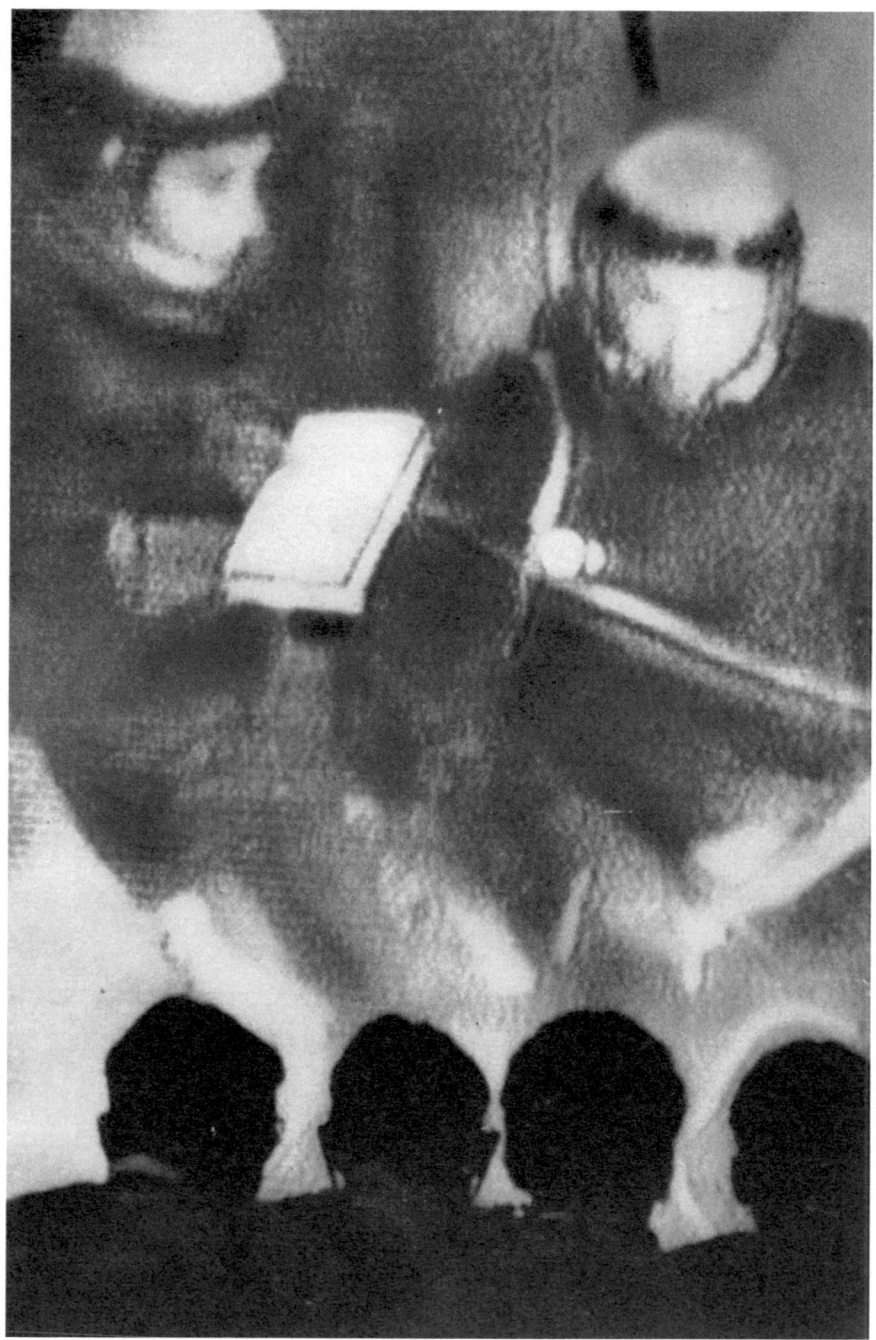

Patsayev (left) and Volkov, the first men on the first space station, seen on the wall screen in the TsUP at Yevpatoriya, with silhouettes of the flight controllers.

now became Mishin's focus. He recalled three of the GOGU members – Tregub, Chertok and Raushenbakh – to Moscow with him, leaving Yeliseyev to lead the specialists in managing the DOS mission, supported by Nikolayev, Gorbatko and Bykovskiy. Generals Kamanin and Shatalov and the other Air Force staff also returned to Moscow. Kamanin's aide, General Goreglyad, was at Baykonur to manage the landing and recovery operation.

On awakening, Dobrovolskiy, Volkov and Patsayev all entered Salyut, which was their new home in space.

Marina Dobrovolskiy recalls of these days: "People were coming and going all the time. The telephone rang. Congratulatory telegrams arrived. The docking was especially important. I remember the flight controllers congratulated mother, saying that the docking was performed excellently and that it was a crucial milestone – the station had begun operations!"

SPACE LABORATORY

In essence, the Salyut space station was a series of cylinders with small, medium, and large diameters. It had a total length of 13.6 metres, a maximum diameter of 4.15 metres and a mass of 18.6 tonnes. It comprised four sections. At the front was the transfer compartment. This was the smallest habitable section. It was 3 metres in length, just over 2 metres in diameter and had a volume of 8.1 cubic metres. It contained the life support and thermo-regulation systems. It also contained the No. 5 control panel for the Orion ultraviolet telescope. On the outside of this section were various masts and antennas, and a pair of solar panels which were identical to those on the Soyuz. The docking cone was on the axis at the front of this section. The hatch on the inward side of the docking system was one of three hatches in the compartment. There was a second axial hatch to provide access to the work compartment, and also a hatch on the outer wall with diameter of 80 cm to facilitate spacewalking, but there were no plans to go outside – indeed DOS-1 carried no EVA suits.

To enter the station, the cosmonauts had first to clear the docking system from the tunnel and then open the hatch to pass through the transfer compartment to the work compartment beyond. This was the largest component of the station and was in two sections. The smaller section (known as the first work compartment) was connected to the transfer compartment via a conical section 1.2 metres long. It was cylindrical, 2.9 metres in diameter and 3.8 metres long. It contained the central control panel, which incorporated a computer – the first on a Soviet manned spacecraft. Facing the panel were seats for two cosmonauts – the commander on the left (as viewed from the rear) and the flight engineer to his right. It was one of seven workstations for controlling Salyut's systems and experiments. The No. 1 station was to control the life support and thermo-regulation systems, and to control the automatic orientation and navigation of the station, but it also included a periscope for manual orientation. From there, actually, the commander could control and fly the station using displays and control handles similar to those of the Soyuz. The central panel consisted of the main control panel and command and signal devices. It provided information on the station's position over the Earth's surface, the number of the current orbit, the times at which the station would enter and exit the Earth's shadow and the periods during which it would be able to establish communication with the TsUP.

The system for orientation and control consisted of the following apparatus:

- ion sensors to measure the orientation of the station relative to its velocity vector;
- infrared sensors to determine the local vertical;
- Sun sensors;
- sensors for the angular speed during the rotation of the station;
- gyroscopes for measuring the angle of the station in three axes;
- an integrator for longitudinal accelerations;
- a stabilisation system;
- a control system for the orientation engines; and
- radio-location rendezvous apparatus.

While firing the manoeuvring engine, small orientation engines would hold the station stable. The system for manual control allowed the crew to align the station towards the Earth, the Moon, the Sun or the stars. While in stellar orientation, they would use a globe marked with the constellations and all stars brighter than the fifth magnitude.

The life support system controlled the gas mixture, eliminated strong smells and filtered out dust. In terms of millimetres of mercury, the pressure was maintained at 760 to 960, the oxygen concentration was 160 to 280, and carbon dioxide was never allowed to exceed 9. The air was cycled through a regenerator which contained an active chemical substance that removed carbon dioxide. Another unit topped up the oxygen. Water vapour was removed by a condensation trap. Special filters absorbed unwanted chemicals released by the materials on the station, the experiments and the crew. The equipment for the air regeneration system was to the left of the No. 1 control station.

The No. 2 station was for manual orientation and navigation. It included the control handles for the orientation of the station, a periscope and a means of stabilising the cosmonaut at his work position. Next was the No. 6 station, which included the flight engineer's seat. To the right, on the side of the compartment, was the No. 7 control panel to operate the scientific apparatus installed externally to analyse the environment around the station.

Aft of the central panel of the No. 1 station was the table for preparing and eating meals. Each cosmonaut had four meals per day, consisting of breakfast, morning tea, the main meal (lunch) and dinner. For the main meal, each cosmonaut had one item (soup or coffee) warmed on a small heater beside the table. They could choose on a daily basis between three types of ration for each of the four meals. For example, ration No. 1 had the following products:

- The 1st breakfast (705–756 calories)
  - o Sausages
  - o Borodin bread
  - o Chocolate
  - o Coffee with the milk

- The 2nd breakfast (600–700 calories)
  - o Russian cheese
  - o Rizhskiy bread
  - o Cookies

- Lunch (798–928 calories)
  - o Green shchi (a type of soup with mixed vegetables)
  - o Chicken meat
  - o Bread
  - o Plum jam with nuts
  - o Blackcurrant juice

- Dinner (593–745 calories)
  - o Caspian roach
  - o Puree
  - o Bread
  - o Honey cake.

The water tanks were located nearby the table and at the aft end of the working compartment. Each man was allowed 2 litres of water per day, but actually they did not use more than 1.2 litres. As on Soyuz 9, silver ions had been added to the water tank prior to launch to keep the water fresh.

Usually, the cosmonauts spent their spare time in this first working compartment, where they had a tape recorder with a selection of pre-recorded music cassettes, a small library and a sketchpad.

Externally, the larger section was 2.7 metres in length and 4.15 metres in diameter. It was joined to the smaller compartment by a short conical adapter. There was no internal distinction, however; the compartment was a single room with total length of 7.7 metres and a volume of 74 cubic metres. Including the transfer compartment, the total habitable volume of the station was 82 cubic metres. The central part of the larger working compartment was occupied by the main scientific equipment (ONA), which took the form of a large white conical unit that rose from the floor almost to the ceiling. It included the OST-1 orbital solar telescope, the RT-2 X-ray telescope, the ITS-K infrared telescope and spectrometer, the OD-4 optical viewer that had a magnification of 60, the FEK-7A photo-emulsion chamber, photographic apparatus and various other apparatus. On the walls around it were three portholes. The No. 3 station to control the scientific apparatus was adjacent to the ONA and included a viewing port. Unfortunately, the protective cover had failed to release when Salyut achieved orbit, and therefore these scientific instruments were unusable. The second control panel of this compartment was the No. 4 station, which was mounted on the adapter between the two sections of the working compartment. It was to control the main medical research equipment, and comprised scientific experiments, a viewing port and a chair.

In the upper corner to one side of the ONA sleeping bags were slung from hooks, but if they preferred the cosmonauts could sleep in the Soyuz orbital module or in the transfer compartment. On the opposite side and in front of the ONA there were exercise devices, including the KTF treadmill, an exercise bike and chest expanders. The crew had special 'penguin' suits designed to stimulate the muscles that would otherwise decay in weightlessness. The Polynom medical apparatus was for general monitoring of the crew's health. A small medical kit, identical to that carried on the Soyuz, provided pain relief, heart stimulation, relief of gastric problems, antiseptics, bacteriostatics and sleeping and stress relief tablets.[11] In fact, during the entire flight there were very few cases when the cosmonauts required medication.

At the aft end of the compartment, behind the ONA and separated from the rest of the working area, was the sanitary and hygienic unit. It had its own ventilators and its surface was a washable material. An airflow drew urine into a collector, where it was separated into its fluid and gaseous components. Solid waste was stored in hermetic tanks. Also at the aft of the compartment were the fridges containing food.

To assist the cosmonauts orientate themselves, the work compartment was painted in different colours – the front and rear were light grey, one wall was green, the other was light yellow and the floor was dark grey.

The cosmonauts had a collection of underwear and sports T-shirts. For cleansing their faces, hands and bodies following experiments, maintenance work or physical exercise they used wet and dry tissues and special towels made of bacteriological materials. From time to time, they were to clean the station using a vacuum cleaner.

Detachable panels on the walls and the floor covered support apparatus, electrical cabling, equipment for operating the station, monitoring the composition of the air, thermo-regulation, radio-links and the main command lines. The cosmonauts could open every panel and check the apparatus mounted on the compartment's structural frames. Hand rails on the walls and floor allowed easy movement in weightlessness. The walls held lockers of food, equipment, documentation, packed clothes, books, hygiene supplies and miscellaneous spare parts for repairs.

The thermo-regulation system had two major elements, one to cool the station and the other to warm it, each with an internal and an external loop. The fluid was based on antifreeze. The external loop ran through radiators with a total area of 21 square metres installed on the surface of the main compartment. The system maintained the air temperature between 15°C and 25°C, the humidity between 20 and 80 per cent, and the maximum airflow at 0.8 metres per second. The temperature and the airflow could be controlled from the central control panel.

An unpressurised section extended the line of the main compartment 1.4 metres to the rear. This was the only section which was inaccessible to the crew. It housed the

[11] Whereas a bactericidal kills bacteria outright, a bacteriostatic is capable of inhibiting the growth or reproduction of bacteria, and so serves to improve the immune system.

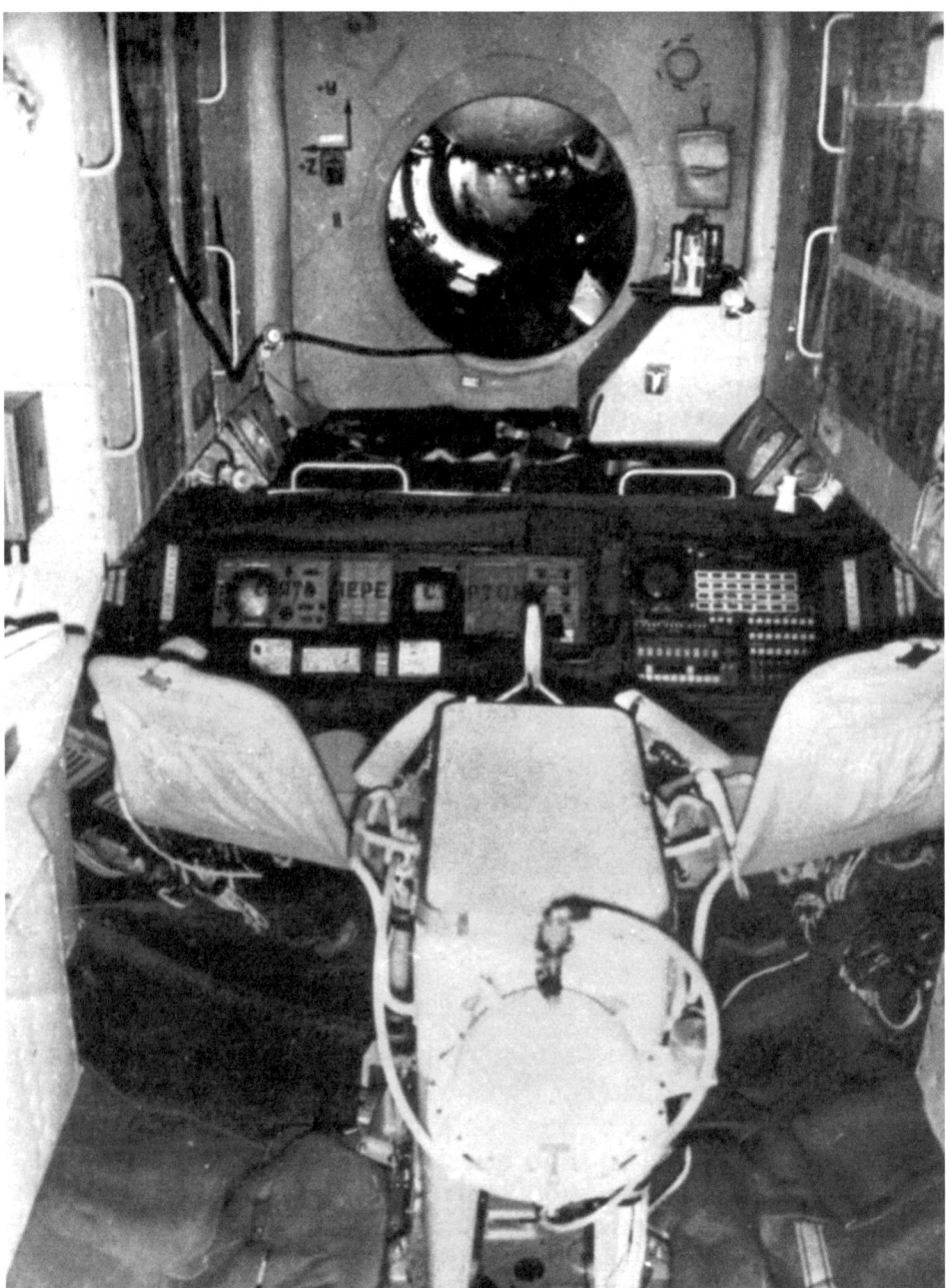

An inside view of the Salyut space station showing the main control panel, the seats for commander (left) and flight engineer, and the open hatch leading to the transfer compartment.

This section of the main control panel on the commander's side shows a globe for navigation and (bottom row, left to right) voltage, current, pressure and temperature, as well as the time, range and approach speed.

KTDU-66 propulsion system comprising a main and a backup rocket engine. It was based on that of the Soyuz, but had larger tanks containing 1,490 kg of propellant (UDMH fuel and nitric acid oxidiser) for a total burn time of 1,000 seconds. At the rear was a smaller cylinder 1.8 metres in length with a diameter of 2.17 metres that housed 32 small orientation engines and had a second pair of solar panels installed on its exterior. Each of the solar panels had an area of 7 square metres, for a total of 28 square metres. In ideal conditions, they had a total output of 2 kW. Because the panels were carried in a fixed orientation on the side of the station, it was necessary to align the station to maximise the illumination of the panels. However, 40 per cent of each orbital period was spent in the Earth's shadow, and at such times cadmium

The flight engineer's side of the main control panel.

accumulator batteries supplied direct (dc) and alternating (ac) electrical currents. A static voltage stabilisation system limited the variation in the voltage to 1.5 per cent. In the docked configuration, the solar panels of the Soyuz spacecraft fed electricity to the station.

In addition to two-way voice and telegraph links, the radio system fed telemetric data to the TsUP. The antennas were on the exterior of the main compartment. The cosmonauts had helmets incorporating headsets. Salyut had four TV cameras: two inside and two outside. One of the inside cameras was static and viewed the area of the central control panel of the working compartment. The other could be set up to record activities anywhere in the station. At launch, one of the outside cameras had documented the separation of the station from the third stage of its Proton rocket. The other had shown the rendezvous and docking operations. The cosmonauts also used them in orienting the station.

Specific references
1. Davidov, I.V., *Triumph and Tragedies of Soviet Cosmonautics*. Globus, Moscow, 2000, Chapter "Полет продожается" (Flight Continues) (in Russian).
2. Kamanin, N.P., *Hidden Space, Book 4*. Novosti kosmonavtiki, 2001, pp. 316–317 (in Russian).
3. Chertok, B.Y., *Rockets and People – The Moon Race, Book 4*. Mashinostrenie, Moscow, 2002, pp. 316–320 (in Russian).
4. Vasilev, M.P., *Salyut on Orbit*. Mashinostroenie, Moscow, 1973, pp. 38–42 (in Russian).
5. Clark, Phillip, *The Soviet Manned Space Programme*. Salamander Books, London, 1988, pp. 56–60.

8

Science and conflicts

EARLY DAYS

The first few days on Salyut were reserved for reconfiguring the station's systems, checking the equipment, starting the scientific investigations, and allowing the crew time to adapt to their new environment. Salyut was considerably more complex than any previous manned spacecraft, with more than 1,300 individual instruments and in excess of 1,200 kg of scientific apparatus.

The Soviet press, television and radio reported enthusiastically this latest success of the manned space programme – the official line was that the Soviet Union had never participated in a race to beat the Americans to the Moon, it was concentrating on space stations to conduct scientific research and benefit the national economy, at which it clearly led the way.

MEDICINE ON SALYUT

Day 3: Tuesday, 8 June
The second day for the cosmonauts on Salyut started at 1 a.m. on 8 June, when the station entered the Soviet communication zone. After breakfast, they checked the life support systems and made a start on preparations for the scientific programme. At 11.02 a.m., the cosmonauts initiated a manoeuvre to raise the orbit to 239 × 265 km with a period 89 minutes. With Salyut's systems confirmed to be in good order the Soyuz was powered down, since its interior would be ventilated by the station's life support system. In operating the complex station for the first time, the cosmonauts made several mistakes. For example, because they forgot to disable the docking regime, they had a problem when they first attempted to reorientate the station.

Daily life on board Salyut involved six major activities:

- the flight programme;
- morning hygiene and toilet;

- physical exercise;
- four meals;
- individual rest time; and
- an 8-hour sleep.

The flight programme included the control and maintenance of the station and its systems, the scientific equipment and investigations (the schedule included almost 140 specific experiments), radio communications and TV broadcasts, photographic sessions, and other tasks for flight operations. Exercise was of crucial importance in weightlessness. In addition to 2 hours per day exercising on the treadmill and with a chest expander, each man was to spend 30 minutes light 'walking' on the treadmill prior to retiring. Many lessons had been learned from the 18-day flight of Soyuz 9 in 1970, and the complex for physical training (KTF) was more substantial than the one available on that mission. The gravitational load imparted by the KTF on Salyut during physical exercise was 50 kg. On 'sports' days, each man had three exercise sessions in a 24-hour period: two of 75 minutes and one of 30 minutes. The flight plan allowed each man 2 to 2.5 hours per day of leisure time, which he could spend as he wished: resting, reading a book, observing the Earth, taking photographs or preparing for a forthcoming experiment. Every seventh day was a 'weekend' for the entire crew. The three men were to follow a phased sleep pattern in order that there would always be at least one man on duty, and at least one resting.

Day 4: Wednesday, 9 June
From 3 p.m. on 8 June to 1 a.m. on 9 June Salyut was out of the communication zone. After their morning toilet and breakfast, for the first time the crew exchanged their flight suits for the ones named 'Athlete' but irreverently known as 'penguin' suits.[1] These suits were designed to impart loads on certain muscles to simulate the forces experienced in everyday life on Earth, in the hope that this would minimise the deterioration of muscles and bones during a long period of weightlessness.[2] The cosmonauts used part of a communication session to demonstrate the suits, and to thank the designers. A system of supports and elasticated straps were attached to the wearer, as it were, by rigid soles and shoulder straps. The plan called for each man to wear his suit only for 40 to 60 minutes, 3 to 6 times per day, while working. They initially had some difficulty in moving their arms and legs while compressed by the elastic, but soon found the suits to be so comfortable that they asked to wear them all day, and later became so used to them that they slept in them as well.

On this day the cosmonauts also began to use the treadmill, but when it was noted that the vibrations which were transmitted through the station's structure caused the solar panels and antennas to 'flap' with an amplitude at their tips of about 5 cm they were asked to use the treadmill only for short periods.

---

[1] The acronym for the Athlete suit was TNK owing to its cyrillic name of Trenirovachniy Nagruzniy Costyum (Training Loading Suit).
[2] On Soyuz 9 Nikolayev and Sevastyanov had tested an apparatus (Athlete-1) intended for this purpose, but it was fixed to the wall of the orbital module and they could use it only at specific intervals.

They started the scientific work by measuring the radiation level inside the station and the flux of micrometeoroids in space around the station. In addition, they tested the wide-angle periscope provided to enable Salyut to be precisely aligned relative to the Sun and the planets. At 10.06 a.m., Dobrovolskiy and Patsayev fired Salyut's engine again to raise the orbit to 259 × 282 km. Although the atmosphere at orbital altitude is rarefied, it can impart a significant drag force that progressively reduces a satellite's orbit, finally causing it to burn up. As the drag was greatest at the lowest point of the orbit, the manoeuvres were designed to raise this altitude. Reducing the rate at which the orbit decayed would extend the interval before another manoeuvre was required.[3] Although the initial engine firings were costly in terms of propellant consumption, in the long term this strategy made sense.

8.29 a.m.

Dobrovolskiy: "Last night I adjusted the orientation prior to stabilising the station; it is easy to control the spacecraft, it responds very well."

Volkov: "I'm doing a rotation according to the programme. The engines are firing smoothly. Viewing through the porthole by the right-hand command post, I can see the red-hot jets. I'm controlling the orientation; the jets are working and everything goes well."

10 a.m.

Volkov: "The engine is switched off. I'm tracking the time."

Zarya: "We understand."

Volkov: "A slight vibration. The machine vibrates."

Dobrovolskiy: "The engine was fired for 73 seconds. The integrator was switched off."

Patsayev: "The engine's parameters are normal."

Zarya to Dobrovolskiy: "We understand, Yantar 1. Telemetry confirms that the engine fired for 73 seconds."

11.44 a.m.

Zarya: "In answer to your question about the 'penguin'. The metal tail should be above your knee. You can regulate its height with the hidden cord in the lower part of your knee. To eliminate unpleasant feelings caused by the tail, move it parallel to the leg."

Volkov: "Yantar 1 is now feeling excellent in his 'penguin' suit."

In his notebook that day, Patsayev wrote up his first astrophysical observations, and made some suggestions for how to improve the design of future stations.

From Patsayev notebook:

> The stars are almost invisible on the daylight portion of the orbit, even when observing through the porthole on the side facing away from the Sun. Only

---

[3] At an altitude of 300 km, the station's orbit would be lowered by about 90 metres per day.

Two bearded cosmonauts on the Salyut space station, Dobrovolskiy and Volkov check instructions for the next scientific experiment in the narrow part of the main compartment. The large white cone in the background houses the main scientific equipment, which could not be used because its protective cover had failed to release following orbital insertion.

> Sirius and Vega can be seen. After sunset, the stars do not twinkle until their line of sight is close to the Earth's horizon.
> Remark No. 1 – Add a protective cover for the button on the control handle.
> Remark No. 2 – Modify the hermetic seal of the rubbish bags.

At 3 p.m. on 9 June, on the 38th orbit with the crew on board, the station left the communication zone.

Day 5: Thursday, 10 June

One of the primary tasks for this first crew was to determine the degree to which the human body (and indeed other organisms) were influenced by long-term exposure to weightlessness.

The crew were to have a detailed medical checkup every five days. This involved taking blood samples and electrocardiograms, and checking the composition of their bone tissue, in particular of their shins. The procedure was more sophisticated than on previous flights. For instance, whereas only the rate of breathing had previously been measured, now this was augmented by measurements of the volume and speed of inhalation and exhalation, and the overall lung capacity. In addition, the arterial blood pressure and the speed of pulsation waves through the arteries were measured by two separate methods. On Day 5 Patsayev took blood samples of all three men for the first time. He was to repeat this several times during the flight. Placed on the surface of filters, the samples were stored at reduced humidity in hermetic probes. After Soyuz 11's return to Earth, doctors determined how the levels of sugar, urine and cholesterol varied in each man's blood during the mission. The sugar level was normal in the blood samples taken during the first and third weeks, but increased in the fourth week just before the cosmonauts left the station. There was an increase in the level of urine in the blood of all three men owing to the manner in which their kidneys adapted to weightlessness. There was no detectable change in the level of cholesterol.

One of the most significant hazards of long-term exposure to weightlessness is the leaching of calcium from bones into the bloodstream, with possible implications for the kidneys. A special instrument was designed to investigate changes in the bones of the cosmonauts. Each day, every crewmember would place a medical belt around his chest. Before doing so, he would smear cream on his skin in order to minimise irritations. The belts had electrodes for vital body functions. During communication sessions with the station, the doctors at the TsUP would receive electrocardiograms, seismo-cardiograms and pneumograms (i.e. breathing activity) in order to monitor the cardiovascular systems of the cosmonauts. In addition, there was the Polynom apparatus to monitor their physiological activity. This could measure 25 different parameters, but only five at any given moment, and it involved two men: one as the test subject and the second to make the measurements, which were recorded for later transmission to Earth. Although more sophisticated than the belts, this apparatus was used only infrequently.

The results of the biomedical tests provided important information on the general health of the three men during their exposure to weightlessness. Dobrovolskiy and Patsayev both had increased hearts rates, increased arterial pressure and an increase in the blood's exchange rate. In contrast, the cardiovascular system of Volkov, the veteran, was more stable.

_0.51 a.m._

"Good morning," called Zarya.

Dobrovolskiy: "Good morning. I report that everything is all right. Yantar 2 just finished exercising on the treadmill. Yantar 3 is resting. During the period between 16.00 and 18.30, ventilation fan No. 7 was buzzing. Obviously something has been drawn into it. We opened the panel. ... Just after 18.30, the buzzing ceased. Can we switch to the second ventilator?"

Zarya: "We understand. Do that. During physical exercise please do the following experiment. During the running period on the treadmill, someone should enter the descent module and look through the portholes to observe the vibration of the solar panels. Monitor the period and amplitudes of any vibrations."

One innovative piece of apparatus on Salyut was the 'Veter' ('Wind').[4] With the 'penguin' suits, it was to help the cosmonauts to overcome the long-term effects of weightlessness. The 'waist' was fastened to the wall by several supporting struts, and the leggings were rubberised. Once a cosmonaut had hermetically sealed his lower body into the apparatus, a pump extracted some of the air from the leggings. The function of this lower-body negative-pressure apparatus (ODNT) was to draw blood into the lower part of the body, just as if the cosmonaut were stood upright on Earth. In weightlessness the feet do not require so much blood, and therefore the cardiovascular system rapidly adapts by transferring 1.5 litres of blood to the upper body – in particular to the chest and head, which is why on their first days in space the cosmonauts felt 'swollen headed'. Over time, most of this excess is removed by

---

[4] This unit is now popular by the Chibis name.

increased urination. The cardiovascular system is greatly stressed on returning to Earth. The reduced amount of blood that is circulating in the upper part of the body drains to the feet, imposing a considerable pressure on the vessels. While in space, the cardiovascular system loses the compensatory function. The doctors call this an 'imbalance'. When a cosmonaut stands up after returning to Earth, his weakened cardiovascular system is unable to supply blood to his head, the brain is temporarily starved and there is a risk of fainting. This is called 'orthostatic intolerance'. The air pressure in the ODNT was reduced gradually to 'train' the cardiovascular system to adapt to a state approximating that of gravity on Earth. The 'vacuum' test had two stages: in the first stage the pressure was reduced to –27 mm of mercury for two minutes and then to –36 mm for three minutes; for a total of five minutes. At the cosmonauts' initiative, the second stage could be extended to –70 mm. Using the ODNT involved two men, one as the subject and the other to operate the apparatus. The 'vacuum' condition was reported to be a pleasant sensation.[5] After each session, the test subject was required to have the parameters of his cardiovascular system measured.

03.54 a.m.

Zarya: "Yantars, today is a medical day, so do not take off your belts."
Dobrovolskiy: "Periodically, I will switch it on."

From Volkov's diary:

10 June. Exercise on a treadmill and with a chest expander. Toilet. I brushed my teeth with real toothpaste. Again, something dropped into the ventilator. This time it was a food bag. When I removed the medical belt there were no red spots on my skin.

Viktor is sleeping in the transfer compartment. His arms are outside the sleeping bag, and float strangely in the air. Zhora is at his position – the left seat of the main control post. He has used the new cream under his medical belt.

I shaved, but not too much – I've decided to grow my beard.

From Patsayev's notebook:

I continued with daily shaving. The razor is specially designed with a setting to collect the hair, but it is not close enough and the hairs fly away.

On 10 June, the cosmonauts began daily participation in TV shows. Wearing their 'penguin' suits, they talked about themselves, reported their activities and showed some details of their home in space. During one *Cosmovision* telecast,[6] Volkov said of Salyut's dimensions: "It's so big that it takes some time to swim from one end to the other."

---

[5] A contemporary Soviet source said that each man was to have two sessions in the ODNT per week, but owing to technical problems this was not feasible, and only two cosmonauts performed the 'vacuum' test, and only once during the mission. One was Dobrovolskiy and the other was very probably Volkov.

[6] The term *Cosmovision* was coined by the journalists for the TV shows from Salyut, not the name of the television programme(s) that participated in broadcasting them.

From Patsayev's notebook:

> We had the first television broadcast. They asked the commander about our work on board the station, and all of us about our first impressions of being in space. It is nice to study geography, astronomy and physics in space with my colleagues. Virtually entire continents, seas, and islands are visible. For example, it is easy to recognise Australia, Crimea and the Mediterranean. In 90 minutes you get a trip around the world!

At 2.40 p.m. the station left the communication zone, and drew to a close the fifth day.

SPACE ASTROPHYSICS

Day 6: Friday, 11 June

The crew began multispectral observations, both of the optical characteristics of the atmosphere and of Soviet territory in order to provide scientists with unique data about certain locations, including lakes.

In addition, the Anna-III gamma-ray telescope was used to make the first such astronomical studies from a manned spacecraft.[7] Volkov aligned the station to point the telescope at its target and then activated the automatic stabilisation system. Then Dobrovolskiy activated the apparatus to measure the energy spectrum of the gamma rays. The instrument consisted of several scintillation counters and one Cherenkov counter for measuring gamma rays, a pair of neon-filled spark chambers equipped with cameras, and a control panel. The gamma-ray telescope had a detector area of 90 cm^2, drew 14 watts of power and was sensitive to radiation at energies exceeding 100 MeV (million electron volts) with an angular resolution of 1 degree, which was twice as good as instruments previously flown on unmanned satellites. Overall, the 45-kg Anna-III apparatus measured 60 $\times$ 40 $\times$ 45 cm, and included a tape cassette with a capacity of 20,000 images.

In effect, the Salyut crew were the first space astronomers. Gamma-ray astronomy had only recently become feasible, and was giving insights into the structure of the universe. Gamma rays are the most energetic form of light. They are produced by fusion reactions in the cores of stars, but are soon absorbed and so stars appear dark in this part of the electromagnetic spectrum. However, they are emitted by violent events such as a supernovas (when a massive star 'explodes') and by the much less dramatic decay of radioactive elements in space. Objects like supernova remnants, black holes, neutron stars and pulsars are all sources of celestial gamma rays. In addition, there are powerful 'flashes' known as gamma-ray bursts which can release more energy in a few seconds than the Sun will emit during its entire 10-billion-year lifetime! The exact cause of such bursts is disputed, and there may in fact be several causes. Thus far it would seem that all of the bursts originate from outside

---

[7] Astronauts on some Gemini missions had previously conducted astronomical photography.

The Anna-III telescope to detect gamma rays.

our own galaxy, but it is conceivable that they might occur in our Milky Way once in every few million years, with one located within several thousand light-years of the Earth once every few hundred million years. By solving the mystery of gamma-ray bursts, scientists hope to develop further insight into the origin of the universe, the rate at which it is expanding, and its size.

The thickness of the Earth's atmosphere is approximately equivalent to 10 metres of water, so gamma rays, X-rays, ultraviolet and infrared radiations from space are absorbed. When the highly energetic atomic nuclei of cosmic rays interact with the atmosphere they generate gamma rays, but these too are absorbed. It is therefore not possible to undertake gamma-ray astronomy at ground level; it must be done at high altitude using instruments on balloons or, better still, on satellites.

The cosmonauts used the Anna-III to:

- determine the telescope's basic operational capabilities;
- investigate how the gamma-ray flux varied with directions in space; and
- correlate such observations with the flux of charged and neutral particles both directly entering the station and as secondary products in the station.

The Anna-III telescope detected gamma rays and charged particles as the station was rotated and stabilised relative to the Sun. In total, it was operated for 20 hours under the control of one cosmonaut.

The main astrophysical experiment on Salyut was the Orion telescope, which was in the transfer compartment. It had two mirrors, one 28 cm in diameter and the other 5 cm in diameter, and a focal length of 1.4 metres. The instrument was designed to make spectrograms of stars in the range 2,000–3,800 ångströms.[8] At a wavelength of about 2,600 ångströms it could provide a resolution of 5 ångströms. The tracking system allowed the telescope to maintain its orientation to within one second of arc. The spectrograms were recorded in the form of photographs on 16-mm tape bearing UFSH-4 emulsion. An airlock and mechanical arm allowed a cosmonaut to replace the film cassettes. The mirrors were coated with aluminium, without protection, to enable them to be re-surfaced if they ever became tarnished by micrometeoroids. To use the instrument, one man (usually Dobrovolskiy) controlled the orientation of the station while Patsayev, who was responsible for this research, aimed the telescope. Patsayev had to operate the system quickly because there was only a 30–35-minute period on each orbit during which observations could be made – this being while in the Earth's shadow. Dobrovolskiy, sitting at the central control panel, oriented the station as specified by Patsayev in the transfer compartment with the Orion. When the target star was visible to the telescope, the station was stabilised and Patsayev started the observation. During the mission he obtained six spectrograms of the star Agena (beta Centauri) in the southern sky and nine of Vega (alpha Lyra) in the north. In fact, Vega is the 'standard star' for spectral analysis of other stars. These stars were selected because of their extremely high surface temperatures (10,000°C in the case of Vega and 24,000°C for Agena). Once an investigation was completed, Volkov used the airlock manipulator to retrieve the cassette of tape and to replace it with another one.

Salyut also had the FEK-7 photo-emulsion camera with a volume of 1.4 litres for detecting the charged particles of primary cosmic rays. The majority of cosmic rays are protons and alpha particles (helium nuclei), but there can also be much heavier nuclei. A precise knowledge of their fluxes as a function of energy was important for several reasons. Interstellar spectra can provide information about how cosmic rays are propagated and accelerated in the galaxy. In principle, this can be derived from measurements made in the upper atmosphere by demodulating the observed solar spectrum. Since protons and helium nuclei have different momenta and kinetic energies per nucleon, the comparison of their spectra provides useful constraints for modulation and acceleration theories.

The FEK-7 camera was designed to search for:

- magnetic monopoles (single magnetic charges; Dirac particles);
- trans-uranium and uranium nuclei in primordial cosmic rays, important for global astrophysics and the determination of the distribution of the sources of cosmic rays; and
- anti-nuclei and trans-nuclei to investigate the symmetry between matter and anti-matter.

Finding such particles would have important implications for theoretical physics. Similar cameras had been flown on the unmanned satellite Cosmos 213, on Zonds 5,

---

[8] An ångström is 1×10^{-10} metre, and is the unit in which spectra are measured. The human eye is sensitive from 4,000 to 7,000 ångströms, running from violet to red respectively. The Orion telescope was designed to observe in the ultraviolet.

The Orion astrophysical telescope.

7 and 8 flying circumlunar trajectories and on the Soyuz 5 mission, but in each case data was able to be collected only for short periods. The FEK-7 on Salyut operated for 17 hours 28 minutes. It was placed in the descent module of Soyuz 11 for return to Earth and analysis by specialists.

Another project was to determine the intensity of charged particles in the altitude range 200–300 km (where the station flew) because this radiation appeared to have been increasing since 1960. It had even been proposed that this region was occupied by clouds of electrons possessing energies as great as 300–600 MeV. When the Sun is active it can suddenly release vast numbers of charged particles, and following a major 'flare' the increased radiation can linger in the inner heliosphere (where the Earth is located) for up to a month. The Earth's magnetic field provides a degree of protection, but even in low orbit a high flux of such particles can cause damage to both electronic and biological systems. During the flight, the crew performed more than 60 operations related to the measurement of charged particles. The instrument used was able to detect protons with energies of 400 MeV and electrons exceeding 8 MeV. The observed electron flows were several hundred times less intense than those previously measured by the Cosmos 225 satellite.

At 1.06 p.m. on 11 June Salyut left the communication zone of the NIP stations, but the ship *Academician Sergey Korolev* in the North Atlantic was able to continue to communicate with it. The final experiment of the day was to investigate optical materials that had been exposed to the space environment. Before the crew retired, Yevpatoriya relayed through a Molniya satellite and *Academician Sergey Korolev* to congratulate them on their successful work so far.

3.47 p.m.

Zarya: "Yantars, the Control Group wishes to thank you for your work during the last days. Have a nice rest, and start the next work day in a good mood."

Volkov: "Thank you. It is nice to hear that. If tomorrow we feel we did like today, then everything will be well."

From Volkov's diary:

> 11 June. A very full programme today. It shouldn't be planned in that way, if you consider adaptation to the conditions aboard the station. The rubbish bags should be redesigned in order to avoid spending so much time opening and closing the hermetic seal.

THE FIRST CONFLICTS

After the excitement of the early days, life on board Salyut settled into a routine. As the new technical flight director at the TsUP, Yeliseyev was in charge of operations, supported by veteran cosmonauts Nikolayev, Gorbatko and Bykovskiy. Reports on how the flight was progressing were submitted to Kamanin several times per day.

Day 7: Saturday, 12 June

At 0.40 a.m. Salyut again entered the communication zone. The cosmonauts began the day by measuring the radiation in the station, then analysed their cardiovascular systems and tested their eyesight in different illumination conditions. Photography of the Earth's cloud cover and various atmospheric phenomena completed the day's scientific work. The crew transmitted another TV show and talked of living in their home in space.

From Volkov's diary:

> 12 June. I woke up. I drank water from the new tank; we finished the first one. After Viktor had prepared the vacuum cleaner, I swam through the compartment cleaning it. Zhora is strapped in his seat and diligently writing something in his flight journal.
>
> Viktor has prepared his sleeping place in the hatch between the descent module and the orbital module. Soon we will be in communication with the Earth, but now, according to schedule, I must exercise.

0.41 a.m.

Patsayev: "We have a suggestion about the medical sensors. It is uncomfortable to wear them all the time. I kept the belt on for three days and the sensors have made indents. Let us make an agreement with you Zarya: tell us when you will be able to receive their telemetry and we'll put them on during that time, but remove them at other times."

Zarya: "We understand. We accept your suggestion."

The flight controllers at the TsUP in Yevpatoriya take a break. Cosmonauts Gorbatko, Yeliseyev and Nikolayev are first, second and fourth in the first row.

2.12 a.m.

Dobrovolskiy: "Now it is time to say something about psychology. I think that the psychologists don't have cause for concern. It is necessary that the three of us take exercise together. In addition, we should do it on a more frequent basis. Firstly, we would be able to encourage one other. ... We should force ourselves to do all of the physical exercise.[9] It is necessary to extend the exercise time to approximately 30 minutes. You should plan this to be done by two or three of us – a minimum of two of us. It is better for the work, too."

Zarya: "About the exercise, all three of you can exercise for 30–40 minutes."

Dobrovolskiy: "All right. Now, about work. All new operations should be planned for the three of us. Only with three of us together could we work with the Polynom sensors and fix problems. It will also be more interesting."

Zarya: "We understand."

Dobrovolskiy: "In addition, it would be easier to repeat the operation."

Patsayev was complaining about the medical belts they had to wear continuously on their chests. Dobrovolskiy was concerned about the general organisation of their

---

[9] Dobrovolskiy had concluded that the enthusiasm of a man exercising alone soon waned; it would be better for the crew to exercise jointly, since then they would be able to encourage one another. As commander, he may have been thinking of Volkov, who had missed several exercises earlier in the mission.

activities. In fact, these complaints marked the onset of psychological tensions – in part irritability arising from the unnatural circadian rhythm, but also due to flaws in mission planning and poor use of the very brief periods of communication with the TsUP.

The plan was for the three cosmonauts to work shifts displaced by 8 hours, and while one man slept his two comrades were to exercise or perform 'silent' work. In general, life on board the station was progressing satisfactorily. During the first two days, they prepared apparatus and started some experiments. The need to exercise and perform medical tests meant that the time available for experiments was limited. In addition, a lot of time was devoted to reading instructions, preparing equipment, placing experimental samples into their containers and chambers, recording results and so on. Consequently, only 4 to a maximum of 5.5 hours per day were available for experiments.

The scientific programme for the DOS-1 station had been agreed only after tense discussions between the TsKBEM managers and the representatives of the various scientific institutions. The station carried much more scientific equipment than any previous manned spacecraft. But if the flight was organised inappropriately, and the time was poorly allocated to the different experiments, then the cosmonauts would not be able to use the equipment in the best manner. One instructor had proposed that the cosmonauts read detailed instructions before each experiment to familiarise themselves with the purpose and methodology, and then, when the experiment was completed, read how they were to record their results. All this reading took up a lot of time.

For Yeliseyev, this was a real challenge:

> The programme was planned in such a way that all important crew activities would be carried out while the station was in range of the tracking stations. This enabled us to check the status of the onboard systems and, if necessary, provide support to the crew. However, due to the timing of the orbit it was impossible to retain the normal terrestrial duration of 24 hours for the crew, and their cycle was 25 minutes shorter. By saying '24 hours', I don't mean the duration of the light and dark times in space, because an orbit lasts only 90 minutes; I mean the sleep cycle of the men – in particular, the time from the start of one morning to the start of the next. We thought that they would soon accommodate themselves to the planned circadian rhythm. However, the physicians saw a serious risk. Alyakrinskiy, a biorhythmology specialist from the Institute of Biomedical Problems, came to the control centre in the hope of changing the programme. He wanted to talk to me urgently. At first, I attempted to avoid him: we were busy, the cosmonauts felt well, and I did not see the need to spend time on medical issues. However, he persisted and I saw him. Our conversation was long and difficult. He really understood the essence of the problem and carefully explained it to me. He asserted that the daily deviations of the rhythm of life from the norm would be very difficult for the cosmonauts, and would cause nervous disruption, if not worse. I did not believe him. In any case, it was not realistic to expect us to rearrange the

programme at this stage. Therefore, I assured him that there was no problem and refused his request. Finally, he gave up and departed.

Nevertheless, as time went by the psychological stresses on the crew worsened.

From Dobrovolskiy's notebook:

> Some days were a nightmare. There was a general absence of everything: no interesting things, no happiness, the monotonous sound of the ventilators, strong smells, numerous experiments. It seemed to me that the TsUP simply wished to test our endurance.[10]

The euphoria of the first days was undermined by the 'ranking' of the crew. They shared a general responsibility for the success of the flight and jointly undertook the programme, but by his enthusiasm Volkov, the only veteran on the crew, threatened the authority of Dobrovolskiy, the commander who was used to the discipline of a military chain of command. Initially minor issues grew into more serious ones. The TsUP sensed that the situation on board was abnormal, and attempted delicately to improve it. This was the first long flight of a 3-man crew, and the first aboard such a large and complex spacecraft. Previous space missions had not been able to study the psychology of a group of people isolated in a craft in a dangerous environment with a biorhythm significantly different to that on Earth and pursuing a schedule of exercise and experiments. The two cosmonauts for the Soyuz 9 mission who spent 18 days in a cramped Soyuz had trained together for more than a year. However, the Soyuz 11 crew had been formed less than four months ago, had not expected to fly so soon, and had a rookie commander and an ambitious flight engineer with little respect for military authority. While Mishin and Kamanin fought for the prestige of 'their' cosmonauts on crews, it was now evident that neither man thought seriously about the psychological issues facing 'mixed' crews on long-duration space flights. In particular, when considering whether to replace Kubasov with Volkov in order to allow Leonov's crew fly this first space station mission, no thought was given to the potential downside of sending Dobrovolskiy's recently formed crew on such a long flight.

3.44 a.m.

Zarya: "Yantar 2, conduct photography, monitor the most visible atmospheric phenomenon and let us know."

Volkov: "Well, now we see a bush fire."

Zarya: "Understood. Another request. If possible, report the porthole conditions.

---

[10] Dobrovolskiy's frustration at the workload was in part because the mission planners had drawn up the schedule of scientific experiments without appreciating the time that it would take to perform them in the weightlessness environment of space – it always took longer than it had during terrestrial training. Even before they could start an experiment, they had to prepare the apparatus, locate and read the instructions, unpack any samples or devices and install them. Naturally, as commander, he worried about his crew's performance. It was demoralising for them to be judged by the TsUP to have fallen behind the schedule. However, in retrospect, it is evident that this crew was inadequately trained to conduct a large scientific programme – they were simply not ready.

Is it possible to see the stars?"

Volkov: "No, it isn't. In sunlight the stars are not visible, but they can be seen just before sunset and [of course] before sunrise."

Zarya: "Understood."

Volkov "The portholes are clean. . . . They are in excellent condition, but some are slightly covered by vapour. The stars are not visible on the daylight side. I made a few observations. Even Jupiter, which is now in the constellation of Scorpio, is not visible."

From Patsayev's notebook:

> 12 June. At night the stars and the Earth are easily visible. We can see the clouds and the illuminated cities – even fires on Earth. We can see the limb of the planet where it occults the stars. During sunsets and sunrises, the long rays of light illuminate high-altitude clouds. Are the stars visible during the days? It depends on the position of the Sun. At angles of less than about 15 degrees we can see the planets and the brighter stars.

8.11 a.m.

Zarya: "Yantar 2. Another question. Could you work with the experiments and at the same time receive information?"

Volkov: "Do you understand, everything depends on the time. Now, for example, I am preparing the Polynom. We spent 1 hour 20 minutes on that."

Zarya: "Understood."

Volkov: "The difficulty is that a man is not fixed in the seat. . . . Everything floats away – as soon as you let go of something it floats away."

Television Report:

Zarya-25 (call-sign of Yevgeniy Frolov, the commentator of the Central USSR TV): "On line is the flight engineer, Vladislav Volkov. We know that for you the station is at the same time a laboratory, your home, even a gymnasium. We would like to hear from you a detailed description while making the first TV tour of Salyut. Now from the Earth we are switching to the portable camera. Did you understand us?"

Volkov: "I understood you very well. I will be pleased to give a tour of the Salyut orbital station. It consists of two segments. The station you can see now, and the Soyuz ship. . . . In the distance, the Soyuz spacecraft is visible, docked with the station. Notice the size of this station! Now the research engineer is swimming here from the transfer compartment."

Zarya-25: "I see him very well."

Volkov: "Now I'll show you the second part of our station. We have our very own sports facility, although admittedly it is not as big as the arena at Luzhniki. Here is a medical seat, the treadmill and handrails. Here is a chamber, some apparatus, the work place of the research engineer and his flight journal and control panel. This is the central control panel – we use it to control the orbital station and the spacecraft at the same time. . . . Now you can see our photographs of Korolev, Gagarin and Lenin. They are always with us in spirit. Now, I'll show you the docking apparatus. Here is the docking place. Do you have questions? Can you see the docking spot?"

Zarya-25: "I can see it very well."

Volkov: "That is the orbital module. This is the transfer compartment. Here is the sleeping zone. Here we rest."

Zarya-25: "We are running out of time. Could Yantar 1 provide a brief summary of the last week?"

Dobrovolskiy: "Zarya, I can hear you very well. In brief, all the systems of the spacecraft are working excellently, and the crew feels well. We're ready to continue with the flight programme."

While the fixed TV camera monitored their activities, the cosmonauts took their exercises, engaged in numerous scientific experiments, and even cast the first votes from space – affirming their support for the Communist Party's policies, of course. Excerpts from the broadcasts from Salyut were repeatedly shown on Moscow TV, and owing to his rugged good looks Volkov soon became an idol for many teenage Russian girls.

The hard working day of 12 June, which began at 0.40 a.m., finished at 2.30 p.m. when Salyut left the communication zone of the Soviet ground stations.

NOTES FROM THE STATION

Day 8: Sunday, 13 June
Salyut entered the communication zone at 0.34 a.m., during its 93rd orbit with the crew on board, but during the next seven orbits its path crossed only a subset of the tracking stations. With the cosmonauts on phased shifts, operations were continuing around the clock. Volkov, for example, had started his working day at 9.30 p.m. the previous evening, Dobrovolskiy joined him at 1.50 a.m., and Patsayev took over from Volkov at 6 a.m.

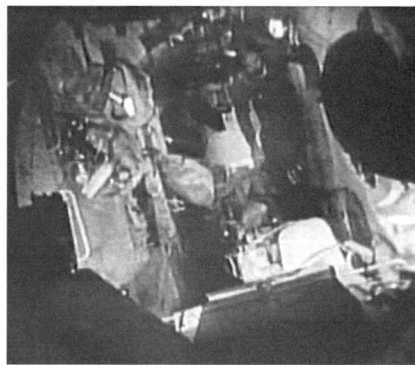

Dobrovolskiy in Salyut's main working compartment.

Most of the seventh day was devoted to biological experiments, both agricultural and genetic. The effect of weightlessness on plant growth was to be investigated by a small hydroponics chamber called Oazis-1 ('Oasis') which regularly fed a nutrient solution to Chinese cabbage and bulb onions. The genetic tests studied mutations in drosophila (tiny fruit flies), tadpole embryos, yeast cells, chlorella and the seeds of higher plants like linen, cabbage and onion. As the degree of mutation of drosophila had been thoroughly studied on Earth, it would be possible to precisely evaluate the influence of the space environment on heredity. Gamma rays were used to stimulate genetic mutations. In addition, Soyuz 11 had delivered fertilised frog eggs, and their development on the station was monitored.

From Volkov's diary:

13 June. The eighth day of the flight. On crossing the equator we started the station's 887th orbit. The other guys are still asleep. Zhora is in the transfer compartment, in a sleeping bag. I cannot see Viktor; he sleeps in my place, on the berth in the orbital module. I've already performed physical exercise, had my breakfast (bacon in the can, blackcurrant juice, plums with nuts and cakes) and drunk water.

Although we're out of radio contact, I will stay on line. After the session, I will perform a medical experiment. I made observations of the starry sky. In the upper region of the night horizon beta Ursa Majoris is clearly visible. At dawn, when the antennas begin to gleam, the stars start to disappear, but not all of them.

In the morning, we cleaned the compartment using the vacuum cleaner. We are currently on the second tank of water, and it appears to be running out already. ...

Two green stalks have sprouted in the Oazis, each about 2 cm long. The guys are still sleeping. I have to awaken Zhora. He should have appeared at 1.30 and it is now almost 2 o'clock. Out of one of the windows there is an antenna brightly illuminated – our next sunrise has begun.

The Earth asked me to put on the medical belt; I did so.

An interesting view: the Earth is still dark, like the sky, but the antenna on the solar panel is brilliant white. The session has started. In my headphones I hear a song from the movie *Fighter Pilot*: 'In a remote landscape my friend flies away.'

Zhora has appeared: "Will you say something good?"

"Greetings to you," I joked.

I checked the strength of my hand using the dynamometer: 35/32, just as previously. It is good. Pulse 52.

From Patsayev's notebook:

13 June: On the porthole opposite to the Sun, frost is visible on the internal surface of the outer glass pane.

Remark No. 1: The bag with instruments has long straps [covering it]. It is better to replace them with slats.

No. 2: The power supply of the vacuum cleaner is too weak. Working in the dim illumination is uncomfortable.[11]

At about 1 p.m., during the jubilee 100th orbit with the crew on board, Salyut left the communication zone. However, during orbits which crossed the eastern part of North America and the Atlantic Ocean the crew were able to communicate with the controllers on *Academician Sergey Korolev*, which relayed the data that it received from the station to the TsUP via a Molniya satellite.

Day 9: Monday, 14 June

Salyut entered the communication zone again at 10.53 p.m. on 13 June, during the 108th orbit in its manned state. By now, its orbit had a low point of 255 km, a high point of 277 km and a period 89.6 minutes.

At a meeting of the Landing Commission at the TsUP, Feoktistov ventured that there were too many long and unnecessary conversations with the crew, which the cosmonauts evidently found irritating. As an example, he mentioned that there was no need to specify each day how to make an emergency return to Earth. The crew could readily obtain such data using the globe on the station's central control panel. Surprisingly, some members of the commission debated this issue, and at the end of the discussion it was agreed that the crew should be consulted and the accuracy of the globe be checked by several brief experiments.

During their eighth day on board, Volkov and Patsayev carried out experiments to improve the station's autonomous navigation system. Patsayev fed this data into the onboard computer to determine the parameters of the orbit.

The scientific work on 14 June included meteorological experiments, a study of atmospheric formations and snow and ice cover. The cosmonauts on Salyut and the unmanned Meteor satellite launched in October 1970 both recorded the cloud cover over the Volga River. The aim was to use the photographs taken by the cosmonauts to improve the interpretation of the TV pictures transmitted by the Meteor satellite. In addition, the cosmonauts studied atmospheric processes related to the formation of hurricanes and typhoons.

As part of the routine medical programme the cosmonauts checked their eyesight by measuring their ability to adapt to the changing lighting outside the station while on the day-side of its orbit.[12]

Later, viewers in homes across the Soviet Union saw a TV transmission in which the cosmonauts talked about their life on the station.

---

[11] Although Salyut had four solar panels and could draw on the panels on the docked Soyuz, it had much more apparatus, and to supply the required electricity the designers had had to reduce the brightness of the illumination in some parts of the station. (Indeed, in pictures taken during the mission it is hard to see the details at the rear of the main working compartment.) This made it difficult to operate the apparatus which was installed in these areas.

[12] Specifically, they measured accommodation and convergence.

3.12 a.m.

Volkov: "Give us more Mayak.[13] We are so bored without it. We can hear it very well over South America, but not elsewhere."

7.56 a.m.

Patsayev: "Can you see us?"

Zarya: "Yes, we can."

Patsayev: "Now, I'll show you our commander. He looks neat and tidy."

From Patsayev's notebook:

> 14 June: We aligned the station to the Sun. The station sometimes oscillated – several feeble lurches, obviously due to the redistribution of the propellant.
>
> Remark: The control panels for the scientific apparatus should be protected by glass safety covers.
>
> Shining particles often accompany the station, flying around in different directions. These are specks of dust.

Half an hour after mid-day Salyut left the communication zone of the ground stations, but while it was in range of *Academician Sergey Korolev* contact with Yevpatoriya was possible via a Molniya satellite.

Day 10: Tuesday, 15 June

The next working day for Salyut began at 10.45 p.m. on 14 June, when the TsUP at Yevpatoriya replied to a call from Volkov, who was on duty. Dobrovolskiy joined him at 3.30 a.m., and Volkov retired when Patsayev awakened.

The cosmonauts used a spectroscope to study areas of the Earth's surface, while at the same time two aircraft made spectroscopic measurements of the same areas for later comparison with the results from space. When the station was passing over the Caspian coast two specially equipped aircraft from Leningrad State University and the Soviet Academy of Sciences flew along the path. An IL-18 airliner operated at an altitude of 8,000 metres and a light An-2 at a mere 300 metres. The aim was to determine the spectroscopic characteristics of the sea and of the soils in the coastal area, and to compare the results from space with those at different levels within the atmosphere in order to identify any distortions that the atmosphere imposed on the readings from space. Once the airborne data had served to calibrate that from space, it would be possible to 'subtract' the atmospheric effects and apply the spaceborne observations to wider areas. Every type of soil, plant and other natural object has its own spectral signature. They can be compared like fingerprints. Thus, the spectral characteristics of soybean plants cannot be mistaken for those of the birch tree, or wheat, larch or lichen. Furthermore, these signatures vary with the age of the plant and the amount of water stored in the soil. Multispectral images provided a valuable new means of monitoring agricultural development and land improvement, and the data was useful to mapmakers, farmers and forest managers.

---

[13] Mayak (Beacon) was a popular radio programme.

Meteorological monitoring, and the study of the cloud cover over the Volga River in parallel with the Meteor satellite continued.

The cosmonauts tested the radiation intensity to determine its effects on biological structures on the station. One goal of this work was to develop an effective means of dosimetry control. In addition, the study of charged particles continued using the FEK-7 photo-emulsion camera.

Then they provided another transmission for Russian TV, this time talking about the medical experiments.

Television Report:

Zarya-25: "Do you hear me? Who is on line?"

Volkov: "Yantar 2 is on line."

Zarya-25: "We have excellent reception. We would like you to tell us about the cardiovascular experiments."

Volkov: "One of our most important tasks is to perform medical experiments. The data will enable scientists to assess the possibilities for long-duration flights of man in space. Today, I would like to show you one of these experiments. I will show it to you now in detail."

Zarya-25: "Please do. By the way, Vladislav Nikolayevich, how are you feeling? How is the entire crew?"

Volkov: "We are feeling excellent. Our training on Earth is largely responsible for that. Now, dear comrades, you see Viktor Patsayev preparing to perform a regular medical examination. Our ship's commander Georgiy Dobrovolskiy is helping him. The experiment is performed using the apparatus you have just seen on your screen. Now Viktor Patsayev is showing the apparatus which he will employ to measure his physiological parameters."

From Patsayev's notebook:

15 June: While the Sun is low (immediately after sunrise or before sunset) the Earth is in a haze. This forms a shroud above the surface, although there is no visible cloudiness. Obviously, some atmospheric layers are lit from the side.

Sometimes there are cloud formations exceeding 1,000 km in length, with a mosaic structure. For example: at 17.40 in the South Atlantic at 50 degrees south and 350 degrees east. Clouds over the ocean looked like foam on the water. The ocean's colour is a delicate blue. The waves are visible usually through the porthole on the opposite side to the Sun, when the Sun is high. The wakes of ships can be seen, as can condensation trails of high-flying aircraft.

As Patsayev made astrophysical and meteorological observations, his colleagues checked the onboard systems and performed essential maintenance. From time to time, they helped the research engineer in the study of atmospheric phenomena by holding cameras up to the portholes (there were more than 20 portholes, and often the cosmonauts had to move from one to another to record specific features). They monitored clouds at different altitudes and times of the day, cyclones and typhoons, ice cover, bush fires and the melting of glaciers. For example, Dobrovolskiy kept an eye on one cyclone that started in the vicinity of Hawaii, moved west until it was a few hundred kilometres off the east coast of Australia, weakened and disappeared.

The TV viewers did not often see Patsayev, since he served as the cameraman and recorded many sequences featuring his colleagues.

In their time off, the cosmonauts read books, listened to music either on the radio or from their cassette player, and sang their favourite songs. The TsUP controllers kept them up to date with the sporting news. Volkov was especially interested in the national soccer championship. Unlike Nikolayev and Sevastyanov, who shaved on a regular basis during their Soyuz 9 flight, Dobrovolskiy and Volkov let their beards grow. As a military pilot, Dobrovolskiy had asked General Kamanin prior to launch for permission to do this. On TV screens and photographs taken on the station, they resemble explorers of remote and unknown places. Patsayev, however, continued to shave.

From Dobrovolskiy's notebook:

> The 907th orbit. We are working against the pressure of time. Despite some problems, we are accomplishing the experiment programme specified down to the minute by Earth. It is extremely difficult to operate the photographic apparatus due to insufficient light. The frame counter is difficult to see. ... We need additional time to prepare and check equipment.

THE NEXT CREWS

As the resources of the station (propellant, air, food and water) were sufficient to continue manned operations until 20 August, the return of Soyuz 11 was set for the last day of June and the launch of Soyuz 12 for between 15 and 20 July. The second crew would depart from the station just before its resources expired. In addition, a review of the resources on Soyuz 11 determined that it was capable of 57 hours of autonomous flight after undocking from the station.

Meanwhile, after a 10-day break on the Black Sea, Leonov's crew returned to the TsPK. There was a debate as to who should replace the ailing Kubasov. In the backup crew were Gubaryev, Sevastyanov and Voronov. Serving as flight engineer on Soyuz 9 Sevastyanov had performed the longest spaceflight a year ago, but the schedule did not provide sufficient time for him to train for Soyuz 12. Although in a short period of time Filipchenko, Grechko and Makarov all joined the DOS group, in mid-June they were reassigned yet again, this time to fly an autonomous Soyuz mission. Then Kubasov passed a detailed medical screening at the Institute for Biomedical Problems, indicating that he had suffered from no more than a simple allergy, which had almost cleared up. Nevertheless, when on 15 June Kamanin recommended that the Soyuz 12 crew should start training to fly the second DOS-1 mission, Mishin nominated Rukavishnikov to replace Kubasov on this crew.[14]

---

[14] If this were to be done, and the Soyuz 12 launch was on schedule, then Rukavishnikov would establish the world record for the shortest interval between successive missions: 101 days.

In addition, Kamanin nominated commanders and military research engineers for three more DOS crews. Later on, Mishin would add his flight engineers to complete them. Using the labels C for commander, FE for flight engineer and RE for research engineer, the assignments were:

- Prime crew (the second crew for DOS-1): Aleksey Leonov (C), Nikolay Rukavishnikov (FE), Pyotr Kolodin (RE);
- Backup crew: Aleksey Gubaryev (C), Vitaliy Sevastyanov (FE), Anatoliy Voronov (RE);
- The third crew: Pyotr Klimuk (C), FE from the TsKBEM, Yuriy Artyukhin (RE);
- The fourth crew: Valeriy Bykovskiy (C), FE from the TsKBEM, Vladimir Alekseyev (RE);
- The fifth crew: Viktor Gorbatko (C), FE and RE both from the TsKBEM.

In general, Leonov and Gubaryev's crews trained for the final mission to DOS-1, and Gubaryev's crew expected to fly in early 1972 as the first to DOS-2, followed by Klimuk's crew. Bykovskiy and Gorbatko's crews were to backup DOS-2. Although Kamanin allowed the last of the four crews for DOS-2 to have two civilian cosmonauts, their chances of flying were low.

Interestingly, one of Mishin's candidates for flight engineer on the DOS-2 crews was Feoktistov who, with Mishin and Tregub's support, approached Kamanin with a view to entering training, but Kamanin's negative attitude towards him remained strong.

The first two crews, prime and backup, were to end their training by 30 June, the day of Soyuz 11's planned return to Earth. In early July they would fly to Baykonur to prepare for a launch less than three weeks later. Although there was only a month remaining before his flight, Leonov asked Kamanin for permission to travel to the GDR (East Germany) to deliver personally to the Dresden Gallery his cosmic water colours. Kamanin refused the request with the following words: "If you do not want to place yourself in a stupid position, then don't tell anyone of this desire of yours. But, know that I will be categorically against your trip to GDR." Kamanin wrote in his diary on 15 June: "It was only a little bit over two weeks left before departure to the cosmodrome, and the commander of Soyuz 12 thinks not so much of the flight in space but about delivering his paintings to the Dresden Gallery."

Specific references
1. Vasilyev, M.P., *Salyut on Orbit*. Mashinostroenie, Moscow, 1973, pp. 21–81 (in Russian).
2. Kamanin, N.P., *Hidden Space, Book 4*. Novosti kosmonavtiki, 2001, pp. 317–320 (in Russian).

9

The fire

"THE CURTAIN"

To the national TV audience, the flight of the Yantars had settled into an established routine with the cosmonauts working to the timetable of scientific experiments, exercises and other activities. The programme was going to plan and the crew were in excellent spirits. There was not even the slightest hint in their transmissions of the clashes between Volkov and Dobrovolskiy. At the TsUP, Yeliseyev, Nikolayev, Bykovskiy and Gorbatko, who were jointly responsible for communicating with the station, worked hard to calm the tensions on board.

Day 11, Wednesday, 16 June
Dobrovolskiy and Volkov performed a test of the various methods for controlling the station. When doing so manually they used the wide-angle optical periscope. In addition, the accuracy of the ion automatic control system was tested. They also checked the intensity of the flashes while the attitude control system's engines were firing. Later, they studied the cloud formations in the upper atmosphere using a radio-mass-spectrometer.[1] During the brief time when all three men were awake, Patsayev performed routine medical tests. In terms of heart rate, Dobrovolskiy had 78 beats per minute and Patsayev 77, but Volkov had just 58; the norm being 60–80. And whereas Dobrovolskiy had an arterial blood pressure of 135/75 and Patsayev 135/85, Volkov was lower at 118/55.

From Dobrovolskiy's notebook:

> 16 June: At the beginning, we did not drink much water. Nor did we eat the assigned amounts. But, like at home, we ate when we felt hungry. However, the days are passing and we are slowly adopting the planned regime.
> "Stupid weightlessness! Another pencil has gone!" yells Vadim.

---

[1] In Russian: *radiochastotniy masspektrometer* (радио-частотный масс-спектрометар).

> Weightlessness is an interesting state. I am writing with Viktor's pencil – I lost mine a long time ago; almost all our pencils have gone.

It appeared that apart from problems with weightlessness and the lost pencils, the mission was progressing normally. But suddenly the situation changed. Just before the start of another communication session, Volkov noticed a smell of smoke from somewhere at rear of the station. As soon as communication was established, he reported: "Aboard the station is 'the curtain'!" The anxiety in his voice was evident. To confuse the Westerners eavesdropping on the station's transmissions, a number of code words had been defined, and 'the curtain' meant something related to fire and smoke. Unfortunately, having forgotten what this code meant, the controllers asked Volkov for an explanation. He furiously shouted in plain language: "There is a fire onboard! We are now entering into the ship!" He meant that they were retreating to the Soyuz ferry. He added that there was also a strong smell of burning electrical insulation. In their haste, they neglected to get the instructions for an evacuation, so he requested assistance: "Read us the instructions for an emergency undocking from the station!"

When the TsUP sought information about the source of the smoke, they were told that it was coming from a panel on the aft wall which separated the habitable part of the station from the propulsion section. The controllers could tell from the agitated voices that the crew were alarmed. While it was logical to evacuate the station, they should not do this while there was any prospect of extinguishing the fire. The first thought that came to mind was that one of the scientific instruments had caught fire. At that time, scientific organisations had yet to develop highly reliable equipment for use in space, and some faults were likely. In the main control room at the TsUP, Yeliseyev and Nikolayev acted to gain control of the situation by telling the crew to switch off all the scientific equipment, try to find the source of the smoke, and then retreat to the Soyuz. But the communication session expired before the cosmonauts could report.

Immediately after the communication session the leaders of the various groups at the TsUP met in the main control room to plan what to tell the crew to do during the next session. Their dilemma was that they did not know the situation on the station. Were the cosmonauts in the ship? Had they sealed the hatch to the station? Might they even have undocked! Since this was obviously no time to engage in a lengthy discussion, everyone was brief and businesslike:

"What should we do?"

"It is necessary to prepare several options."

"Explain."

"Let's begin with the worst case: that they have undocked the spacecraft from the station."

"We'll need several orbits to determine the status of the station. If they remained nearby, will they have enough fuel and life support to dock again?"

"We will have to calculate that."

"Ask your specialists."

"Okay."

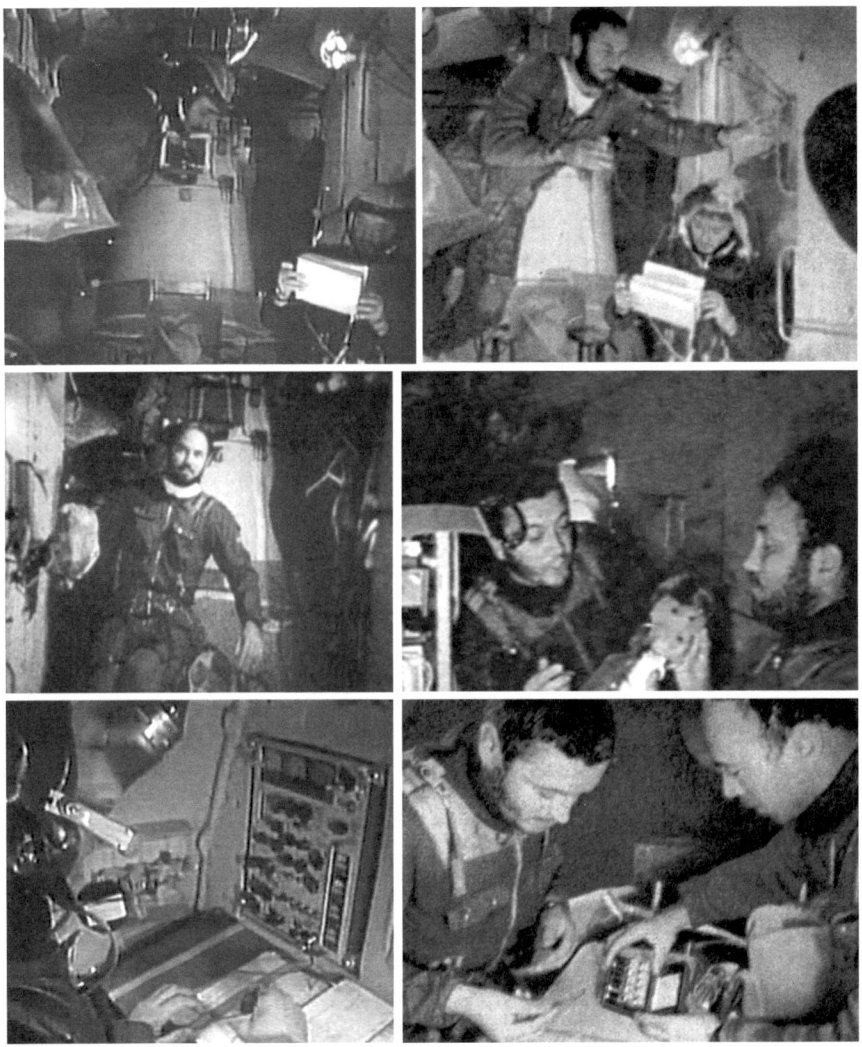

Life on board Salyut. In the two upper photos Dobrovolskiy and Volkov are in the working compartment, wearing their 'penguin' suits. Dobrovolskiy relaxes (middle left) after the fire on board the station. Dobrovolskiy (right) and Volkov discuss the flight programme (middle right). At times they were in conflict over how best to proceed. Patsayev can be seen working the Orion telescope (bottom left), and with Volkov taking blood samples (bottom right – note also the shoulder strapping of Volkov's 'penguin' suit).

"It is important to know whether they closed the station's hatch before undocking, because if they didn't then we've lost it!"

"That is clear. If they have undocked, then there is no urgency about the ship. We must focus on checking the station: first, the composition of the atmosphere and the power supply system."

"We should switch on the internal camera and assess the situation for ourselves."

"Agreed. Analysis Group, see to this."

"What if the cosmonauts are still on board the station?"

"We must question them. But first we must calm them down. They will probably have switched off the faulty instrument, but what if this had no effect?"

"Then the situation will be urgent."

"Let us prepare two additional plans: one for an urgent evacuation of the station, and the other a normal evacuation that returns the systems to the automated regime. Planning Group, this is for you."

"Good."

"If the faulty apparatus is switched off, the first step is to identify it, as otherwise we won't be able to reactivate the other instruments. Today's programme of work is already lost. Let us form a working group to find the problem. Representatives of Planning, Analysis and Experiments will participate, with the latter in charge."

"Agreed."

"We will have to remove the smoke from the station."

"The Analysis Group should prepare proposals."

"A longer period of communication will be required. We urgently need to connect all the command-measuring sites and arrange additional [telemetry] communication channels from the Ministry of Telecommunications."

"Okay, good. Now get to work. We will reconvene five minutes before the start of the next session and coordinate our efforts."

After the specialists had dispersed for their assignments, Minister Afanasyev rang from Moscow to ask what was happening on Salyut; as did Kerimov and members of the Central Committee. Yeliseyev explained only that a scientific instrument had caught fire, the cosmonauts had switched off all of the instruments, and specialists at the TsUP were studying a number of options to overcome the problem.

Yeliseyev also called Mishin, who immediately convened Bushuyev, Semyonov, Tregub, Feoktistov and Chertok. As Mishin told the TsKBEM team: "Yeliseyev has just reported that there is a fire on the DOS. The crew is preparing for an emergency landing. We must alert Kamanin to prepare the recovery team. Tregub must initiate work with the Ballistics Group to determine the best orbit on which to undock to ensure that the landing will be on our territory." It was decided that Tregub should go to assist Yeliseyev – although because a flight to Crimea would take five hours it was entirely possible that by the time he arrived the cosmonauts would themselves be back on Earth. The others would remain in Moscow and monitor the situation via internal channels. If it proved possible to continue the programme as planned, then in five days Chertok and Raushenbakh would join Tregub at the TsUP for the final phase of the mission.

"THE SMOKE ISN'T BEING PRODUCED ANY MORE"

The controllers met again five minutes before the next communication session and Yeliseyev's team prepared brief instructions appropriate to each of the three options. Just in case, he invited Eleonora Krapivina, who had spent a lot of time studying the crew in training and could evaluate their capabilities in an emergency situation. For Yeliseyev, it was important to have someone on hand to assist him in providing the most important instructions to the cosmonauts in the brief time available during the communication session.

When radio contact could be expected, Yeliseyev called: "Yantar! This is Zarya! On line!"

Instead of the station commander, who was responsible for reporting on incidents as serious as this, the response came from Volkov.

Volkov: "Zarya, this is Yantar. We hear you well."

Zarya: "Where are you?"

Volkov: "In the station."

Zarya: "Report what is happening."

Volkov: "The smoke isn't being produced any more, but there is still smoke in the station. We have headaches."

It was evident from Volkov's voice that he was tired, almost exhausted, but there was no sign of the previous anxiety. The smoke had come from the control panel of the scientific apparatus (PUNA) located on the wall at the rear of the main working compartment. This suggested that the problem was simply the failure of one of the science instruments. The controllers were greatly relieved. The instructions for this situation were very simple – to switch on the filter to cleanse the atmosphere.

Flight director Yeliseyev and Colonel Gorbatko played an important role in calming the Salyut crew after a fire broke out on the station, convincing them to continue the mission.

For reassurance, Yeliseyev explained the procedure for abandoning the station: "The order of the steps for an emergency evacuation is printed on pages 110 to 120. It lists what you should do after your transfer into the descent module. After transfer, prepare the spacecraft according to the instruction on 7K-T, pages 98a and 98b.[2] To undock, read pages 133 to 136. However, return only on command from the Earth. Don't hurry. With the panel switched off, the smoke should cease. If you choose to depart, leave the filter on. Take tablets for your headaches. The telemetry indicates that the carbon dioxide and oxygen concentrations are normal. The commander will take the decision about transferring to the ship and undocking from the station."

As commander of the station, Dobrovolskiy understood that it was time for him to take control of the communication: "Zarya, I am Yantar 1. We understand. There is no hurry. PUNA is switched off. Now two of us will be on duty, one will rest. Don't worry, we want to continue working."

Zarya: "Yantar 1, this is Zarya. We have analysed the onboard systems and we believe our recommendations will restore the situation. We hope you will be able to continue the flight according to the plan. The smell of the smoke will disappear. We suggest that you rest tomorrow, then resume the normal regime. Later, after you have left the communication zone of the ground stations, the ship *Academician Sergey Korolev* will contact you."

General Kamanin, who was planning to fly to Yevpatoriya on the afternoon of the same day, had been informed of the problem by General Shatalov in Zvyozdniy.[3] When Kamanin arrived at Yevpatoriya, Colonel Bykovskiy informed him: "The situation has improved. There is no longer smoke, just the smell of soot. But in the last six hours the crew has been so busy that they have not had dinner, and therefore are in need of rest."

During the emergency Volkov had become extremely nervous and, as the veteran, had usurped Dobrovolskiy's role and attempted to resolve the situation by himself. When he used expressions like "I decided" and "I did" in later conversations with Yeliseyev, Nikolayev and Bykovskiy it became clear that he was too emotional and independently minded to realise or acknowledge his errors.

In one of his last interviews, published in 1989, Mishin recalled: "I had a complex conversation with Volkov. He declared himself to be in command. When the cable burned, they lost their heads and wanted to depart the station. I calmed them down." In addition, Mishin ordered Volkov to respect the commander: "Everything must be solved by the crew commander; carry out his orders." But Volkov had replied: "The whole crew decides things together. We will sort out how to proceed by ourselves."

---

[2] Recall that 7K-T was the model of the Soyuz in use at that time.

[3] Kamanin was already planning to fly to Yevpatoriya on 16 June at 4 p.m. Prior to his flight, he went to the TsPK and met Popovich (on the eve of the latter's trip to Paris with Sevastyanov), Khrunov (about to visit the United States) and Volynov (who again asked Kamanin to be included in one of the forthcoming crews). Around 1 p.m., when Kamanin was having lunch, Shatalov approached him with the news of the fire on the station. Kamanin went straight to the airport and at 2.05 p.m. his Tu-104 departed for Yevpatoriya.

The tracking ship *Academician Sergey Korolev.*

At 10.30 p.m., the station entered the communication zone on its 155th orbit with the crew on board. Dobrovolskiy and Patsayev had calmed Volkov and sent him to rest, and he had fallen asleep. Kamanin conversed with Dobrovolskiy and Patsayev. After recounting the sequence of events on the station and describing the health of the crew, Dobrovolskiy judged the situation to be "almost normal". Although it was clear that they were exhausted by the day's events, he concluded: "We'll probably be able to continue the flight."

In his diary Kamanin added: "Prior to the launch of Soyuz 11 we agreed with Georgiy Dobrovolskiy that in describing the status of the station and the crew, if he had no doubt about continuing the flight he should say 'outstanding' or 'good', and if he had doubts then he should say 'satisfactory'. But the station commander forgot this." Kamanin was also dissatisfied that Dobrovolskiy appeared to have deferred to Volkov, who, after reminding everyone that he was the most experienced member of the crew, had dominated the communications with the TsUP.

A few orbits later, *Academician Sergey Korolev* made contact with the station and then informed the controllers at the TsUP that the situation on board was improving: Dobrovolskiy and Patsayev had eaten a meal and Volkov was still asleep.

The sudden emission of smoke in the station had strained the relationships between the members of the crew to the limit. During the crisis the cosmonauts continued to make entries in their notebooks. One remark by Dobrovolskiy clearly indicates his concern: "If this is harmony, what is divergence?"

Day 12, Thursday, 17 June
The next day, while the controllers analysed the telemetry received from the station, the crew visually inspected the locus of the fire, identified the faulty apparatus, and isolated it from its power supply. It was the fan to cool the panel for controlling the orientation of some of the scientific equipment. When the fan seized, the motor had continued to try to drive it, and the winding of the stator had overheated and issued a dense smoke. Although there had been no flame, as such, this was the first case of a 'fire' on the manned space mission.

On the recommendation of the TsUP, the cosmonauts reactivated the instruments one by one until all the scientific equipment was again operational.

Although the filter removed the smoke, the crew remained concerned about the composition of the atmosphere.

4.26 a.m.
   Zarya: "During the 955th orbit, perform a functional test with the apparatus."
   Dobrovolskiy: "What is the composition of the atmosphere?"
   Zarya: "It is normal."
   Dobrovolskiy: "Are you watching the oxygen?"
   Zarya: "The oxygen is normal. We are watching it for you."

With this assurance, Dobrovolskiy retired for some much-needed sleep. Patsayev was already asleep. Volkov, now sounding less anxious, was on duty.

7.31 a.m.

Zarya: "Yantar 2, please remind Yantar 3 that on the 957th orbit he is scheduled to do a stabilisation."

Volkov: "I won't awake them, they're so tired."

Zarya: "It isn't necessary now. Let them rest."

11.56 a.m.

Zarya: "We have a question. How many times did each of you use the vacuum unit – how often and for how long? You can reply tomorrow if you don't have the details to hand." This was an enquiry about the Veter lower-body negative-pressure apparatus.

Dobrovolskiy: "Understood. The vacuum unit is good. During one test, I reduced the pressure to –70 mm [of mercury] and felt excellent. The loads aren't like those on Earth, they are much less, and it is possible to increase the vacuum level without risk."

Later in the day, General Agadzhanov advised members of the State Commission that the situation was satisfactory, and since the crew were in no immediate danger there was no reason to curtail the flight. When the issue was raised of whether a fire might occur in another system, it was decided that the cosmonauts should switch off all the scientific apparatus until Chertok and his team could determine the status of the station's electrical system and assess the potential of another fire.

How did the families of the cosmonauts react to these dramatic events? Svetlana Patsayeva was at a Young Pioneers' camp and hence was personally unaware of the problem in space: "But," she recalled, "for my mom these days were very difficult. She wrote a diary during the entire mission. I read these very personal records only following her death. Her diary clearly shows how much she worried about the crew. She knew the dangers. Indeed, she worked at an enterprise near dad's and actually knew the complexities of a space mission.[4] Mom was present during the periods of communication with the crew, and was up to date with what was happening on the station. I didn't know of the fire, but mom knew from the conversations of the crew with the Earth. And she knew how serious it might become."

Marina Dobrovolskiy was also unaware of the incident: "The technical side of the flight did not greatly concern me. I thought about dad, how he was feeling, what he was doing, and when he would return. But I was always sure that my father would find the correct solution and skillfully overcome even a very difficult situation. So it was with the fire. Indeed whether or not the station would continue in operation was dependent on a command decision."

In the meantime Western observers picked up Volkov's unencrypted transmission that there was a fire on board the station, and, their suspicions aroused, observed an intriguing change to the daily routine on 17 June – the Soviet press reports made no mention of either scientific work or a TV transmission, they referred instead to the

---

[4] Vera Patsayeva worked at the Central Scientific Research Institute of the Academy of Sciences (TsNIIMash), which was adjacent to the TsKBEM's main building in Kaliningrad.

cosmonauts having carried out "minor corrective work", explaining that there were tools, spare parts and safety devices on the station.

SPACE BIRTHDAY

Day 13, Friday, 18 June

On the second day after the fire the mood of the cosmonauts noticeably improved. Volkov and Patsayev even gave a 5-minute TV report in which they showed some of the scientific equipment – in particular demonstrating how the enormous bulk of the solar telescope dominated the compartment. They also talked about monitoring the Earth from the station. Of course, by this point they had removed all evidence of smoke, and at no time did they refer to the fire. Watching the broadcast, Kamanin noted a "discordance" between the tired unshaven faces of the cosmonauts and the impressive background of the solar telescope.

On direction from Chertok, the controllers read to the crew the plan for switching on the apparatus. One by one all the medical equipment and several of the scientific instruments were reactivated. All worked normally. The scientists were keen for the remaining apparatus to be reactivated in order to resume the scientific investigations but Yeliseyev and Tregub, who was now at the TsUP, with the support of Chertok in Kaliningrad, decided to await a comprehensive analysis as to which of the various investigations should be continued.

After a brief medical check, Patsayev resumed using the Orion telescope to study stars in the ultraviolet.

7.24 a.m.

Patsayev: "Can you hear me well? The Orion experiment was performed at the planned time: the second regime, the second star on map No. 3. I started working at 6.34 and the timer was started at 6.45. Exposures of 10, 30, 90 and 270 [seconds] were made. I will determine the rewind time and let you know later. All indicators changed colour: green, orange and white. As for the rest, it was all normal. Report finished."

Zarya: "Yantar 3, continue. The Control Group appreciate you resuming normal work."

Dobrovolskiy: "Yes, he is happy too."

Zarya: "Yantar 3, a reminder: don't forget to check the No. 5 panel for the control of the scientific apparatus before the start of the second part of the experiment."

Patsayev: "I won't forget. I prepared it earlier. I cleaned the porthole and the glass of the visor. All is normal."

Zarya: "Excellent, excellent. According to preliminary results, everyone is happy with the first part of the experiment."

Patsayev: "Yes, it wasn't bad. We are happy too. The object was held stable, and the automatic system worked well."

1.21 p.m.

Patsayev: "I'm reporting the situation with the Orion. The third star went well except that we could not finish the final exposure planned for 810 seconds owing to the sunrise; it was just 720 seconds. I had to cut it short because of the glare from an illuminated antenna. The remainder of the operation was normal."

Zarya: "Understood, Yantar 3. Thank you for the information."

Patsayev: "You're welcome."

Zarya: "Yantar 3, as we aren't going to be on duty tomorrow, Happy Birthday from the Control Group."

Patsayev: "Thank you, thank you."

Zarya: "We wish you all the best. The others are preparing greetings for tomorrow. Most sincere greetings to you."

Patsayev: "Thank you."

Zarya: "All of you should have a tube of juice tomorrow, to make a toast."

Patsayev: "But there is no glass."

Zarya: "We hope Yantar 2 will find something for this occasion. See you later."

At 2.30 p.m. Salyut began its 182nd orbit in its manned regime.

It had been intended that on 20 June the cosmonauts would observe the launch from Baykonur of the third N1 lunar rocket in order to test the Svinetz instrument that the military had designed to detect intercontinental ballistic missiles, but on 18 June the launch was delayed, initially to 22 June, then later to 27 June, which ruled out the possibility of the cosmonauts making the observation since on that day the station would not pass over the cosmodrome. However, Kamanin confirmed that the plan for them to observe night launches of solid-propellant missiles from Baykonur would proceed. Meanwhile, Mishin had flown to the cosmodrome to supervise the preparations for the N1 launch. The postponement meant that he would not be able to fly from Baykonur to Yevpatoriya until a few days before the planned end of the Soyuz 11 mission.

10.19 p.m.

Volkov: "I have just awakened. I slept for about seven hours. I slept well, and feel well. The others are now resting."

Zarya: "Where are the others sleeping? In the working compartment?"

Volkov: "Yes. Again on the floor, next to the filters. They support themselves on both sides by their legs, and sleep."

From Patsayev's notebook:

> Here, we don't need legs – we swim like fish in an aquarium. Loose objects will float out of reach if you don't attach them to something. Untidy people are not welcome in space! There are interesting differences to being on the Earth – for example to drink water, to eat, in movement, with clothing, and with sleeping. Here, we must learn everything afresh. ... Weightlessness is both good and bad. It is good when working with instruments, and facilitates easier movement. But the return to gravity will be difficult. In the future we will need spacecraft with artificial gravity.

Although the Soviet press resumed their familiar routine of reportage on 18 June, the radio monitors at the Kettering Grammar School in England detected telemetry on the Soyuz 11 frequency – the first such transmissions since 9 June – indicating that for some unannounced reason the cosmonauts had powered up their ferry.[5]

Day 14, Saturday, 19 June
The birthday of Viktor Patsayev on 19 June further relieved the stress on the crew. They performed medical examinations and operated such scientific apparatus as the TsUP permitted. In fact, the Control Group had decided to scale down the scientific experiments, prohibit the communication of unnecessary information, and gradually increase the physical exercise regime in preparation for the return to Earth.

Several times during the day, the cosmonauts were moved from one experiment to another. They were physicians, biologists, astronomers and meteorologists, and the scientists back on Earth were keen for the results. Astronomers wanted observations from above the atmosphere of cosmic radiation that would provide information on how the universe was structured. Physicians and biologists wanted to know how the human body and other organisms reacted to long-term exposure to space – both in terms of weightlessness and the radiation environment. Physicists wanted to know how various materials behaved – fluids, for example, display interesting properties in weightlessness. Technologists wanted to know if it would be possible to create a whole new range of materials possessing unique attributes. When Salyut was being designed the Academy of Sciences had suggested that a scientist be included in the crew, the logic being that only a scientist could analyse the results of an experiment in space and suggest a procedure to follow up an interesting observation. However, because a commander and a flight engineer were required to operate the station, and the Soyuz could accommodate a maximum of three cosmonauts, there was room for only one researcher on the crew. It was therefore decided that the third member of each crew should be a professional cosmonaut who had been trained as a researcher and investigator – which is why Patsayev's role on the crew was 'research cosmonaut'. The scientific programme was developed by the scientists, who spent a great deal of time explaining how to use the apparatus and how to analyse the results. In addition, senior representatives for each scientific investigation were permitted access to the TsUP, and there was a special radio channel between the scientists and the station's crew to enable the scientists to discuss the performance of their experiments and to offer the cosmonauts advice. Although the scientists and the crew worked together closely, the cosmonauts never spoke the surnames of the scientists on the radio for security reasons. When an experiment was successful, the contented scientists were often able to exit the TsUP with graphs and tables. If a problem developed, then the scientists would retire to attempt to understand the failure and devise a remedy for the next opportunity. Despite some frustrations, the experience gained in

---

[5] In fact, Soyuz transmissions continued until 21 June, then nothing more was heard until 24 June.

attempting to undertake a scientific programme on an orbital station was priceless. Sometimes a modification of the apparatus or a revised operating procedure was suggested for a future flight. In some cases, it was concluded that the work would be better done by an automated satellite – for example, once a telescope had been precisely aligned on a celestial source, the observation could be marred by the vibration of the station in response to the cosmonauts moving around. But, on the other hand, there was merit in testing new apparatus on a manned station to ensure that it worked properly prior to assigning it to an automated satellite.

Volkov and Patsayev conducted a multispectral optical study of the atmosphere at sunrise and sunset – each of which occurred in orbit every 90 minutes, providing a wealth of data. During this experiment they established the diversity in colour of the atmospheric upper layers during sunset and sunrise, and its correlation with aerosol particles. At 2.58 p.m., when the station flew over the northwestern coast of Africa, the cosmonauts saw a vast sand storm. In addition, they placed the station into solar orientation and checked the accuracy of the gyroscopes after such a prolonged time in space.

Continuing their medical programme, the cosmonauts made further measurements of their cardiovascular systems and bone density. They also assessed the ability of their eyes to differentiate colours in order to determine the degree to which the eye is affected by weightlessness.

Conditions inside the station were normal: the temperature was 22°C, the pressure was 880 mm of mercury, and the smell of smoke had cleared.

7.13 a.m.
Zarya: "We all send Happy Birthday greetings to Viktor Ivanovich. We wish him successful work."
Patsayev: "Thank you."
Zarya: "We hope that the commander will organise a party."
Dobrovolskiy: "We offered him a day of rest apart from physical exercise, but he has so much technical work."
Then, after a pause, Dobrovolskiy reported: "We performed photography of the twilight horizon. When the Sun appeared, a small part, less than half of its disk, was visible. We took pictures of it."

10.19 a.m.
Zarya: "Yantar 3, again we all send our greetings on your birthday – we wish you a successful flight and happiness in your life. Your family sends their most sincere wishes."
Patsayev: "Thank you for your greetings. Although you are far away from us, we always feel your support."

Patsayev's 38th birthday was the first birthday to be celebrated in space. Knowing the recent crisis and tensions between the members of the crew, the psychologists at the TsUP had prepared a special programme. Patsayev's wife and children were in the communication centre in Kaliningrad, watching the TV signal from the station.

With them was the famous TV anchor, Yuriy Fokin. The communication officer at the TsUP was Nikolayev.

Nikolayev: "How is the table prepared?"

Patsayev: "The table is prepared excellently: cold veal, cookies and blackberry juice in tubes."

Nikolayev: "Did you find the bottle?" He was referring to the traditional bottle of celebratory champagne.

Patsayev (laughing): "No, we didn't. We looked for the bottle everywhere, but we couldn't find it. The delicacy was the onion, which was a present from Vadim. We sliced it into three parts and shared it. Zhora's present was lemon."

The onion and lemon were smuggled on board by Volkov especially for the first birthday in orbit. The TV viewers could see the table set with tubes of juice, cheese, fruits, nuts and cans of veal. The items were held in place by tapes across the table. Patsayev sat at the table and Dobrovolskiy and Volkov floated in the background, smiling happily. Not having champagne glasses, they toasted loudly with the plum juice tubes. Patsayev said that of his presents he most enjoyed the onion, which was the first 'fresh' food that any of them had tasted since entering space.

His son Dmitriy recalls: "We were invited to the communication centre and had a chance to talk with dad, but only for a brief time. I don't really remember what we said – I was 13 years old and there were many interesting devices in the room that distracted me. However, I do remember that Dobrovolskiy and Volkov presented an onion to him."

Svetlana Patsayeva, even younger, remembers the visit only by what her mother

"Did you find the bottle?" asked General Nikolayev (right) as Patsayev celebrates his birthday in orbit.

told her: "The crew were given congratulations from their friends and relatives. And there was music too. Our friends especially asked that our father's favourite song be played. Previously, the people on the ground had recorded the congratulations from our family. They brought me home from Young Pioneers' camp for this. Someone came to our home with the equipment to record our words, me playing the piano (in fact, I was just starting to learn music, and my playing was not very good), and the sounds of our parrot."

After the communication session Patsayev's friends gathered in his apartment in Moscow, where his wife, Vera, had prepared a celebratory lunch. It was an unusual birthday party, as the person being honoured was not present. Someone had a bottle of French champagne, but it was decided to defer opening it until Viktor was home. A note was affixed to the bottle bearing the signatures of all the attendees, together with the message: 'Vitya, you were searching for this bottle in space, but it was here on Earth.'

From Dobrovolskiy's notebook:

> 19 June. Viktor's birthday. His wife sent a greeting letter with the words: "Mum has arrived, she is feeling well." Viktor was so impassioned.

From Volkov's diary:

> 19 June. Today is Viktor's birthday. We laid the table. The onion was a real delicacy. Zarya gave him their greetings. The Earth asked for a report.
>
> 21.30. Start my duty. I will be the first to see the globe instrument indicate '1,000 orbits'. This historic event will occur during my time on duty. Simply unbelievable.
>
> I slept at the new place, which is similar to the roomette in a wagon.[6] For the last two days I have slept well – about eight hours. Tomorrow we expect the radio programme *With Good Morning* to be transmitted.

So, with the stress of the fire behind him, Volkov was once again sleeping well.

ONE THOUSAND ORBITS

Day 15, Sunday, 20 June

At 2.14 a.m. the Salyut space station completed its 1,000th orbit since its launch on 19 April. It was in the communication zone at the time, and cosmonaut Gorbatko was the communication officer at the TsUP. He pointed out that the crew had been on board for 206 orbits, and joked that perhaps they should remain for an additional thousand orbits. In accordance with the flight plan, 20 June was a rest day in space. They made a TV report showing off the station's various sections and its equipment. While in radio contact, they reported observations of the Earth and its atmosphere

---

[6] On Russian trains which travel for many days and nights, some wagons provide roomettes in which passengers can have privacy.

that they had made in recent days, including the African sand storm they had seen the previous day.

From Volkov's diary:

20 June. The third week of our work in orbit has started, but the station has been in space for two months, making 1,000 orbits. Because the commander Dobrovolskiy and research engineer Patsayev are sleeping, I'll be on duty at the time of the 1,000th orbit. In the sleeping bags, I can only see their heads. In these 'beds' you get so comfortable that sometimes you grow reluctant to get up.

There is only one orbit until the 1,000th circle. It has just started the 999th. In a few minutes, Zarya will call me. Through the static I hear:

"Yantar! Here is Zarya. On line!"

"Zarya! I am Yantar 2, I hear you excellently!"

"Yantar 2, how is it going?"

"How is it going? Normal. My crewmates are asleep. With no one to talk to, I don't feel so cosy in this huge space home. It is a feeling that is familiar to anyone who, as the sailors say, has duty on the ship's bow. As I speak to you I feel as if I am at home. I know that the weather below isn't very good, being cloudy, windy and rainy. Up here, away from the portholes, the Sun is blinding and the Earth is covered with the clouds."

"Don't you have rain?" Zarya asked in jest.

"No, we don't have rain. Nothing Earth-like is in this vicinity. Just the real splendour of space!"

"Here, they are preparing the *With Good Morning* radio programme."

"That is good news."

We have heard the pre-recorded selections of music on our tape recorder so often that they are no longer our favourites! We are therefore eager for the promised radio programme, in particular our music requests.

It is interesting how the commander and I look with bearded faces on the TV screens. My beard reminds me of a Tatarian-Mongolian man. Honestly, I don't tend to it any more.

The Earth asked: "How do you hear the short waves?"

"It is good – especially in the western hemisphere. It is so pleasant to hear words in your own language while passing over South America."

Next a question about our plants: "Do you look after the shrubs?"

"Of course! In fact, more often than planned in our flight programme. We have a special love for our greenery. We feel that it links us to the remote – yet so close – Earthly realm. We devote great attention to our 'little cosmic garden'. The vegetables grow well."

The communication session is over. The next will be on the 1,000th orbit. How long will my two crewmates sleep? Will I alone see the number 1,000 appear on the display of the globe? No, the crew commander will be with me. I'll awaken him in half an hour. He will take the duty, and communicate with Zarya.

We know that the 1,000th orbit will start at 00.44.44 on 20 June. In these final minutes, the only thing that I do is watch the onboard clocks. Yes! The first seconds of the jubilee orbit have begun.

When journalists at the TsUP asked Yeliseyev, the flight director, about the crew, he said: "Each man has a different character and, of course, during communication sessions this is very noticeable. Patsayev doesn't speak very often, we almost never hear him. He will just let us know which experiment he has finished, or ask details about his work. Volkov speaks the most. He also expresses his emotions the most. He not only talks about the flight programme and the investigations, he also asks us about soccer scores and weather conditions. He sends his regards. This is totally in accordance with his spirit and nature – on the Earth he was also so communicative. In terms of emotions, Dobrovolskiy is somewhere in the middle. He always speaks calmly and certainly."

00.59 a.m.
Volkov: "The 1,000th orbit is a working orbit. Although today is a rest day, we decided to devote it to the Earth – we photographed the cloud cover, the oceans and the landscape for geological studies and issues relating to the national economy. In general, we are doing work which is usually assigned to working orbits. We want to spend every minute of our 'leisure' time maximising the results for return to Earth."

Zarya: "We send you our warmest regards – there are so many greetings."

Volkov: "About 4 o'clock, when Viktor wakes up, we'll do our physical exercise and then do what I have already said – photographing and monitoring the Earth."

Zarya: "If you have the information to hand, please tell us what you have done for the last 24 hours in terms of medicine."

Volkov: "We are doing all the experiments required by the physicians."

Zarya: "Understood. Thank you. Well done!"

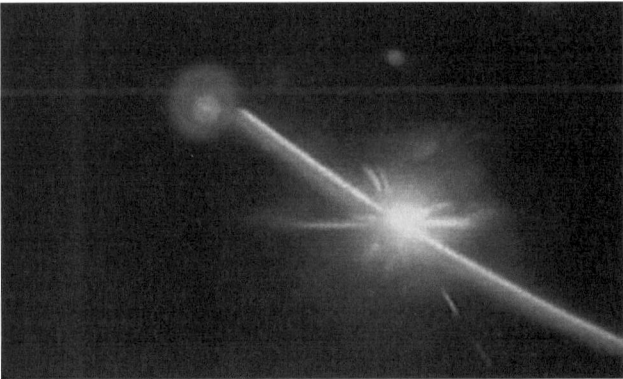

One thousand orbits around the Earth. A sunrise recorded from the Salyut station.

Volkov: "I carefully log our food and water consumption. Tell our comrades there who are responsible for this that I am logging it all. We have written reports on the operation of all the systems.[7] On board this ship, we are sharing our duties – each of us has a different area of responsibility. Everything is as planned."

3.57 a.m.

Zarya: "Yantar 3. Firstly, we are happy with your Orion work. Tomorrow we also plan Orion work. Did you perform two sessions with Orion with one or two stars?"

Patsayev: "Two stars."

5.30 a.m.

Dobrovolskiy: "Yesterday at 14.58.00 above the northwestern coast of Africa, I observed a sand storm at 344 degrees longitude and 17 degrees latitude."

8.36 a.m.

Zarya: "One request: please water the plants twice a day – at the start of the day, and at the end."

Patsayev: "In the instructions it says to water once only."

Zarya: "Understood. However, it is necessary to do so twice. Report the general conditions of the shrubs, and in particular the development of the first real foliage. Report on it daily."

For most of 20 June the cosmonauts rested and monitored the Earth, its clouds and ocean, and made observations of the stars. In addition, Dobrovolskiy provided a TV report for viewers on Earth – the request for which was probably an attempt by the TsUP to highlight his role as the station's commander.

Television Report:

Zarya-25 (TV reporter): "Yantar 1, as the first commander of the Salyut orbital station, do you have any impressions?"

Dobrovolskiy: "I have great impressions. I am lucky that my first space flight is to this station. It is composed of two spacecraft: the station itself and the transport ship docked with the station. It is a large complex. It allows us to conduct a great deal of scientific work. The designers, engineers and diligent workers did an excellent job of providing comfortable living conditions for the crew."

Zarya-25: "We understand that you have controlled both the Soyuz spacecraft and the Salyut station. Obviously, they have different characteristics. Can you speak of the differences in flying these vehicles in space?"

Dobrovolskiy: "I can tell you that our training enabled us to master the techniques required. We have no difficulty. It is very easy to control the transport

---

[7] In assuring the controllers that he was taking meticulous notes, Volkov, who was well aware of how he caused problems for his colleagues, flight controllers and even his boss Mishin, may have been trying to make amends as the mission approached its conclusion.

ship, and the entire orbital station is very responsive – easily controllable. In general, it is just as each of us dreamed flying in space would be like."

Zarya-25: "Understand. Yantar 1. Specifically, what have you done as the station commander?"

Dobrovolskiy: "As a matter of fact, my first task was one of the most interesting operations – docking. We wanted so much to conduct it in the best possible manner. As for work, the station is so large and there are so many possibilities for work that each member of the crew has specific responsibilities. It is a complex issue. On the flight to the station we had some discomfort [adapting to weightlessness], but after entering the station we began to work at full strength and soon it was as expected."

Zarya-25: "Thank you very much, Yantar."

Despite the fact that the general health of the cosmonauts was acceptable, in their two weeks in space they had spent considerably less time on physical exercise than planned owing to the following reasons:

- when the load-bearing 'penguin' suits were worn during exercise, they tore, and their function was greatly reduced once the elastic sections had become damaged;
- some of the supporting struts of the Veter lower-body negative pressure unit were damaged early on, and thereafter the cosmonauts rarely used it; and
- use of the treadmill was restricted because the noise was sufficient to disturb anyone attempting to rest, and because it transmitted vibrations through the station's structure which caused the solar panels and antennas to oscillate and the propellants to slosh.

TRACKING SHIPS

As noted, the mission of the first Salyut station was controlled from the TsUP in Yevpatoriya, Crimea, supported by several tracking ships of the Soviet Academy of Sciences.

In March 1971 *Academician Sergey Korolev* had relieved *Cosmonaut Vladimir Komarov* in the North Atlantic, near Sable Island off the Canadian coast. Its first task had been to support Soyuz 10 in April. Now it was supporting Soyuz 11. Most of the crewmen of *Academician Sergey Korolev* originated from Odessa, the city in which Dobrovolskiy was born. They also had fond memories of Volkov, who had visited the ship in December 1970 and attended its launch. It had all the apparatus needed to control the most complex operations of the Soyuz-Salyut orbital complex, including orbital manoeuvres. It could communicate with the TsUP via a Molniya satellite. When the station's path took it over the eastern region of North America or the North Atlantic, *Academician Sergey Korolev* would be able to communicate with it for up to 12 minutes, and two or three communication sessions were possible each day.

An older and less sophisticated ship was stationed in the equatorial region of the Atlantic Ocean. This was *Bezhitsa*, which was on its fifth voyage since its launch in

February 1967. It had taken its station at 13 degrees west and 1.5 degrees south in March 1971 to support the Salyut mission. It could communicate with the crew of a spacecraft and receive telemetry, but did not routinely transmit to the TsUP – this required the use of the internal channels of the Soviet Navy. Also in the southern Atlantic Ocean was *Kegostrov*, which was another of the smaller vessels launched in 1967, and also on its fifth voyage. It had sailed in February 1971 and taken its station at 24 degrees west and 22 degrees south. Like *Bezhitsa*, it was equipped to receive telemetry from the spacecraft and communicate with its crew. Depending on the schedule decided for the return of Soyuz 11, one or other of these two ships was to monitor the critical braking manoeuvre.[8] Several other communication ships were located in the South Atlantic to assist with the operation of Salyut: *Morzhovets, Borovochi, Nevely* and *Ristna*.

Specific references

1. Yeliseyev, A.S., *Life – A Drop in the Sea*. Aviatsiya and kosmonavtika, Moscow, 1998, pp. 77–79 (in Russian).
2. Siddiqi, Asif A., *The Soviet Space Race with Apollo*. University Press of Florida, 2003, pp. 778–780.
3. Vasilyev, M.P., *Salyut on Orbit*, Mashinostroenie, Moscow, 1973, pp. 81–107 (in Russian).
4. www.ski-omer.ru (in Russian, about Soviet tracking ships).
5. Kamanin, N.P., *Hidden Space, Book 4*. Novosti kosmonavtiki, 2001, pp. 320–325 (in Russian).

---

[8] Specifically, *Kegostrov* was to monitor the braking manoeuvre if this were scheduled for the second orbit after Soyuz 11 undocked from Salyut, and *Bezhitsa* would do so if it occurred on the third orbit.

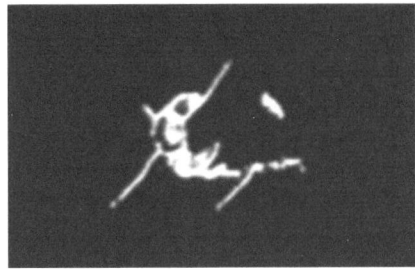

10

Drawing away from the station

FINAL DAYS

In their first fortnight on board the Salyut station the cosmonauts had performed a large amount of scientific work and accumulated results to be returned to Earth for analysis by specialists. As the mission drew to a conclusion, the crew were in high spirits.

Day 16, Monday, 21 June
Work resumed with the Orion astrophysical observatory, this time with stars in the constellation of Serpens. Volkov was in charge of navigation. He spent a lot of time 'sitting' by a porthole on the station's floor, 'hunting' for landmarks on the Earth and in the sky. Once Volkov had selected a landmark, Dobrovolskiy would orient and stabilise the station to enable this to be viewed. During the next orbit, Patsayev controlled the two telescopes of the Orion system, one on the exterior of the transfer compartment and the other affixed to a porthole inside it, to simultaneously record spectrograms of a single star in different sections of the ultraviolet spectrum.[1] The cosmonauts also continued measurements of gamma rays, the electrically charged nuclei in cosmic rays, and the intensity of free electrons in the orbital environment. At 2.21 p.m. Patsayev, who had started his career in meteorology, sent a greeting on behalf of the Salyut crew to the attendees of the National Meteorological Congress in Leningrad.

From Patsayev's notebook:

> 21 June. The Moon looks the same as when viewed from Earth. Sometimes a round rainbow 'spot', or halo, is visible through a porthole on the opposite side to the Sun.
> The boundaries of clouds can be determined by their shadows. Thicker

---

[1] He observed the star Vega (alpha Lyra).

clouds are moving away in regular order, and cloud belts on the night side are visible in moonlight. ...

The can openers are inadequate, often creating shards while opening the can. The seal of the rubbish bags is unsuitable, letting the stench out. ...

It is essential to have a work site for performing repairs, a workbench with instruments. ...

The station lights are inadequate. The inscriptions on the push buttons for switching on the food heater and the vacuum cleaner are barely visible. It is too dark at the work sites, especially at No. 3 [which was located adjacent to the large conical module housing the main scientific equipment].

"THE GREEN CORNER"

Day 17, Tuesday, 22 June
When Salyut entered the communication zone of the ground stations Dobrovolskiy was on duty, continuing studies of the physical properties of the atmosphere using a manual spectrograph. For this work he observed the horizon immediately before the Sun rose above the horizon, continuing until it had risen, and then he repeated the sequence in reverse at sunset. At the same time Patsayev measured the polarisation of the sunlight reflected by the Earth surface. Volkov and Dobrovolskiy performed meteorological observations, in particular of a cyclone near Hawaii. Meanwhile the gamma-ray measurements continued.

2.01 a.m.
Dobrovolskiy: "We saw a big cyclone at 168 degrees longitude and 30 degrees latitude, and photographed it."

A few hours later, they monitored the development of a cyclone near the eastern coast of Australia.

6.41 a.m.
Volkov: "Zarya, I am reporting. At 6.36 hours we observed and photographed a cyclone at 125 degrees longitude, near Australia."

Then the cosmonauts made another telecast.

Television Report:

Zarya-25: "Today I ask you to tell us about your biological experiments related to the effects of weightlessness on the growth and development of higher-order plants. Can you start your report?"
Dobrovolskiy: "Zarya, this is Yantar 1. I will ask Yantar 2 to swim over to one of the containers of scientific apparatus, and he will show you the objects we study."
Zarya-25: "We see you, excellent."
Volkov: "Comrades, we continue the introduction to our station and its extensive programme of scientific work. In the time available, we will explain the complexity

A cyclone seen from Salyut.

of the biological investigations that we are performing. I will now show you the special section where our mignonettes are situated – nine plants.[2] However, for that I have to swim."

Zarya-25: "Go ahead, we will observe you."

Dobrovolskiy: "I would like to show you the special section with the container for the plants. The name of this container is Oazis."

Zarya-25: "We can see it excellently on the TV screen."

Dobrovolskiy: "This container holds nine bags, each with the seeds of different plants brought from Earth."

Zarya-25: "Which ones?"

---

[2] Mignonette is the common name for a small family of herbs and shrubs that inhabit arid regions.

Dobrovolskiy: "It is difficult to tell exactly. I am reluctant to move them, because they have not yet grown sufficiently to be recognisable."

Zarya-25: "Are they growing?"

Dobrovolskiy: "Here are the sprouts of the plants. You should be able to see them. The first one appeared just two days after the start of operations using this container. The second sprout, this one, is actually higher than the first, and even has four little leaves. Can you see? Next, the sprouts in bags No. 2 and No. 1 appeared."

The commentator asked how they tended the plants.

Dobrovolskiy: "We continuously observe the plants. It is a real pleasure to watch how they grow. Several times per day, we look into our 'green corner'. The plants are developing in normal conditions. We water them twice each day using a special solution. They are illuminated by three lamps. Beside the Oazis is another unit with seeds of other plants, as well as water bacterium, flies and chlorella."

Zarya-25: "Thank you very much. We would like to continue this discussion, but your station is leaving the communication zone. We will take our leave of you, and wish you a successful flight. All the best."

In the unit adjacent to the Oazis, shown by Dobrovolskiy, were tadpole embryos. The fertilised spawn was brought to the station by Soyuz 11. The embryos had been put into storage on 10 June, after several days of development, so that they could be studied on Earth for any deviations from normal development.

From Dobrovolskiy's notebook:

> 22 June. Today Viktor has decided to sleep in the orbital module. Previously, he was sleeping in the same place as Vadim.
>
> All the time we are busy with work: changing the water tanks, activating scientific apparatus and adjusting it, taking pictures, controlling the station's systems, making the day's TV programme, communicating, etc. Vadim relaxes by reading Pushkin or Lermontov. Viktor uses the Era a lot,[3] working either with the cassettes or the cine or photographic cameras. . . .
>
> From *Academician Sergey Korolev*, the crewmembers from Odessa have forwarded me greetings in lyrics. . . .
>
> From the perspective of its construction technique, the station reminds me of home. Although it has four 'rooms' – the descent and orbital modules of the ship and the transfer and the working compartments of the station – here everything is optimal for work and rest. We have the engineers, technicians and workers to thank for this. However, people usually think of home as a place to relax after returning from work. Here, it is impossible. On the Earth, a home is where you are surrounded by relatives and friends. Here there are only the three of us. And it cannot match the air, the sea, the Russian fields, the snow or the wind – everything that our minds associate with 'home'. . . .
>
> Watching the Earth, you can see the direction of the station's flight, but on

---

[3] The Era investigation, which began on 16 June, was to detect high-energy electrons at orbital altitude.

rapidly moving away often you cannot.[4] We try to wear the 'penguin' load suits all the time – we even sleep in them, although at first that was not so pleasant. ... We often work with the vacuum cleaner.

Meanwhile, at the TsUP the Landing Commission met and then recommended to the State Commission that Soyuz 11 should return on 30 June, making the landing on the third orbit after undocking from Salyut. The recovery zone had to a generally flat unpopulated region without major rivers, lakes or forests. They selected an area of the steppe some 150–200 km southwest of Karaganda in Kazakhstan.

Day 18, Wednesday, 23 June
With the end of the mission imminent, Adamik Burnazyan, the Deputy Minister of Public Health, called the 'stars' of aerospace medicine, including Oleg Gazenko and Abram Genin, to the TsUP. On 23 June they expressed their confidence that by the use of the treadmill and elastic expanders for exercise, wearing the 'penguin' suits to condition their muscles and bones, using the Veter lower-body negative-pressure unit to sustain their cardiovascular capacity and routine monitoring by the Polynom apparatus, the Salyut crew would be in much better shape on their return than were the Soyuz 9 cosmonauts. But Dobrovolskiy, Volkov and Patsayev had not exercised as much as intended. Although it was evident from an analysis of the medical data and discussions with the crew that they were tired, tense, and lacked concentration, the physicians attributed this to poor organisation of the flight, an overly ambitious work programme, and the unfamiliar daily rhythm of the operational schedule of the station, which was shorter than the 24-hour norm. The Air Force felt that the crew would be able to finish the flight as planned, but would have difficulty readapting to gravity. Kamanin suggested that Volkov would have the greatest difficulty because he had exercised less often, was drinking insufficient water, had refused to eat meat, had often complained about problems with the physical training equipment, and had shown the greatest tendency to make mistakes. This was rebutted by the doctors of the Ministry of Public Health, who said that Volkov ought to readapt more rapidly because of the three men he was the keenest sportsman and because he had flown in space previously. It was true that his first flight in October 1969 had lasted just five days and that on his return he had initially felt faint, but after sitting for several minutes and drinking some water he had been able to stand. In addition, the doctors pointed out, Volkov was the most active member of the Salyut crew, continuously 'swimming' back and forth within the station. In fact, they were sure that even although he had not exercised as much, Volkov was the strongest of the trio.

Volkov was on duty at the start of 23 June. Six hours later, Dobrovolskiy joined him and began a very busy working day. Then at about 8 a.m., before Volkov went to sleep, Patsayev awakened and joined Dobrovolskiy.

---

[4] Dobrovolskiy was referring to the speed at which the human eye adapts to rapidly changing lighting conditions. Although he was able to determine the direction of the station's motion after watching the surface of the Earth for a while, if he quickly switched his attention to another area that was differently illuminated then it took a while to perceive the motion which he knew to be occurring.

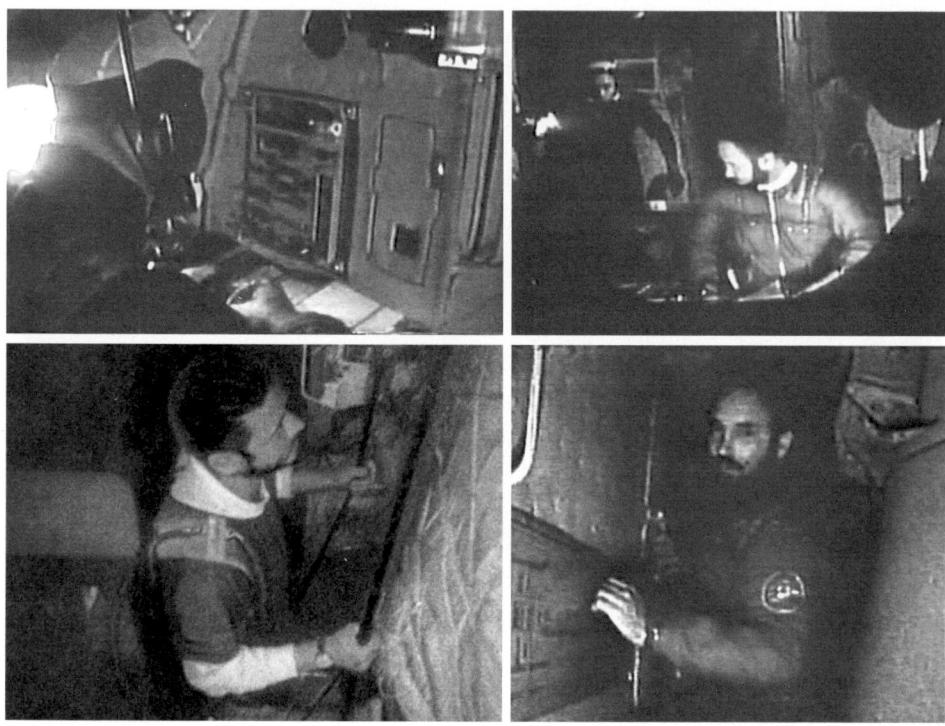

The final phase of the mission. Patsayev in the transfer compartment (top left), and with Dobrovolskiy (top right). Volkov exercises on the treadmill (bottom left), and Dobrovolskiy (bottom right) in between the housing for the scientific equipment (to his rear) and a wall panel.

From Volkov's diary:

> 23 June. I didn't take my leisure time. Instead, I could not resist spending it photographing Earth. I began by recording the mountains of Europe covered with fantastic patterns of snow (Mont Blanc) and the Persian Gulf. It was simply unbelievable work! How could I turn away and rest? Above all, there was minimal cloud cover.

Dobrovolskiy tested the optical characteristics of the station's wide-angle visor by checking the diffusion levels of its various different projection screens. In addition, supported by Volkov's navigation measurements, he experimented with the 3-axis orientation of the station. During these tests, Patsayev studied how the gases of the thrusters affected the optical coating of the portholes. They continued to monitor and photograph the Earth, in particular regions in central Kazakhstan and the Pamir Mountains. Starting at 5.48 p.m., in concert with a meteorological satellite, Volkov photographed a cyclone in the Indian Ocean at a longitude of 60 degrees east and a latitude of 45 degrees south. The working day was finished by a medical inspection, and at 9.40 a.m. the station exited the communication zone of the ground tracking stations.

Day 19, Thursday, 24 June
Dobrovolskiy and Patsayev successfully performed one of the most important tasks of the military programme. This involved using the Svinetz apparatus to observe the night launch of two solid-propellant ballistic missiles, one from a silo at Baykonur and the second from a mobile launcher. In addition, the cosmonauts continued to test Salyut in different regimes of manual and automatic orientation, with different angular rates. They also resumed astrophysical observations and took pictures of the Earth for the purposes of geology, geodesy and cartography.

3.09 a.m.
   Volkov: "At 1.06 a.m. I studied a typhoon or cyclone at longitude 29 degrees and latitude 50 degrees."

From Patsayev's notebook:

   24 June. I observed bright particles before sunrise. These were of differing sizes, at distances of between 1 and 10 metres, moving at differing speeds in differing directions. Some of them were variable in brightness.

7.33 a.m.
   Zarya: "Has your appetite changed, and how much food do you eat daily?"
   Dobrovolskiy: "No loss of appetite. We eat everything."
   Zarya: "Do you require anything else related to physical exercise?"
   Dobrovolskiy: "In general, it would be good to run for half an hour. We use every free minute to perform physical exercise."
   Zarya: "Understood. Could you tell us the most difficult physical exercise, and the reason for this?"
   Dobrovolskiy: "I'll ask the guys. . . . There are no such exercises. We actually want to overload ourselves."
   Zarya: "Do you wear the 'penguin' suits continuously?"
   Dobrovolskiy: "Yes, we do – even while sleeping."

On the evening of 24 June Dobrovolskiy, Volkov and Patsayev broke the duration record set by the Soyuz 9 cosmonauts Andriyan Nikolayev and Vitaliy Sevastyanov, who, upon their return, had not been able to stand upright and had found gravity to be so severe that they were concerned they might die whilst asleep. While in space their cardiovascular systems had adjusted to weightlessness, and had then been slow to readjust to the terrestrial environment. At first there had been concern that their hearts would never recover, but by the 6th day they were back to normal. However, even prior to the flight of Soyuz 9, physicians had begun to consider how to reduce the effects of weightlessness. Their strategy was to prevent the heart from becoming accustomed to working in the lightly loaded regime, in order to make recovery after returning more rapid and less stressful. Although the Salyut crew had not exercised at the start of the mission as much as intended, they were making up for it now that they were in their final week. The breaking of the endurance record marked a major milestone. Yeliseyev and Gorbatko called from the TsUP to congratulate them, then passed on the advice that when they landed they should remain in their couches and await the physicians.

7.29 p.m.

Zarya: "I congratulate all of you on exceeding the flight endurance record. In two orbits the 19th day will be over and the 20th day begun. Hold on, and keep going."

Dobrovolskiy: "Understood, understood. Thank you."

Zarya: "Well done guys, hold on! How is the physical exercise going?"

Dobrovolskiy: "We use everything accessible on board."

Zarya: "We wish you the most successful end to your task."

Dobrovolskiy: "Thank you. We will complete it. We feel well – more or less."

Zarya: "Well done! According to all the data, everything on board is excellent."

Dobrovolskiy: "Yes, everything is normal. Thank you, and we send our greetings to you all."

In a medical check, Dobrovolskiy had a pulse of 72 beats per minute, Patsayev 74, and Volkov 88. Dobrovolskiy had a respiration rate of 20, Volkov 12, and Patsayev 18. The arterial pressure in the case of Dobrovolskiy was 110/78, Volkov 110/70, and Patsayev 130/75.

With regard to the recommendation of the Landing Commission that the descent should occur on the third orbit after undocking, Tregub tried to convince Kamanin to bring this forward to the second orbit in order to reduce the time that the tired men would spend in the tiny ship. But Kamanin refused, because a second-orbit descent would mean a night-time recovery. A third-orbit landing just 24 minutes before sunrise would give sufficient illumination to speed the recovery operation and facilitate any medical intervention.

"DO NOT WORRY"

Day 20, Friday, 25 June

The measurement of the distribution of high-energy electrons at orbital altitude that was started nine days ago, was continued. This used the Era apparatus, which could detect charged particles in the space through which Salyut passed. Patsayev used it to study how the ionosphere varied along their orbit. He also measured the electron resonance of special antennas designed with a different configuration.

2.57 a.m.

Zarya: "Are you feeling well?"

Dobrovolskiy: "Yes, everything is normal. We feel well. Tell the managers that everything is going to plan. We are even doing some experiments which you did not plan."

Zarya: "Understood, but do not miss your rest periods."

3.32 a.m.

Dobrovolskiy: "We have noticed that over this last 24-hour period our eyes have become tired. We move from bright light into the shadow. And it is dark in the ship. To be honest, the illumination inside is inadequate. ... We have just observed a very large cyclone at 12 degrees north and 128 degrees east.

Zarya: "Received. One minute to the end of communication."

Dobrovolskiy: "Understood. End of communication."

As the mission drew to an end, the cosmonauts became more tired and emotional. At the same time, the physicians recommended that they intensify their exercises to improve their ability to readapt to the Earth's gravity.

7.20 a.m.

Volkov: "Today when I was doing physical exercise I overloaded myself, and so I am tired. However, I liked it."

Zarya: "That is good. The physicians are very glad that you exercise so much."

Volkov: "I tried to do everything as you recommended, but tired myself out."

Zarya: "Now you can see how good that is."

Volkov: "I don't know if it is good or bad."

Zarya: "It is good, it is good. The physicians said it is good."

Later in the day, the cosmonauts made their penultimate *Cosmovision* telecast.

Television Report:

Dobrovolskiy: "We are wrapping up a mission that will last just over three weeks. We are packing equipment, documentation and some of the scientific apparatus, and placing it in the descent module for return to Earth. We will return with a great deal of interesting materials. The scientists, engineers and technicians are eager for them. To be honest with you, we are impatient too, because we have grown a little bored."

Zarya-25: "We can see you excellently. Please, could you explain what are you doing at the moment?"

Dobrovolskiy: "Now? Well, Yantar 2 is going to sleep earlier than normal. Next it will be me, and finally Viktor Ivanovich Patsayev. Then we'll wake up and exercise to strengthen ourselves ready for departure."

Zarya-25: "You know, dear comrades, we are watching your unprecedented flight with the greatest interest. We are delighted with your heroism and magnificent work. We wish you ... a successful end to the flight and a soft landing."

Dobrovolskiy: "Thank you very much. We will see you later on Earth."

Zarya-25: "Indeed, see you later on Earth."

Dobrovolskiy: "Do not worry. Everything will be just fine with us."

Zarya-25: "We are sure of that. Have a happy flight and a successful landing."

Dobrovolskiy: "Thank you very much."

At 10.30 p.m. the Salyut crew finished their 315th orbit and exceeded by almost 50 hours the previous endurance record. According to Kamanin, observations of the crew showed that they looked tired and had a low attention span. Furthermore, they tended to provide evasive answers to questions about their health.

In the evening, the Landing Commission met again and confirmed the plan to descend on 30 June on the third orbit after undocking, but the landing point was relocated (without explanation) to 200–250 km southwest of Karaganda. The current weather forecast in the recovery zone was favourable. Nikolay Gurovskiy, one of leading aerospace physicians, reported that the medical group would be prepared for all possible situations. The physicians emphasised that the cosmonauts should remain as still as possible following landing, and await the arrival of the doctors in the recovery team. Gurovskiy again stated that it was the opinion of the Ministry of Public Health that the Soyuz 11 cosmonauts would adapt to conditions on Earth more readily than had Nikolayev and Sevastyanov after their 18-day flight.

Day 21, Saturday, 26 June
At 8.04 a.m. on 26 June Dobrovolskiy, Volkov and Patsayev started their 21st day in space. Their task was to conclude the scientific and technical experiments. Using apparatus mounted outside the station, they finished measurements of the flows of high-energy particles and the flux of micrometeorites – there were sets of sensors for micrometeorites outside the transfer compartment and the larger part of the working compartment. In addition to the radiation in the station, they measured the intensity of the heavy nuclei in cosmic rays and electrons in the 300–600 MeV energy range, all of which was to be correlated with the level of solar activity.

The positions of the micrometeoroid detectors on Salyut's exterior.

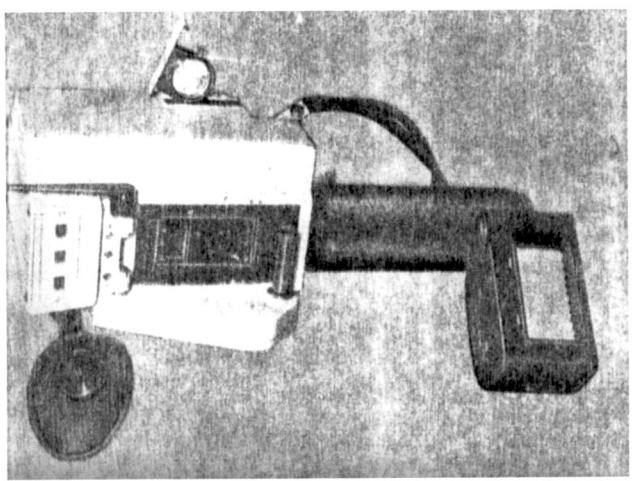

A manually operated instrument to measure the radiation inside the station.

From Volkov's diary:

26 June, 14:00. The 21st day has started. Zarya congratulated us on breaking the world record for the longest flight in space.[5] Their greetings were most welcome. ... We were deeply touched. Our eyes were watery with emotion. The guys were sleeping when I received these greetings on my regular duty. I did not awaken them, but they somehow perceived the news and emerged from their sleeping bags.

Our sleeping bags remind us of a beehive – small holes which we enter at the sleeping time and swim out when we hear the wake-up command (that is, when the man on duty awakens you by shaking your shoulder, or sometimes your head).

By the way, something about the sleeping time. For some reason, the last two nights I slept very little – perhaps three hours in total. I could not force myself to sleep. Last night, I even tried to read Yevgeniy Onegin just before bedtime. I spent an hour reading, to no effect – even the book did not help.

On my previous flight, I did not have dreams. Now, I have as many as I want – even more than on Earth.

When the air inside the station was tested the temperature was 22°C, the pressure was 880 mm of mercury and the composition was normal. The station's systems were performing extremely well.

10.14 a.m.
Zarya: "Yantar, this is Zarya. Why do you complain?"

---

[5] Note that although the Soyuz 9 record was 18 days, the International Astronautics Federation required an endurance record to be exceeded by 10 per cent to recognise it as having been 'broken'; hence the delayed congratulations.

Dobrovolskiy: "I complain because of the 'torture' of the medical sensors. Oh my God! ... Oh, oh, oh! These doctors ... Oh! Right hand, left leg!"

From Volkov's diary:

> 26 June, 17:00. The working day is finishing. Tomorrow is Sunday. Before bedtime, we changed the tank of cooling-drying aggregate in the sanitary-hygiene facilities.
> I have checked my ['penguin'] flight suit for landing.

6.41 p.m.

Dobrovolskiy: "Of which investigation you are talking?"

Zarya: "The medical one. What you have not completed today, you must precisely complete tomorrow. Also, we ask that you time your work involving the Polynom."

Dobrovolskiy: "We are trying to work as on Earth, but here the conditions are different. The amount of work is the terrestrial one, and that is why we are short of time."

From Dobrovolskiy's notebook:

> 26 June. Volodya Shatalov read to me a clipping from the *Pravda* newspaper. At a session of the Odessa City Council, I was elected an honorary citizen of the city.
> Earth has provided us with a forced physical exercise regime.
> Soon will be landing time!

After finishing the scientific programme, the final days of the flight were devoted to intensive physical training, medical examinations and the other preparations for returning to Earth. In concert with controllers at the TsUP, they had already started to prepare Salyut to resume operating in its unmanned regime. They were to check

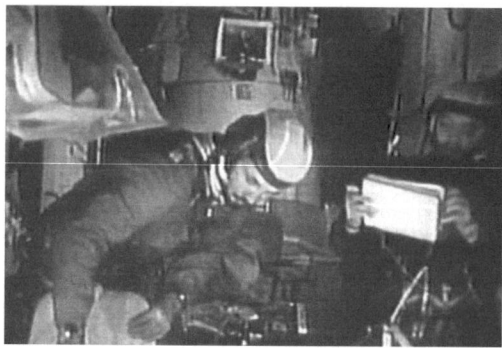

Left: Dobrovolskiy and Volkov check instructions. Right: Dobrovolskiy controls the flight programme, as Volkov (in the background) exercises on the treadmill.

The Soyuz 11 cosmonauts were very popular among the Soviet public, who followed the flight of the first space station crew with the great interest, in this case in the newspaper *Izvestia*.

and switch off all equipment that would not be required. The quality of the supplies of water, food and other consumables that would be needed for the next crew had to be checked. In parallel, they prepared the Soyuz, which had been powered down for more than three weeks. The scientific materials to be returned to Earth were stowed in the cramped descent module in such a way as not to alter its centre of mass or to overload it. The crew were permitted to bring back to Earth only items specified by special instruction. Bags of rubbish were loaded into the orbital module, and would be discarded with that module.

As the cosmonauts were packing up their things on that 26 June, Aleksey Isayev, General Designer of OKB-52 (Himmash) and one of the pioneers of Soviet rocketry, suffered a lethal heart attack. He was 63. Isayev led work on the development of the primary and backup engines for all Soviet manned spacecraft, including Salyut. The KTDU-1 braking engine for Vostok and Voskhod and the KTDU-35 for Soyuz had successfully de-orbited all Soviet cosmonauts. Immediately after Isayev's death the Kremlin issued an announcement that identified him by name for the first time.

Day 22, Sunday, 27 June
On the next day, 27 June, the Soviet Union suffered another severe blow when the third launch of the N1 lunar rocket from Baykonur failed. The flight began well, but after 57 seconds a stabilisation problem caused the automatic control system to turn off all the engines of the first stage and the 3,000-tonne rocket crashed not far away from the launch pad.[6] This was a serious loss for Mishin, because it undermined his ambition to send cosmonauts to the Moon in the near future.

As the world's first space station, Salyut was the last hope for the Soviet manned space programme. The Soyuz 11 crew had proved that the DOS design was capable of sustaining long-duration missions. In conjunction with the daily telecasts that had enabled people right across the nation to participate in the excitement of living in a space station, the research they undertook demonstrated what flying in space was all about. The Americans had landed on the Moon. So what! Soviet cosmonauts were the masters of Earth orbit, which was where the true benefits were to be gained.

In the meantime, the Salyut crew devoted their 22nd day in space to the increased exercise regime and medical tests.

2.32 a.m.
Dobrovolskiy: "We all have normal blood pressure: Yantar 3 is 115/75, Yantar 1 is 120/70 and Yantar 2 is 115/60. After exercise, our pressure and pulse went from 140/55 to normal in about a minute's time ... different from conditions on Earth."

The physicians rescheduled the rest times so that the cosmonauts would be fresher for the landing. However, this meant that for the first time all three men would sleep at the same time. Thus far, at least one man had been on duty at all times. Although

---

[6] The situation could have been worse – the catastrophic N1 launch failure of July 1969 had destroyed the launch pad! Fortunately, two N1 pads had been constructed.

the cosmonauts accepted this new regime for the remainder of the mission, they did not like it.

8.27 a.m.

Dobrovolskiy: "I have a question about the sleep schedule. It says that Yantar 3 is to go to sleep at 12.40, that Yantar 2 will be awakened at 14.00, and that during this time Yantar 1 will rest."

Zarya: "Correct. We will realign you slowly. Do you understand?"

Dobrovolskiy: "The logic of this alignment is understood. Can the station remain without anyone on duty?"

Zarya: "It is the decision of the Control Group. Did you understand me correctly?

Dobrovolskiy: "I understood. However, we are not happy with it."

Zarya: "Follow the programme. It will be alright. The station is in good order. Do not complain, just do it. The Control Group says the new plan is necessary."

Dobrovolskiy: "Understood."

Zarya: "It is necessary to follow the new schedule. We will monitor the telemetry, and if necessary we will awaken you. Do not worry. ... Don't forget that your task now is to rest."

Volkov: "We plan to nap on our leisure days, because there is not enough time for this on working days."

Although busy with physical exercise, medical tests and preparations to return to Earth, the cosmonauts periodically took time to observe the Earth.

1.42 p.m.

Volkov: "We observed a cyclone over South America at 22 degrees east and 46 degrees south."

Zarya: "Logged."

On 27 June the cosmonauts made their seventh and final *Cosmovision* telecast. By now they were the best-known cosmonauts since Gagarin, Titov, Teryeshkova and Leonov. Surprisingly, this time the 'star' was the most reticent member of the crew – Viktor Patsayev. Interestingly, although the preparations to return to Earth were well underway, the subject was the food that they had been eating during their record-breaking stay is space.

Television Report

Zarya-25: "Many television viewers and radio listeners would like to know: how do you eat?"

Patsayev: "Our food is either in cans or in tubes. We also have small packages of desserts such as prunes and cookies. The food is stored in two freezers – which are very large units. We keep tubes and juices in special containers. Some food can be heated – we have two heaters."

Zarya-25: "You have been in space for 22 days. Has your weight changed?"

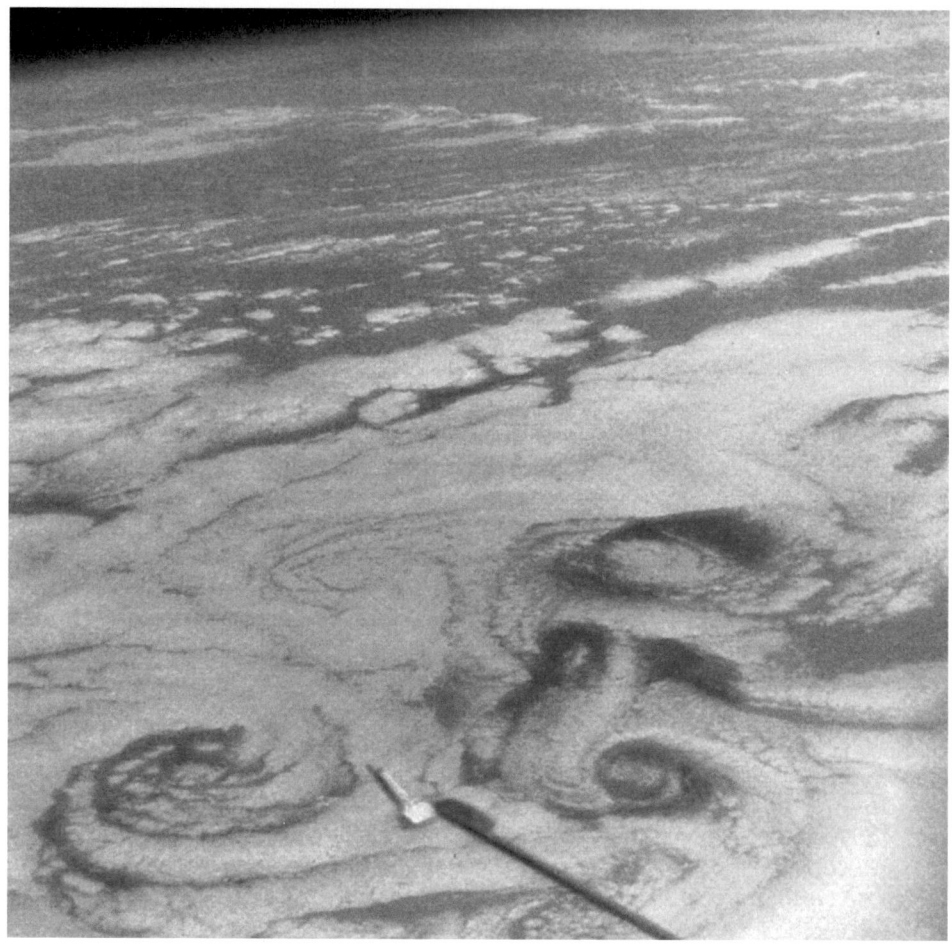

As the mission drew to an end, the cosmonauts continued to monitor terrestrial meteorological phenomena.

Patsayev: "I don't think so."

Zarya-25: "What do you do in your rest time?"

Patsayev: "We don't have much leisure time, but when we do we read – we have a small library with books by Lermontov, Pushkin and Tolstoy. And we also listen to music on our cassette player."

Day 23, Monday, 28 June

Their penultimate day on Salyut began on the morning of 28 June. At 12 noon the station completed its 342nd orbit with a crew on board. While the cosmonauts made their preparations to return to Earth, the landing support team at the TsUP kept up to date on the meteorological forecast for the dawn period in the recovery zone. The most important factor was the wind speed. If the descent module were to land

At Yevpatoriya, the flight controllers were happy with the progress of the mission, and were eager for the crew's return. In the first row (left to right) are Feoktistov, Nikolayev, Kamanin (with Yeliseyev behind him), Kerimov, Agadzhanov and Chertok. (From the book *Rockets and People No 4*, courtesy www.astronaut.ru)

On the eve of Soyuz 11's return to Earth, members of the State Commission arrived at the TsUP in Yevpatoriya from Moscow and Baykonur. Seated in the first row (left to right) are Raushenbakh, Chertok, Agadzhanov, Nikolayev, Mishin, Afanasyev, Kerimov, Bugayskiy (with Semyonov behind) and Shatalov.

on its side, as often happened, and there was a strong wind, then it might roll after landing, and even on a flat surface this would be unpleasant for the men inside, especially if they were feeling weak. In the worst case, if the wind speed exceeded the permitted maximum the module might be damaged on impact and the crew injured – perhaps even fatally. However, the forecast was still favourable. The Landing Commission prepared two sets of instructions for the cosmonauts: the first for the primary landing site and the second – to be used only if the first attempt were to fail – for the reserve site.

Having realised that the cosmonauts were tired, the TsUP worked with them step by step in the process of preparing Salyut to operate in its automated regime in the weeks between the departure of its first crew and the arrival of its second crew. As a result of this close supervision, which was feasible only during the periods when the station was in communication, the effort took much longer than expected. The same procedure was adopted for preparing the Soyuz spacecraft. As part of the process of 'mothballing' the station, it was thoroughly cleaned and the rubbish was stowed in the orbital module of the ferry for disposal.

With the landing imminent, experts from the TsKBEM and Himmash arrived at the TsUP. Headed by General Kerimov, the expert group included Boris Chertok, Boris Raushenbakh, Yuriy Semyonov and Viktor Bugayskiy. As on the occasion of the docking three weeks previously, many off-duty controllers again came into the control centre. And of course Very Important People flew in simply in order to take part. As all the preparations for the descent were well in hand, most of the guests took advantage of the delightful weather and passed the time by walking along the beach. Despite the recent launch failure of the N1 rocket, everyone at the TsUP was happy with the progress of the Soyuz 11 mission and was confident that tomorrow's undocking would go well and that the extraordinary crew would land safely.

"THE HATCH IS NOT HERMETICALLY SEALED!"

Early on 29 June Mishin, Minister Afanasyev and Academician Keldysh flew from Baykonur to Yevpatoriya. No one at the TsUP wished to talk about the N1 failure – in part because Mishin was not in the best of moods, but also because most of the people present were firmly of the opinion that the real future of the Soviet manned space programme was operating orbital stations.

Day 24, Tuesday, 29 June
Shortly after 8 a.m. the cosmonauts began their 24th day in space, but they were all asleep at that time.

4.49 p.m.
  Zarya: "Hello."
  Volkov: "Good morning."
  Zarya: "How are you feeling?"

Volkov: "Good."

Zarya: "And your mood?"

Volkov: "As always. We are on your schedule. We will put on our 'penguin' suits now. Everything is in order. The systems of the Soyuz are normal."

Dobrovolskiy: "What is the weather like in the recovery region?"

Zarya: "The weather is excellent. All is ready. We are waiting for you."

The State Commission met at 7.30 p.m. and confirmed the landing parameters. General Nikolayev reported that everything on the station and the ferry craft was as it should be. Re-entry was to take place on the third orbit after undocking from the station, with the landing timed for 2.18 a.m. on 30 June, approximately 100 km east of Dzhezkazgan in northern Kazakhstan. The crew were not to open the hatch, they were to await the recovery team led by General Leonid Goreglyad and the physician Colonel Anatoliy Lebedyev, who expected to arrive within 20–30 minutes in order to assist them out of the capsule.

When the communication session started at 7.45 p.m. Dobrovolskiy and Volkov reported that the 'mothballing' of Salyut had been finished, all items that were to be returned to Earth had been stowed in the descent module, and the cosmonauts were wearing 'penguin' suits and were ready to depart as planned. Yeliseyev pointed out that telemetry indicated that Volkov had forgotten to switch on Salyut's noxious gas

Dobrovolskiy towards the end of the mission, re-entering the station after checking out the Soyuz 11 spacecraft. On his left shoulder is the TsPK patch.

filter. Volkov initially argued that the TsUP had actually recommended leaving this switched off, but when the log of the previous day's communication was reviewed he accepted his error and returned to the station to activate the filter.

Finally ready to exit, they closed the hatches: first the hatch between the working compartment and the transfer compartment and then, after they had passed through the tunnel into the ferry, the hatch with the passive docking unit. Next was the hatch in the orbital module with the active docking unit. First Volkov, then Patsayev and finally Dobrovolskiy passed into the descent module.

A hermetic seal of the final 60-cm-diameter hatch was of key importance, because when the orbital module was jettisoned this hatch would separate the men from the vacuum, extreme temperatures and radiation of the space environment. As the last man in, Dobrovolskiy closed the hatch, which was on a single 127-mm arm and was sealed by rotating a large grip. But the Hatch Open indicator on the display panel remained lit – without a hermetic seal, the air would leak from the descent module when the orbital module was jettisoned. For the crew, who did not possess pressure suits, this would be fatal.

The TsUP heard Volkov's strained voice: "The hatch is not hermetically sealed! ... What can we do? ... What can we do?"

Yeliseyev calmly advised: "Don't be disturbed. Open the hatch and turn the grip fully to the left, then close the hatch again and turn the grip six and a half times to the right." He also directed that while the hatch was open they should use a tissue to swipe the ring of the hatch to see whether something had become lodged inside and was precluding a hermetic seal. Volkov and Dobrovolskiy carried out this operation, but the indicator remained illuminated. They repeated the procedure several times, but to no effect. After assessing the situation, the TsUP told the cosmonauts to inspect the sensors which sent the open/closed signal to the display panel.

Yeliseyev recalled of this dramatic time: "We asked the cosmonauts to verify the operation of the sensors that sent signals to the display panel. The sensors are in the form of buttons – just like a door bell. As the hatch closes, it pushes the sensors and they produce signals. All the sensors were in working order. But the guys found that the hatch hardly touched one of the buttons, with the result that it did not push down sufficiently to send the signal. We asked them to verify this repeatedly, and this was confirmed. We requested that they verify visually whether the hatch closed tightly, and they reported that it did. Because the automation would not permit carrying out further operations unless it received the correct signal from the hatch, we decided to generate the signal artificially – we simply asked them to apply a strip of insulating tape to hold the button in the correct position and then to shut the hatch. They did so, and visually confirmed that the hatch was correctly closed."

Once Dobrovolskiy had taped the problematic sensor, he closed the hatch and the Hatch Open indicator went out.

"It turned off! The indicator turned off! Everything is in order!" Volkov joyfully informed the TsUP.

During the 20 minutes that it had taken to resolve the problem, the mood both on board the spacecraft and in the TsUP had been tense.

Left: The hatch between the descent and the orbital modules. Right: Yeliseyev tells the cosmonauts how to circumvent the warning indication and hermetically close the hatch.

In the second half of the 15th orbit of the day, the pressure in the orbital module was reduced to 160 mm of mercury to verify the seal of the descent module's hatch; it proved to be airtight. By the 16th orbit of 29 June Soyuz 11 was finally ready to undock from the station.

9.25 p.m.
Patsayev: "The Hatch Open indicator is off."
Zarya: "All clear. Go ahead and undock."
Patsayev: "The Undock command was issued at 21.25.15."
Volkov: "Separation achieved. Separation achieved."
Volkov: "I watched the undocking visually. The station moved left of us, during a turn."
Zarya: "The landing will occur ten minutes before sunrise."

At 9.35 p.m. the cosmonauts reported through the ground station in Yeniseysk in Siberia that they had achieved a normal separation. Having sufficient propellant to manoeuvre, Dobrovolskiy drew to a halt at a range of about 35 metres and then turned his spacecraft to enable Patsayev to take photographs of Salyut through his porthole in order to document its condition.

Day 25, Wednesday, 30 June
The crew of Soyuz 11 had two full orbits to make the preparations for their descent. With two hours remaining to re-entry, Kamanin (call-sign 'No. 16'), his retirement imminent, made one of his rare calls.

00.16 a.m.
Kamanin: "Yantar, I am No. 16, how do you hear me?"
Dobrovolskiy: "No. 16, I hear you excellently."
Kamanin: "Here are the landing conditions. Above the territory of the USSR it is

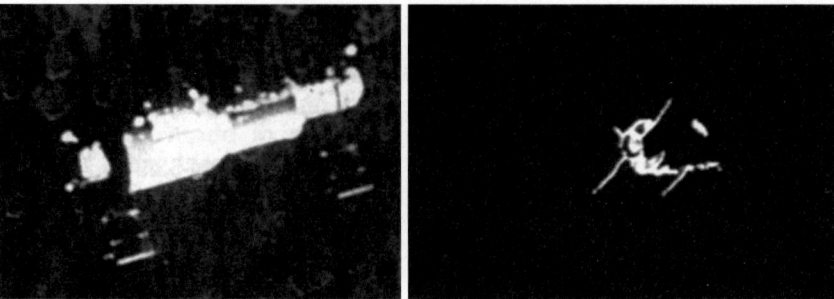

After undocking, Patsayev snapped these pictures of the first Salyut space station. (Courtesy Mark Wade)

slightly cloudy: 3–4 marks. In the landing area it is clear with a visibility of 10 km, the wind is 2–3 metres per second, the temperature is 16°C, the pressures at ground level is 720 mm of mercury. During your descent, constantly report by short-wave and VHF on all antennas – especially those under the hatch of the descent module and on the parachute. After landing, follow your instructions: don't open the hatch, don't make any rash movements, await the medical team. I wish you a soft landing. See you soon on Earth!"

Dobrovolskiy: "Understood: the landing conditions are excellent. Here everything is in order, the crew is excellent. We thank you for your help and good wishes."

And then a few moments later, Dobrovolskiy: "We are following the programme. The Earth will appear shortly. I am starting orientation. To the side is the station. Splendid, it is a beauty. Now, I am starting orientation."

Patsayev: "I can see the horizon in the lower part of the porthole."

Volkov: "The 'Re-entry' indicator is blinking. The SOUD indicator is blinking. It is normal." The SOUD was the system for orientation and control.

Zarya: "Yes, it is."

Dobrovolskiy: "Systems checked. Everything is normal. The horizon has already appeared. The station is above me."

Zarya: "Good-bye Yantars, until the next communication session."

As the crew of Soyuz 11 began their journey back to Earth, the Salyut station on which they had lived for so long receded to a tiny speck gleaming against the dark background of space.

Specific references

1. Vasilyev, M.P., *Salyut on Orbit*. Mashinostroenie, Moscow, 1973, pp. 107–155 (in Russian).
2. Yeliseyev, A.S., *Life – A Drop in the Sea*. ID Aviatsiya and kosmonavtika, Moscow, 1998, p. 81 (in Russian).
3. Kamanin, N.P., *Hidden Space, Book 4*. Novosti kosmonavtiki, 2001, pp. 325–332 (in Russian).
4. Harvey, Brian, *The New Russian Space Program*. Wiley-Praxis, 1996, pp. 278–279.

11

Cosmonauts dead on landing

SOYUZ LANDING OPERATIONS

The most critical and dramatic phase of a manned space flight is the return to Earth. For a Soyuz mission, it starts with the orientation of the spacecraft for the braking manoeuvre and ends approximately 90 minutes later with the landing of the descent module on the Kazakh steppe and the evacuation of the crew. This phase involves a sequence of twelve specific actions, the successful completion of which is vital for the safety of the crew. Indeed, to date, the worst accident in the history of the Soviet manned space programme – the death of Vladimir Komarov – occurred during the return from orbit.

The OKB-1 designers based the return operation on the presumption of excellent visibility in orbit for the orientation and braking manoeuvres, as well as on Earth for the landing. Traditionally, the in-space activities were done on the daylight part of the orbit so that the crew could confirm the orientation of their spacecraft relative to the illuminated horizon, and landing was timed to occur at dawn. Setting up for re-entry is crucial, as even a small misalignment of the braking engine in relation to the direction of travel could result in the descent module missing the landing site by hundreds of kilometres. In addition, if the entry angle were too shallow, the descent module might only 'graze' the atmosphere and remain in an extremely low orbit which, although it would soon decay, would likely not do so before the crew ran out of air. The orientation and control system (SOUD) developed in Department No. 27 of the OKB-1 under the leadership of Boris Raushenbakh was used to orientate the spacecraft with its main braking engine facing in the direction of travel. Normally, the braking manoeuvre to initiate the descent trajectory occurs 25 minutes after the completion of the orientation manoeuvre, while travelling northeast at an altitude of about 250 km over the Gulf of Guinea towards the coast of Africa. The KTDU-35 had a single combustion chamber, and was designed by Isayev's OKB-2 bureau. It delivered a thrust of 417 kg, and could be fired up to 25 times for periods between one and several hundred seconds, accumulating a total time of at least 500 seconds. It was this engine that performed the manoeuvres of the rendezvous with Salyut. An almost identical engine with a thrust of 411 kg served as a backup for the braking manoeuvre. The propulsion module contained four tanks (two for fuel and two for oxidiser) containing approximately 900 kg of propellant.

At the onset of the braking manoeuvre the cosmonauts feel a gentle jolt, followed by uniform deceleration. Depending on the ballistics of the descent, the engine fires for between 145 and 194 seconds to reduce the speed from the 8 km/s required for orbit by 100–120 m/s to initiate the descent. In passing over the Mediterranean at an altitude in the range 110–150 (usually 130) km, the spacecraft adopts an orientation in which its longitudinal axis is more or less perpendicular to the direction of travel, with the orbital module 'on top' and the propulsion module 'beneath' so that when the three modules are separated, aerodynamic drag cannot cause a collision with the descent module. At the time of separation, less than ten minutes after the braking manoeuvre, explosives simultaneously jettison the orbital and propulsion modules and discard from the descent module all unnecessary elements such as its antennas and periscope. Only the descent module is equipped with the shielding required to survive the thermal stress of entry into the atmosphere; all the discarded items burn up. Owing to mass limitations and the relatively short time of its autonomous flight – about 30 minutes – the descent module is not equipped to issue telemetry. Instead, at all stages of the descent following separation the commander loudly calls out the progress of the automated sequence of operations and on conditions in the descent module, and this commentary is encoded in the form of Morse code and transmitted by a small VHF antenna on the outer part of the hatch at the top of the capsule – the

The main breaking engine KTDU-35 visible at the rear of a Soyuz spacecraft.

one which had provided access to the orbital module, and had thermal protection on its exterior. In addition, telemetry from various systems on board is recorded by the 'Mir-3' device, which has a duration of 76 minutes.[1]

In contrast to the spherical Vostok and Voskhod capsules, the descent module of the Soyuz is capable of controlling its path through the atmosphere. This phase of the descent starts over eastern Turkey, 16 minutes after the braking manoeuvre and about 6 minutes after separation. The module has six 10-kg thrusters positioned on its sides which draw their propellant from tanks located in the base, directly behind the couches. The flight control system fires these thrusters as necessary to maintain the broad base facing the direction of travel. In addition, because the module has an offset centre of mass to generate aerodynamic lift, the thrusters can roll the capsule to steer left or right and upward or downward so as to aim for a given landing point. Furthermore, an aerodynamic flight subjects the crew to a lesser g-load than does a ballistic path. The entire module is coated with an ablative material for protection against the heat of re-entry, but the base, which is subjected to the most extreme thermal stress, is covered by a thick shield of azbetextolite material. The maximum thermal and deceleration forces occur while over the Caspian Sea. The Kazbek-U couches enable the cosmonauts to return with their backs facing the direction of travel and in the optimal body-position to endure the deceleration.[2] At this time, the module is sheathed by a hot plasma which, being opaque to radio waves, inhibits communication. The module bounces and shakes in response to the aerodynamic forces of its passage. It is a very noisy time. After the time of greatest thermal stress, the incandescence of the surrounding plasma fades to show blue sky. As the module continues to slow down, the strong vibrations cease and there is a welcome silence.

The parachute deployment begins at an altitude of about 9.5 km. First a cover is jettisoned to allow a small pilot chute to pull out a drogue chute with a canopy area of 14 square metres. This is designed to stabilise the module, and it is released after 17 seconds to initiate the deployment of the main chute at an altitude of about 7 km. This chute is stowed in an egg-shape container behind the heads of the crew that has a volume of just 0.27 cubic metres. It deploys in several stages to produce a canopy of 1,000 square metres by an altitude of 5 km. Small VHF and short-wave antennas on the shrouds transmit signals to the recovery helicopters. By 50 seconds after the start of the deployment of this chute the rate of descent ought to have been reduced to 6–8 m/s. If the rate of descent exceeds the maximum permissible value, the main chute will be jettisoned and the reserve chute of 570 square metres deployed. This is stowed in a separate container adjacent to that for the main chute, with a volume of 0.17 cubic metres. If used, the reserve chute will deploy at an altitude of 4.6 km and achieve the minimal landing speed of 10 m/s.

---

[1] When Vladimir Komarov's capsule struck the ground at high speed, the 'black box' was destroyed by a combination of the shock and the subsequent fire. The design had been strengthened in order to survive a recurrence of such an event.

[2] When the direction of the force is from the feet toward the head, the body is exposed to the maximum load. The optimal position is when the force acts at an angle of 10–15 degrees to the chest-to-backbone direction (known as 'breast-spin') because this minimises the component from the head to the feet.

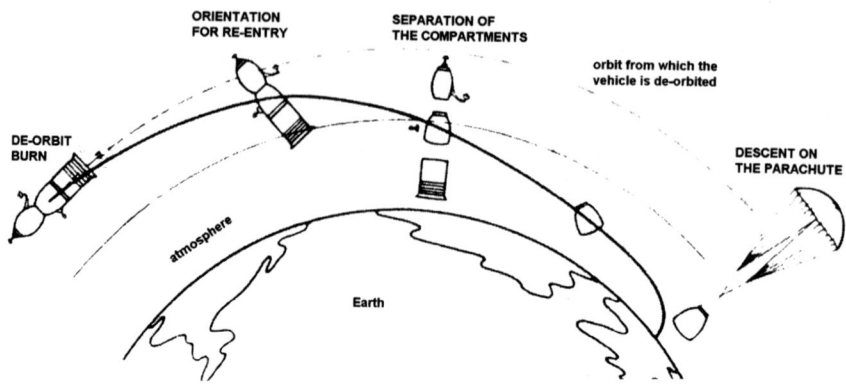

Immediately after the braking manoeuvre, the Soyuz spacecraft separates into three modules to enable the descent module (in the middle) to re-enter the atmosphere on its own. (From the book *Soyuz - A Universal Spacecraft*, courtesy Rex Hall)

The normal deployment of the Soyuz parachute system: (1) the pilot and drogue chutes deploy in turn; (2) on the drogue chute; (3) jettisoning the drogue deploys the main chute; (4) while on the main chute, two ventilation valves open; (5) the base heat-shield is jettisoned; (6) the harness of the main chute is repositioned for landing; (7) retro-rockets fire 1 meter above the ground to soften the impact; and (8) the descent module lands and the chute is jettisoned.

Three last operations of the previous graphic are shown by this collage of pictures of a Soyuz descent module landing. The final pre-landing operations proceed as the capsule descends on its main chute (top left); dust is raised as the retro-rockets fire (top right), and the cloud of dust continues to obscure the capsule as the parachute is jettisoned.

After the deployment of the parachute, a pair of valves on the top of the module automatically open to allow the internal pressure to match that outside. At this time, in preparation for landing, each cosmonaut makes sure that his body is comfortable in his contoured couch and shock-absorbing rods elevate the couches from the floor. At an altitude of 3 km (75 seconds after descending through an altitude of 5.5 km), the basal shield is jettisoned to expose four solid-propellant retro-rockets (DMP) to be used to cushion the landing. With the heavy shield discarded, the rate of descent slows. On a nominal descent, there is ten minutes remaining to landing. Each retro-rocket has 22 jets arranged in two rings near the edge of the module's base. They are fired simultaneously by an altimeter at a height of 1–1.5 metres over the ground to reduce the impact speed to 2–3 m/s, with the shock being absorbed by the couches.[3] Once the Soyuz is on the ground, the parachute is jettisoned in order to preclude this from dragging the module across the surface if there is a strong wind.

The landing area is on the flat Kazakh steppe. The 'landing window' usually starts three hours before dawn and ends just before sunrise. In addition to enabling the in-orbit manoeuvres to be made in daylight, this schedule permits the recovery team to observe the descent module without being blinded by the rising Sun. If the descent is on target, the recovery helicopters will soon settle close alongside. If the recovery crew is unable to arrive quickly, the spacecraft commander will open the hatch and exit. Because the hatch swings into the cabin towards the flight engineer's side, the research engineer is second to exit, after which the flight engineer transfers to the central couch prior to exiting.

As cosmonauts are under stress during the descent they suffer an 'adrenalin rush', and even when everything functions as intended they can be taken by surprise. For example, the crew of Soyuz 7 were initially confused when, after the deployment of the main chute, they felt fresh air rush into the cabin through the valves designed to equalise the pressure. And on Soyuz 4 the crew were surprised when the shock-absorbers raised their couches just before the landing. The first mission to return in abnormal conditions was Voskhod 2 in March 1965, with cosmonauts Belyayev and Leonov. When the automatic orientation failed, Belyayev did so himself and landed 400 km off-target in a snowy forest, and they had to spend two nights in the frozen capsule surrounded by wolves and bears. Another serious incident occurred in January 1969 during the return of Soyuz 5 with cosmonaut Volynov, when the propulsion module failed to separate and blocked the heat shield as the spacecraft entered the atmosphere. Volynov was alarmed by the rise in temperature and smell of soot in the cabin, but fortunately at an altitude of 80 km the connections between the modules melted, the propulsion module was torn away by the atmospheric drag, and the descent module stabilised. However, it was off-target and the landing was so hard that Volynov suffered broken teeth. Of course, the worst accident occurred during the return of Soyuz 1 in April 1967. Owing to the lax technical discipline of the people who applied the thermal treatment to the descent module, the volumes of the main and reserve parachute containers were reduced, with the result that when

---

[3] The height sensor is a gamma-ray altimeter ('гаммалучевой высотомер').

The recovery team opens the hatch to help the cosmonauts out of the capsule. On some occasions the capsule comes to rest upright, but here it is on its side, which can be uncomfortable for the crew.

the parachutes were inserted they were packed too tightly.[4] At an altitude of 9.5 km the hatch of the main parachute container was jettisoned, as planned. This drew out the pilot chute, which deployed the drogue chute. Unfortunately, the drogue was not able to pull the main chute from its container. Seventeen seconds later, the hatch of

[4] The root cause of Komarov's death was the thermal treatment of the descent module and the placing of the parachutes into their containers. Because the parachute containers of both the Soyuz 1 and Soyuz 2 descent modules did not have hatches when they were sent for the application of their thermal treatment, the technicians decided not to ask for the hatches to be supplied and instead 'closed' the openings using improvised covers that did not form a hermetic seal. During the treatment, some molecules of the thermal protective material penetrated the containers and coated their walls, thereby both reducing their volumes and making the smooth interior surfaces rough. When the treatment was finished, the technicians tried to put the parachutes into their containers and, on finding that they would not fit, opted not to inform their managers but instead (according to Mishin) to use some kind of tool to force them in. It is ironic that the early problems suffered by Soyuz 1 led to the cancellation of the launch of the second spacecraft for this joint mission, as otherwise both crews would almost certainly have been killed.

the reserve chute jettisoned and pulled out the reserve chute. What happened next is disputed: one account says that the reserve chute was in the so-called aerodynamic shadow of the drogue; another says that it became twisted with the other lines. But both accounts agree that the parachute was unable to deploy fully.[5] In any event, the module struck the ground at a speed exceeding 50 m/s, causing the main instrument panel to break free and crush the chest of Komarov, killing him.

THE SILENCE OF THE COSMONAUTS

Dobrovolskiy, Volkov and Patsayev knew well the risks of the return operation, but on the third orbit after undocking from Salyut they were in excellent spirits and impatient for the landing. At 1.10 a.m. on Wednesday, 30 June, while out of radio contact over the Pacific Ocean approaching Chile, Dobrovolskiy, assisted by Volkov, oriented Soyuz 11 to position its main engine facing the direction of the flight.

One of many disputed issues concerning the final phase of this mission is the time of the last words from the crew.

The last officially published communication from Soyuz 11 was at 00.16 a.m., when Kamanin in the TsUP spoke to Dobrovolskiy, who reported that they were in the process of preparing for the orientation manoeuvre. At that time they could still see the Salyut station. Then the controller signed off with: "Good-bye Yantars, until the next communication session."

The official sources do not give a chronology of the last conversations with the Soyuz 11 crew, or between the cosmonauts.

In his 1971 book *Soviets in Space*, Peter Smolders cites the following words from Dobrovolskiy as the last communication received by the TsUP: "I am beginning the descent procedure."

Yeliseyev's book offers the following account of the final words received by the TsUP: "The last communication session is ending. Immediately before leaving the zone of radio visibility, Volkov managed to call loudly to say: 'Prepare cognac, see you tomorrow!' ..." However, owing to the phrase "see you tomorrow" the time of this reported communication is unclear – was it on 29 June or 30 June. Nevertheless, the words "Prepare cognac" would be a typical final message prior to an imminent reunion. It may well have been that immediately before the loss of communication Dobrovolskiy said he was "beginning the orientation" and then in the final seconds Volkov managed to add his remark.

Between 1.22.00 a.m. and 1.31.25 a.m. Soyuz 11 passed over South America and then set off across the Atlantic Ocean. As noted, for optimal visibility at the landing site the braking manoeuvre was to be made on the third orbit after undocking from the station. This was why Soyuz 11 had a different re-entry trajectory than previous missions. One circuit of the Earth lasted on average 89 minutes. During this interval

---

[5] The recovery team found the pilot, drogue and reserve parachutes at the landing site; the main chute was destroyed inside its container by the fire that followed the crash.

the planet rotated through 22.2 degrees, so Soyuz 11 was north of the equator at the moment that the engine fired, somewhat to the north and west of the typical braking position for a Soyuz descent. The engine was fired automatically at 1.35.24 a.m., as planned. At that time, Soyuz 11 was over the Atlantic between the northeast coast of South America and the coast of Africa. The engine fired for the planned duration of 187 seconds and was automatically switched off after reducing the speed of the spacecraft by the requisite 120 m/s. Another interesting detail – in contrast to most of the previous flights, in this case the braking manoeuvre was made during the descending portion of the orbit – i.e. after the ship had passed the apogee point. Following the braking manoeuvre, the automated control system would reorient the vehicle for the separation of the modules, perform the separation, control the path of the descent module through the atmosphere in order to aim for the target, manage the parachute deployment sequence, jettison the heat shield, fire the retro-rockets and jettison the parachute. The crew were not required to participate in any of these critical operations.

Did the tracking ships in the Atlantic Ocean detect signals from Soyuz 11 during the braking manoeuvre? Chertok's memoirs and Kamanin's diary, two of the most widely cited sources, offer contrary accounts.

Chertok wrote:

> After undocking from the station, two orbits are allowed to prepare for the descent. The crew will conduct manual orientation while out of our visibility

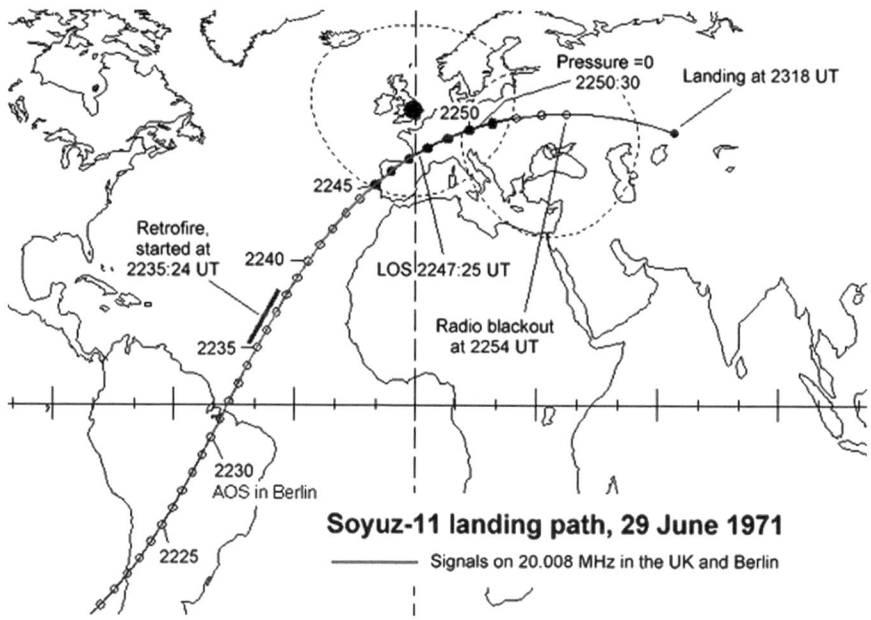

Soyuz 11's descent track. (Courtesy Sven Grahn)

zone and pass control to the gyro instruments. The command for the start of the descent activity will be emitted from NIP-16, with NIP-15 as the reserve. The KTDU will fire for braking at 1 hour 47 minutes on 30 June. ...

All indications on the panel were normal, and the cosmonauts reported the achievement of all operations on time. ... Everything went according to the timetable. The tracking ships received information as the spacecraft passed above, and reported to the TsUP that the braking engine had operated for the estimated duration and was switched off by the integrator [when the correct velocity had been attained]. The control-measuring complex and the GOGU were satisfied with the control of the spacecraft on the landing orbit.

After engine cut-off, the spacecraft exited the communication zone of the tracking ships in the Atlantic. The orbital module and the propulsion module were jettisoned from the descent module while passing over Africa.

Based on this, we can conclude that the TsUP had information from "the tracking ships" that the braking engine was fired and shut off as expected, and that Soyuz 11 then re-entered as planned. Also, Chertok implies that *several* ships were involved in tracking this particular re-entry! Furthermore, he said that Soyuz 11 left the radio zone of the ships when the main engine switched off, which is a point also made by the official TASS report (see the next chapter). However, he was mistaken in giving the time of the braking manoeuvre as 1.47 a.m. (this was the time that the modules were separated) and incorrect in saying that the separation occurred above Africa (it was the typical scenario for the previous Soyuz missions, but not in this case).

Another author, Colonel Ivan Borisenko, the 'Sporting Commissar', has said that communication was briefly established with Soyuz 11 about this time, then lost at the moment of the separation of the modules.

However, in his diary entry of 30 June General Kamanin says:

According to the re-entry programme, the KTDU must start at 01.35.24 and should turn off after 187 seconds. We impatiently waited for a report of the braking manoeuvre. Shatalov repeatedly called Yantar on line, but there was no response from the crew. ...

At 1.47.28 the separation must occur, ... but there are no reports about this. We did not know whether Soyuz 11 had begun the descent, or had remained in orbit. The period of communication calculated for the case of the ship not leaving orbit (01.49.37–02.04.07) began. There was an oppressive silence in the room. There was no communication with the crew or any new data about Soyuz 11. Everyone understood that something had occurred aboard the spacecraft, but no one knew what. The minutes of expectation passed terribly slowly.

So, according to Kamanin, no one in the TsUP knew whether the main engine had fired on time or if the braking manoeuvre had been completed. He did not mention receiving the information from the tracking ships in the Atlantic that Chertok cited. There was no response from the spacecraft to Shatalov's calls. The silence from the spacecraft shortly before, during, and after the braking manoeuvre, which was about ten minutes before the separation of the modules, is another interesting detail. With the exception of Kamanin, no other source (Chertok, Yeliseyev, Feoktistov, Rebrov, and others) spoke of the silence of the crew in the braking period – while Soyuz 11 was passing over the tracking ships. Yeliseyev, who was in the TsUP with Kamanin, Chertok, Feoktistov and others, did not refer to tension in the control room owing to uncertainty concerning the braking manoeuvre. He wrote nothing about the tracking ships and signals they might have received from Soyuz 11; only of data from the radar stations which detected the descent module after its path had carried it onto Soviet territory.

So what really happened? Let us consider the tracking ships in the Atlantic. Due to the position of Soyuz 11 during the braking manoeuvre, only a ship located in the equatorial region could have received a transmission during this time. *Bezhitsa* was at its operating station near the coast of Africa in the Gulf of Guinea, at 1.5 degrees south, 13 degrees west, until 29 June. From this station, it would have had two or three opportunities each day to monitor the success of the braking manoeuvre. But it had been at sea for four months, and was low on provisions. It was to sail to Las Palmas in the Canary Islands in early July for replenishment. Since this station was of crucial importance to monitoring Soyuz 11 during its braking manoeuvre, it was decided that *Kegostrov*, in the South Atlantic at 22 degrees south, 24 degrees west, should move to relieve *Bezhitsa*. On 29 June *Bezhitsa* received an unexpected order to leave its station. Amazingly, it left before *Kegostrov* arrived to replace it! On the morning of 30 June, local time, when it was realised that *Kegostrov* would not be in position before Soyuz 11's braking manoeuvre, the head of the Soviet Naval Fleet personally ordered *Bezhitsa's* captain to urgently return to his previous station so as to monitor the braking manoeuvre – not just the telemetry but also the commentary from the crew. However, it was apparent that *Bezhitsa* would not be able to resume its former station in time.[6]

Why was Soyuz 11 allowed to proceed with the undocking and return to Earth if a tracking ship to monitor the braking manoeuvre was absent? As noted, in planning the mission there were discussions about whether it should be for 45, 30 or 25 days. Finally, guided by the ballistics, Mishin had decided to accept the '25'-day duration and shorten it by one day, with the landing on 30 June instead of 1 July. This is the first important detail to consider when pondering the reasons for Soyuz 11's return without a tracking ship in this key position. It would appear that in the final stage of the mission the usually excellent co-ordination between the TsUP (in fact, the State Commission) and the Soviet Naval Fleet failed, causing *Bezhitsa* to leave its station prior to the arrival of *Kegostrov*. In addition, there had been a dispute between the Air Force (Kamanin) and the TsKBEM (Tregub) about whether Soyuz 11 should return on the second or the third orbit after it undocked from the station. A return on the second orbit would have taken the familiar route across Africa, but would have meant landing in darkness. During the additional orbit, the eastward rotation of the Earth displaced the longitude at which the spacecraft would perform its

---

[6] While sailing towards the assigned station, the crew of *Bezhitsa* heard the terrible news of Soyuz 11 on Radio Moscow.

The tracking ship *Bezhitsa* was unable to monitor Soyuz 11's braking manoeuvre.

northward crossing of the equator 22 degrees to the west.[7] The descent trajectory for Soyuz 11 was therefore different to the one with which everybody was familiar – as indicated by the mistake in Chertok's account. Instead of firing the main engine while passing above the Gulf of Guinea, where *Bezhitsa* was to have been, the braking manoeuvre started at 10 degrees north, 40 degrees west, and was concluded at 29 degrees north, 32 degrees west. At Soyuz 11's altitude, the communication zones of *Bezhitsa* and *Kegostrov* were about 15 degrees in radius, but beyond about 10 degrees the signal was weak. In fact, not only was *Bezhitsa* off-station when the spacecraft performed its braking manoeuvre, that fact that it was sailing at maximum speed in an effort to resume its station meant that it did not even attempt to listen. And *Kegostrov*, being even further away, could not have received a signal from Soyuz 11 at the vital time. This is why (as Kamanin noted) no one in the TsUP knew whether the spacecraft had made the manoeuvre. And, of course, even if one of these two ships had been in position, neither was equipped to relay the VHF transmission from the spacecraft to the TsUP, which is why the control room did not hear the cosmonauts' voices, only "silence". *Academician Sergey Korolev* and *Cosmonaut Vladimir Komarov* were equipped to relay signals from a spacecraft to the TsUP, but only when a Molniya satellite was conveniently positioned, and in this case *Komarov* was out of service and *Korolev* was in the North Atlantic and too far away to receive signals during the spacecraft's braking manoeuvre.

CODE '111'

At 1.45 a.m., almost seven minutes after finishing the braking manoeuvre, Soyuz 11 crossed the coast of Portugal. Shortly thereafter the automated system rotated it through 90 degrees in order to position the orbital module on top and the propulsion module facing down. At 1.47.28 a.m., while passing over France, twelve explosive

---

[7] A further complication was that owing to the difficulty in achieving a hermetic seal of the hatch prior to undocking, the cosmonauts were initially 20 minutes behind the flight plan.

charges jettisoned the orbital module and six more jettisoned the propulsion module. Because the main radio transmission equipment was in the propulsion module, this terminated all signals from the descent module except those from the VHF antenna incorporated into the descent module's hatch. Shortly thereafter it came within range of the antennas at Yevpatoriya, but the controllers still did not know that the braking manoeuvre had been achieved and that, consequently, the descent module was on its way home. If everything was going to plan, then by now they ought to have picked up the VHF transmission. Although Kamanin ordered Dobrovolskiy to report, there was no reply. If the braking manoeuvre had *not* been performed, then the spacecraft would be in communication between 1.49.37 a.m. and 2.04.07 a.m., and when this session opened Shatalov, who was responsible for communications during re-entry, made repeated calls to no effect.

Just like everyone else in the TsUP, Yeliseyev, the technical flight director, was surprised: "We had asked Dobrovolskiy to make continuous reports as soon as the descent module entered our communication zone, but he was silent. It was strange that Volkov was silent too – he had been very talkative in the recent sessions."

As time passed without news, the anxiety amongst the people in the main control room rapidly increased as they realised that something must have happened. In fact, no one could have imagined the terrible event that had overwhelmed the crew in the cramped descent module.

Soyuz 11 flew over Germany and Poland and onto Soviet territory. At 1.54 a.m. the Soviet tracking radars reported that they had detected it north of the Black Sea at an altitude of about 40 km and 2,200 km from the aim point. It was sheathed by plasma, and hence temporarily out of radio contact. The radar detection was good news, because it confirmed that the spacecraft was on its way home. The controllers in the TsUP assured one another that the silence from the crew must be the result of a radio system failure. The tracking radars reported the reducing range: "Distance 1,800 ... 1,000 ... 500 ... 100 ... 50 km from the planning landing site."

The small drogue parachute deployed on time. Then, at 2.02 a.m., at an altitude of about 7 km, the main chute deployed. During the 15 minutes or so of the descent on the main chute the crew were to make radio contact with the recovery team via the VHF and short-wave antennas built into the shrouds of the parachute, but there was no word. The basal heat shield was automatically jettisoned. At 2.05 a.m., with 13 minutes remaining, the recovery crews on an IL-14 aircraft and four Mi-6 and Mi-8 helicopters reported to the TsUP that they could see the module swinging on its red-and-white main chute and that they had detected signals from it, although there was still no word from the cosmonauts.

The manager of the recovery team, General Kutasin (call-sign 'No. 52'), who was in one of the helicopters, reported directly to the TsUP. The clarity of this radio link was excellent. According to Yeliseyev, beaming smiles came to the faces of the controllers upon hearing that a transmission had been received from the antennas on the main chute – the first signals received from Soyuz 11 since it departed from the communication zone during preparations for the orientation manoeuvre above the Pacific Ocean: "Finally, we heard a report from a helicopter in the planned landing area that they could see the parachute. It was wonderful! ... Then, the report from

The recovery team spotted Soyuz 11 descending on its main parachute (top left). It landed on its side (top right), and a few minutes later the recovery helicopters landed alongside (bottom).

No. 52: 'It has landed. Our helicopters are landing nearby.' Well, it seemed that was all. Next, they would report the general state of the crew, and with that we would finish our work. Only a few minutes more."

Colonel Ivan Borisenko, the 'Sporting Commissar', who was actually the member of the recovery team responsible for officially logging the landing parameters, has written: "There was no radio contact with the cosmonauts. ... From the Mi-6 in which I was flying we saw the descent module slowly descending, swinging under the large canopy of the parachute. The soft-landing retro-rockets fired correctly, the module almost stopped for a moment in the air, then settled onto the ground."

The four small rockets automatically fired at a height of 1 metre in order to soften the landing, in the process raising a cloud of dust. At 2.16.52 a.m., Soyuz 11 landed 202 km east of Dzhezkazgan, having overshot the target by 10 km. Exactly 23 days 18 hours 21 minutes 43 seconds had elapsed since it lifted off from Baykonur. At almost the same time, the helicopters landed nearby.

The TsUP awaited General Kutasin's next report, but the radio remained silent. Yeliseyev recalls the dramatic wait:

> Five minutes passed by; 10; 15. ... No news from No. 52. ... How strange. Usually, someone remains in the helicopter to report on the radio the events as they happen. ... One hour has passed. ... No. 52 is still silent. ... It means that something has happened. ...
>
> Suddenly, using an internal channel, Kamanin asked me to come. He was alone in the room used by the State Commission. He never called someone without a reason. As I ran to him, he looked darkly at me and said: "Now they have given me the code '111', which means that they have all perished. We agreed a code: '5' means that their general state is excellent; '4' means good; '3' means there are injuries; '2' means severe injuries; '1' means that a man perished; '111' means that all three perished. It is necessary for us to fly to the landing site, I have ordered the plane."
>
> Kamanin, Shatalov and I were immediately driven to the airport, where an aircraft was ready. I can no longer remember the airport at which we landed. We transferred to a helicopter and were flown to the landing site.

Kamanin did not mention the '111' code in his diary, but he wrote that for at least the first 30 minutes whenever he asked for a report from the landing site the reply

The Soyuz 11 recovery operation was handled on site by Kamanin's aide, General Leonid Goreglyad.

was always: "Wait." Then he received the following message: "General Goreglyad has flown from the landing site to Dzhezkazgan and reported via [short-wave] radio that the outcome of the space flight is the most tragic one."

"DOBROVOLSKIY WAS STILL WARM"

When the State Commission was informed of the terrible news, Afanasyev, Mishin, Kerimov and others refused to believe it, and asked for confirmation. About an hour later, General Uglyanskiy reported from the landing site that within a few minutes of the module landing, members of the recovery team, led by General Goreglyad, had opened the hatch and found the cosmonauts inert and without any signs of life.

Interestingly, Chertok has a different account of events in the TsUP immediately after the landing. In the absence of reports from the landing site, General Kerimov had thought that Marshal Kutakhov, the Commander in Chief of the Air Force, and as such in overall command of the recovery team, wished to have the privilege of informing the Kremlin of the successful conclusion of the historic mission. In fact, this report should have been made by Kerimov, who, as the Chairman of the State Commission, was responsible for reporting to Moscow; specifically to Ustinov and Smirnov. After 30 minutes without a communication from the landing site Kerimov decided that he really should call Ustinov to complain about the breach of protocol. But then he learned the truth. Pale, Kerimov gave the tragic news:

> Two minutes after the landing, members of the recovery team ran from the helicopters to the descent module, which was laying on its side. Outwardly, there was no damage whatsoever. They knocked on the side, but there was no response from within. On opening the hatch, they found all three men in their couches, motionless, with dark-blue patches on their faces and trails of blood from their noses and ears. They removed them from the descent module. Dobrovolskiy was still warm. The doctors gave artificial respiration. Based on their reports, the cause of death was suffocation. There were no strange smells in the cabin. The procedure for evacuating the bodies to Moscow for analysis has been accepted. Specialists from Podlipok and the TsPK have set off for the landing site.

The stunned silence in the crowded control room was broken when someone said that the spacecraft must have suffered a decompression that had exposed the crew to the vacuum of space.

When the recovery team had run from their helicopters to the descent module, it was believed that the silence from the crew was simply the result of a radio failure. The team included Air Force doctors to assist the cosmonauts – who must surely be debilitated by their return to gravity after three and a half weeks in weightlessness. When the crew failed to respond to loud banging on the side of the module, they urgently opened the hatch and were shocked to find the men inert, as if asleep or unconscious. But the fact that their bodies were limp and there were trails of blood indicated that they were injured; even though the cause was not apparent. Normally, the recovery team would simply assist the cosmonauts to emerge from the 60-cm-diameter hatch. It would be more difficult to extract their inert bodies. The task was complicated by the fact that the module had come to rest with the couches stacked one above the other. One man reached into the cramped cabin, released the belt on Dobrovolskiy's couch and drew him out. Patsayev's couch was higher up. Owing to the manner in which the hatch swung into the cabin, it was more difficult to reach Volkov. As each body was retrieved, the doctors applied manual cardiopulmonary resuscitation. The activity was recorded by a film camera brought to document the joyous return. Furthest away was Dobrovolskiy. His body was still warm and limp. His bearded face was lifeless, his mouth was open and there was a dark patch on his right cheek. His rescuers valiantly tried to revive his heart using chest compression and lung ventilation. To the right, military medics tried to revive Volkov, with one positioned on the body to exercise the chest while the other knelt to give ventilation. Volkov's right sleeve had been rolled up in order to attempt a transfusion. Nearest the cameraman was Patsayev, with his body oriented in the opposite direction to the others, and with a civilian medic to either side of him, attempting resuscitation by artificial respiration.

It would later be determined that when the recovery team pulled the cosmonauts from the module they had been dead for in excess of 30 minutes. Furthermore, they had spent 11.5 minutes exposed to vacuum. Humans and experimental animals had sometimes suffered rapid decompression in terrestrial laboratories or on scientific balloons at high altitude, but the Soyuz 11 crew were the first humans to suffer the vacuum of space at an altitude in excess of 100 km. Cardiopulmonary resuscitation is only likely to be effective if given within six minutes of the cessation of the heart, since after this the brain is permanently damaged. The rescuers had stood no chance of reviving the cosmonauts.

There is only one film record of the rescue effort. It shows two medics tending to each body. In addition to manual chest compression and lung ventilation, they had heart-lung and defibrillation (electroshock) apparatus. The effort was observed by a number of military officers, some standing close by and the others waiting beside the helicopters.

As there are no official reports available from the people directly involved in the effort to resuscitate the crew, the details remain unknown. Colonel Borisenko only briefly reported: "We ran to the landing point. The recovery team opened the hatch and pulled out Dobrovolskiy, Volkov and Patsayev, who had no indications of life. The doctors did everything possible, but it was too late. Based on the preliminary examination by Dr. Anatoliy Alexandrovich Lebedyev at the landing site, the crew perished from the rapid decompression of the cabin of the ship."

One of the doctors, and one of very few witnesses to the drama at the landing site, was Levan Stezhadze: "For more than an hour we tried to resuscitate them with the heart-lung machine. The heart reanimation lasted over an hour. We tried using the defibrillation equipment. It was good apparatus. . . . However, there were no signs to show that revival was possible. For example, when I inserted a needle into the heart of one cosmonaut, instead of blood there was only air."

Drama at the landing site. Top left: Medical workers try to revive Dobrovolskiy. Top right: Medics attend to Patsayev (foreground) and Volkov (in the middle). Bottom left: After conceding that the cosmonauts were dead, their bodies were draped with white blankets. Bottom right: Specialists begin the inspection of the descent module.

"IT IS INTOLERABLY PAINFUL!"

As Kamanin, Yeliseyev, Shatalov, Mishin, Feoktistov, Afanasyev, Kerimov, Karas, Vorobyev, Severin and others travelled to the landing site, their route took them to Aktyubinsk, the hometown of Patsayev. On arriving at Dzhezkazgan's airport, they were told that General Goreglyad had already organised the transport of the bodies to Moscow. From Dzhezkazgan the group flew in two helicopters to the landing site on the steppe, arriving at 4.00 p.m., whereupon the members of the recovery team recounted the day's events.

Yeliseyev emotionally described the scene:

> The module was on its side with the hatch open. The guys had already been transported away. One of the doctors reported that it was clear that there had been a decompression, and their blood had boiled. The doctors attempted to transfuse blood, but to no effect. When they opened the hatch, the guys were

still warm, but gradually ... hopes faded. ... It is intolerably painful. What an absurdity! A flat field, excellent weather, the module in good condition, and the guys dead. And suddenly something struck me as an electric shock: was it the hatch? Might this be my fault? But they had checked the seal! Might it be something that they had not seen? ... I will not try to describe what I felt at that moment.

Shatalov and I went to the descent module to fill in the form describing its state on landing. The module was immediately surrounded by the military to prevent anyone approaching it without permission. The first thing I observed was the fountain pen. After my flight I had presented Viktor Patsayev with my pen "for good luck". Now it was lying on the sand – evidently it fell out of his pocket when they pulled him out. In my head flashed a recollection of how we arrived at my home with Vadim and Victor after the meeting of the Military-Industry Commission which established their crew. We were happy, and sang songs. When saying goodnight I gave Viktor my pen. ... And here it is – the end of the dreams and the plans. ...

We inspected everything, inside and outside, and wrote our observations: everything was normal. Then they took from the descent module the tape recorder on which were the parameters measured during the descent. They sealed this in a special container for transport with the escort to Moscow. It would explain the tragedy. We flew in the same aircraft.

Kamanin provided a less emotional but more detailed account of the visit to the landing site. Although Yeliseyev wrote of the module that "everything was normal", Kamanin noted the unusual position of one of the valves: "Prior to nightfall, we had time to conduct only a general inspection of the ship, cabin, seats, parachute system, etc. Judging by the results of this inspection, Soyuz 11 performed a soft landing – there was no significant external damage. In the cabin, all the transmitters and all the receivers were switched off. The shoulder straps of Volkov and Patsayev were unfastened, and Dobrovolskiy's belts were ... tight only at the waist. The shutter of one of the two air valves was inverted to 10 mm. There were no other deviations in the cabin."

About an hour after the group from Yevpatoriya arrived at the landing site, they were joined by specialists from the TsKBEM and the TsPK who flew from Moscow. With this group was Aleksey Leonov, commander of the original Soyuz 11 crew. In his book *Two Sides of the Moon*, he has written: "When the rescue forces reported that the crew was dead, I was instructed to fly to the landing site immediately with ... Vitaliy Sevastyanov.[8] We were appointed members of the government committee dealing with the aftermath of the disaster and our main task was to secure the spacecraft and take photographs of the scene. It took us about 3 hours to reach the site, by which time the bodies of the crew had already been removed. Their blood-soaked seats and signs that attempts had been made to resuscitate them, were the only evidence of the tragedy."

---

[8] By mistake, here Leonov wrote Yeliseyev's name.

On the morning of 1 July another group of specialists arrived from Moscow with equipment to test the hermetic seal of the descent module. They closed all openings, including the valve set in the unusual position, and increased the internal pressure above ambient by 100 mm of mercury. When there was no indication of a leak they increased the pressure first to 150 and then to 200, and waited 30 minutes, but again the pressure remained constant. Having established that the decompression was not the result of a meteoroid puncturing the shell of the module, the module was flown to Moscow later that day for a thorough investigation.

Specific references

1. Hall, Rex and Shayler, David J., *Soyuz – A Universal Spacecraft*. Springer-Praxis UK, 2003, pp. 67–74.
2. Grahn, Sven, 'Salyut-1, its origin, flights to it and radio tracking thereof'. web site, www.svengrahn.pp.se/trackind/salyut1/salyut1.html
3. Yeliseyev, A.S., *Life – A Drop in the Sea*. ID Aviatsiya and kosmonavtika, Moscow, 1998, pp. 81–82 (in Russian).
4. Kamanin, N.P., *Hidden Space, Book 4*. Novosti kosmonavtiki, 2001, pp. 333–338 (in Russian).
5. Chertok, B.Y., *Rockets and People – The Moon Race, Book 4*. Mashinostrenie, Moscow, 2002, pp. 338–341 (in Russian).
6. Shayler, David J., *Disasters and Accidents in Manned Spaceflight*. Springer-Praxis, 2000, pp. 397–405.

12

Farewell

THE ANNOUNCEMENT

The tragedy was revealed to the world in a message released by the Soviet national news agency at 6 a.m. on the 30 June:

> *TASS reports the deaths of the crew of the spaceship Soyuz 11, Lieutenant-Colonel Georgiy Timofeyevich Dobrovolskiy, Flight Engineer Vladislav Niko-layevich Volkov and Research Engineer Viktor Ivanovich Patsayev.*
>
> *On 29 June 1971 the crew of the Salyut orbital station fully completed the flight programme, and was directed to make the landing. The cosmonauts transferred the results of their scientific research and logs to the transport spaceship Soyuz 11 for return to Earth. After completing the transition, the cosmonauts took their seats in the Soyuz 11 spaceship, checked the systems and prepared the spaceship for undocking from the Salyut station.*
>
> *At 21.28 the Soyuz 11 spaceship separated from the Salyut station, and continued its flight separately. The crew of Soyuz 11 reported to Earth that the undocking operation had occurred normally, and that all their systems were functioning normally.*
>
> *In order to make the descent to Earth, at 01.35 on June 30, after orienting the Soyuz 11 spaceship, the braking engine was fired. This functioned for the required duration. Once the braking manoeuvre had been concluded, all communication with the crew ceased.*
>
> *In accordance with the automated programme, after aerodynamic braking in the atmosphere the parachute system was operated and the soft-landing engines were fired before landing. The flight of the descent module ended in a smooth landing in the preset area.*
>
> *A helicopter-borne recovery team landed at the same time as the Soyuz 11 spaceship, and upon opening the hatch found the crew of the spaceship in their couches without any signs of life. The causes of the crew's deaths are being investigated.*

By their selfless work in the testing of sophisticated space equipment – both the first manned orbital station Salyut and the transport ship Soyuz 11 – Dobrovolskiy, Volkov and Patsayev have made a tremendous contribution to the development of manned orbital flights. The exploits of the courageous cosmonauts Georgiy Timofeyevich Dobrovolskiy, Vladislav Nikolayevich Volkov and Viktor Ivanovich Patsayev will for ever remain in the memory of the Soviet people.

On Moscow TV, the reading of this announcement was followed by portraits of the cosmonauts and the continuous playing of solemn music. It was announced that the space heroes were to be given a full State funeral. The nation was stunned. The deaths of the Soyuz 11 crew shook Moscovites even more than the death of the first man to fly in space, Yuriy Gagarin, in 1968. People wept openly in the streets. For over three weeks the record-breaking flight had been featured on both radio and TV. Dobrovolskiy, Volkov and Patsayev were not seen as just the latest cosmonauts, but as a crew that had accomplished something really new, had broken records, and had unquestionably demonstrated the Soviet lead in the development of orbital stations. Yet, at the final stage, the victory had been transformed not merely into failure but into an overwhelming tragedy.

THE FUNERAL

The post-mortems were conducted in the Burdenko Military Hospital in Moscow by 17 physicians. All three cosmonauts had suffered brain haemorrhages, subcutaneous bleeding, damaged ear-drums and bleeding of the middle ear. Nitrogen was absent from the blood; it, together with oxygen and carbon dioxide, had boiled and reached the heart and brain in the form of bubbles. The formation of gas in the blood was a symptom of rapid depressurisation. The blood of all three men contained enormous amounts of lactic acid, fully ten times the norm, which was an indication of terrible emotional stress and anoxia.

On Thursday, 1 July, the bodies of the cosmonauts were delivered to the Central House of the Soviet Army on Spaskiy Street, where they were laid in open coffins on a catafalque with sombre drapes and multicoloured military banners. Garlands and wreaths were arranged around the coffins. Dobrovolskiy was the nearest to the entrance, Volkov was in the middle and Patsayev was furthest. All three had been dressed in dark civilian suits and bore on their chests Gold Stars to signify that they were Heroes of the Soviet Union. Dobrovolskiy and Patsayev had been awarded the nation's top honour posthumously, and Volkov, who had already received one after his first space flight in 1969, gained a second star.

The only one to display any sign of an injury was Patsayev, who had a dark mark similar to a bruise covering most of his right cheek. Dobrovolskiy and Volkov were said by journalists to look uninjured. But for General Kamanin, who was himself in a state of deep shock, only Volkov looked "as alive"; the faces of Dobrovolskiy and Patsayev were "almost unrecognisable".

Cosmonauts (right to left) Kubasov, Filipchenko, Gorbatko and Teryeshkova form a guard of honour for their fallen colleagues.

In the eight hours in which the cosmonauts were on display, tens of thousands of people filed past to pay their respects. Among them were the First Secretary of the Communist Party Leonid Brezhnyev, Premier Aleksey Kosygin, President Nikolay Podgorny, members of the Politburo, senior members of the military, academicians, spacecraft designers and cosmonauts, and foreign leaders and ambassadors. The three-man military guard of honour was exchanged every three minutes. For a time they were joined by members of the cosmonaut corps.

The family mourners were in the front part of the room: Lyudmila Dobrovolskiy with daughters Marina (12) and Nataliya (4); Lyudmila Volkova with son Vladimir (13); and Vera Patsayeva with son Dmitriy (14) and daughter Svetlana (9). With them stood Valentina Teryeshkova, who been the person who informed them of the tragedy. Behind, in black suits, were the cosmonauts' parents: Mariya and Timofey Dobrovolskiy, Olga and Nikolay Volkov, and Mariya Patsayeva, together with their siblings. After several minutes spent standing in silent tribute, Brezhnyev and his colleagues went to the families to express their personal condolences. At one point, Brezhnyev covered his face with his hand and started to cry.

An emotional farewell to the Soyuz 11 crew. Top: Patsayev (left), Volkov (centre) and Dobrovolskiy (right) lie in state in the Central House of the Soviet Army. Middle: Party and government leaders form a guard of honour. Bottom left: Cosmonaut Teryeshkova presents Brezhnyev and Kosygin to the mourners. Bottom centre: Brezhnyev covers his face in grief. Bottom right: Of the three cosmonauts, only Patsayev showed any visible sign of injury, in the form of a dark mark covering most of his right cheek.

At 10 p.m. the Central Army House was closed to the public. At 1 a.m. on 2 July the bodies were cremated. At 10 a.m. the urns containing the ashes were returned to the hall, and for two hours the room was reopened to the public.

Shortly before noon, the American astronaut Colonel Thomas P. Stafford arrived in Moscow to attend the funeral as President Nixon's representative. He flew there from Belgrade, where, with cosmonaut Pavel Popovich, he had been attending an exhibition entitled *Space for Peace*. "Before I reached Belgrade, I heard the news that the Soyuz 11 crew had died on their return to Earth. My first worry was that the stress of a long-duration flight had killed them, and I wondered what it would mean to our Skylab crews." The call from the American embassy in Belgrade to urgently pack his bags and travel to Moscow came as a surprise. When Komarov was killed in 1967 Washington had asked to send astronauts Alan Shepard and Frank Borman to the funeral, but the request had been refused. On landing in Moscow Stafford rode with cosmonaut Beregovoy, his host, to the Central Army House, where he paid his respects. While there, he was introduced to Aleksey Leonov, unaware that Leonov was the original commander for the Soyuz 11 mission.[1]

Colonel Popovich had also returned to attend the funeral. He had hastily called the *Space for Peace* organiser to explain why he must curtail his visit: "The guys have died! This weightlessness will kill all of us."

At noon the Central Army House was closed to the public, in order to enable the family mourners, close friends and members of the cosmonaut corps to prepare for the procession to Red Square. Each urn was decorated with a large looped garland and mounted on a rectangular metal cradle that had two long carrying handles. The urns were taken to individual carriages that were drawn by armoured cars. The pallbearers for Dobrovolskiy's urn included Leonov, Shatalov, Nikolayev and Stafford.

As the cortege made its way slowly to Red Square with the carriages side by side, military officers walked ahead, some with portraits of Dobrovolskiy, Volkov and Patsayev and others carrying cushions bearing their decorations. A guard of honour marched alongside. And Brezhnyev, Kosygin, Podgorny, members of the Politburo and the government, friends, relatives and other cosmonauts followed behind with the mourners. An accompanying military band played solemn music. The route had been closed to normal traffic. Despite the hot and humid day, hundreds of thousands of people stood in line. Buildings along the route flew their flags at half-mast and displayed black-framed pictures of the dead cosmonauts.

As the procession turned into the cobbled Red Square, thousands of people stood behind barricades around its periphery to observe the final farewell in front of the Lenin Mausoleum. The party on the reviewing platform included national leaders and senior military officers.

The main speech was read by Andrey Kirilenko, a member of the Politburo and head of the State Funeral Commission which was formed on the day of the tragedy, whose membership included Ustinov, Smirnov, Afanasyev, Keldysh and Shatalov.

---

[1] In addition, neither Stafford or Leonov knew that in 1975 they would command the two spacecraft of the joint mission involving an Apollo and a Soyuz spacecraft.

Pallbearers carry the urn with Dobrovolskiy's ashes. On the near side are Leonov and Stafford. Cosmonauts Nikolayev and Popovich are partially visible behind Stafford. On the opposite side, are Kirilenko and Shatalov. (Courtesy NASA)

Members of the public join the funeral procession in Moscow's Red Square.

Chertok and Semyonov among the mourners in Red Square. (From the book *Rocket and People, Book No 4*, courtesy www.astronaut.ru)

"They died at their post, as heroes die". The urns with the cosmonauts' ashes during the final part of the funeral.

In addition, he was the coordinator of the special commission created to investigate why the cosmonauts had died.

"Together with the entire Soviet people and our friends abroad," Kirilenko began, "the Central Committee of the Party, the Presidium of the USSR, and the Soviet government deeply mourn the loss that befell our country. ... To the last second of their lives Georgiy Dobrovolskiy, Vladislav Volkov and Viktor Patsayev stayed at the controls of their ship. They died at their post – as heroes die. They were full of vigour, fully confident of fulfilling the assignment from the Party and the people. And they fulfilled that assignment. The results of their observations are

The final farewell was in front of the Lenin Mausoleum.

priceless for science, for the future of cosmonautics, for mankind. ... It wasn't idle curiosity that drew them into space, but the need to unravel more and more of the mysteries of the universe for the good of men. We will continue this difficult but necessary work."

Mstislav Keldysh, the head of the Academy of Sciences, was the second orator. He agreed that the Soyuz 11 mission had been a major step in the development of Soviet cosmonautics: "The Salyut-Soyuz 11 flight heralds the start of a new stage in exploring outer space, namely using long-term orbital stations in near-Earth orbits."

Generals Nikolayev and Shatalov represented the cosmonaut corps. Shatalov read an open letter written by their colleagues: "We know that our road is a difficult and thorny one but we do not doubt the correctness of our choice, and are always ready for the most difficult flight. ... We express our firm confidence that what occurred must not halt ongoing development and perfection of space engineering and man's striving for space. ... Today, we pay a final tribute to our talented and courageous comrades, but there is not just grief in our hearts, there is also pride in what they did for their country in space."

Finally the urns were taken behind the Lenin Mausoleum to the Kremlin's wall, to be interred alongside those bearing the ashes of cosmonauts Vladimir Komarov and Yuriy Gagarin. As the urns were inserted into their niches, cannons fired in salute. Each niche was sealed with a black plate that bore the name of the cosmonaut and the dates of his birth and death. Their photographs and decorations were placed on pedestals alongside, and the families and friends moved in to pay their final respects.

The whole world shared the grief. The Soviet newspapers were full of tributes and messages of condolence from foreign leaders. Among many who sent messages of sympathy to the Soviet people were Queen Elizabeth II, the Pope, Presidents Nixon and Pompidou and Premiers Chou En-lai and Indira Gandhi.

In a letter to Podgorny the Queen wrote: "My husband and I were shocked to hear of the deaths of your three cosmonauts. We extend our sincerest sympathy to you and to the Soviet people on the occasion of the sad loss of these intrepid men."

On behalf of the United States, President Nixon wrote to the Soviet leaders: "The American people join in expressing to you and the Soviet people our deepest sympathy on the tragic deaths of the three Soviet cosmonauts. The whole world followed the exploits of these courageous explorers of the unknown and shares the anguish of their loss. But the achievements of cosmonauts Dobrovolskiy, Volkov and Patsayev remain. It will, I am certain, prove to have contributed greatly to the further achievements of the Soviet programme for the exploration of space and thus to the widening of man's horizons."

President Pompidou wrote: "All Frenchmen, like me, admired their extraordinary exploits."

In the Vatican, Pope Paul interrupted a general audience to announce the deaths. He expressed sadness for "this unexpected and tragic epilogue", and offered prayers to the families of the three men.

The Chinese Prime Minister Chou En-lai sent a telegram to express sympathy to the Soviet people for their "deep grief" over the deaths of the cosmonauts, and to "convey heartfelt condolences to the bereaved families".

Brezhnyev and Kirilenko help to carry Dobrovolskiy's urn to its final resting place in the wall of the Kremlin.

The ashes of the three cosmonauts have been interred in the wall of the Kremlin.

Indian Prime Minister Indira Gandhi said that the three men had "died as heroes on behalf of science. Their achievement in the exploration of space [was a major] contribution to progress."

In the wake of the successful Apollo 11 lunar landing, NASA and the Soviets had begun to consider the possibility of a joint manned space mission. In January 1971 George Low, NASA's Deputy Director, had led a group of specialists on a visit to Moscow to explore the options, and they met several cosmonauts. Low now sent a letter of condolence.

Valentina Teryeshkova comforts Dobrovolskiy's daughter Marina at the wall of the Kremlin. Dobrovolskiy's mother Mariya stands in the background together with cosmonauts Feoktistov (with glasses) and Gorbatko. (Courtesy Peter Pesavento)

Patsayev's family (left to right): daughter Svetlana, wife Vera and son Dmitriy. Behind is Viktor's mother Mariya. (Copyright Svetlana Patsayeva)

The popular writer Konstantin Simonov wrote in *Pravda*: "Warriors know that the most difficult aspect of a reconnaissance mission is to return across the front line to one's own position. The front line in space reconnaissance, in the struggle to reveal the mysteries of nature, is re-entering the Earth's atmosphere; the final step before landing. It was precisely at this final step that the crew of the Salyut orbital station perished."

Mikhail Rebrov, a special correspondent of the newspaper *Krasnaya Zvezda*, and a close friend of many of the cosmonauts, summed up to the overwhelming feeling at this tragic, yet triumphant, moment in the history of the Soviet space programme: "We know the road to space is difficult and dangerous. But once having embarked upon this road we must continue, for no difficulty or obstacle can turn a man away from his chosen path. The cosmonauts have told us: 'As long as our hearts beat, we will continue to explore the universe.' Wonderful and brave people are now dead. Their names will illuminate the arduous road into outer space like stars."

Kenneth Gatland, vice-president of the British Interplanetary Society, wrote: "The entire space community today mourns the three space heroes whose ashes are being buried in the wall of the Kremlin. Before the tragedy that befell them, they opened a new era of space conquest by occupying the world's first space station. Their epic flight will stand as a landmark in space history."

The writer and broadcaster Patrick Moore said: "Certainly, the uppermost thought in my mind is sadness at the deaths of these three brave men. They will never be forgotten. Unfortunately, nothing can bring them back, but the sympathy of the whole world will go out to their relatives, to their countless friends, and to all the people of the USSR."

One of the last sites to record the three cosmonauts alive was the amateur satellite tracking station at Kettering Grammar School in England. Its leader, Geoffrey Perry, said that they received signals from Soyuz 11 as it was passing 200 km above the island of Madeira in the Atlantic, off the northwest coast of Africa. "At that time we were certain that all three men were still living. After you have been listening to three men's heartbeats for 24 days, it is difficult to put into words your feelings on discovering that they are dead. We are all very upset."

The leaders of the Soviet space programme were quick to reaffirm that manned missions would continue.

Writing in *Pravda* on 4 July Academician Boris Petrov, who was the chairman of the Interkosmos Council, spoke of the conquest of space as a "difficult path", then repeated Brezhnyev's statement, made prior to the launch of Salyut: "Soviet science considers the creation of orbital stations with replacement crews to be the highway to space." Petrov argued that platforms in "near-Earth space" would enable man to make comprehensive studies of the Earth and of astronomy. He said that "the 1970s will see the development and application of long-term manned orbital stations with replacement crews, making it possible to switch from occasional brief experiments in space to regular work by scientists and specialists in space laboratories." He went on: "The experience of the Soyuz 11 crew has shown that the Salyut station is well designed for experiments in orbital flight conditions. Such stations offer broad prospects for the continuation and development of the research that was undertaken by the first Salyut crew. ... In due course larger and more complex multipurpose and specialised space stations will be built. But the significance of the work carried out by the first crew of the first manned orbital station ... will never fade." Speaking of the tragedy, he said: "Soyuz ships have already made several space flights, and have safely returned cosmonauts to Earth. When such complex machinery is being tested and mastered, accidents can never be ruled out."

The disaster overshadowed the Congress of Soviet Writers' hosted by the Kremlin, where the famous poet Yevgeniy Yevtushenko read a memorial poem:

Two-way Link for Ever

In Kamchatka and in Arbat,
Above the Angara rapids
The sorrowful expiration: guys have perished,
As the requiem above the country

None – no matter how it was crowned –
Will not return to its house finally
To three hearts, large, human
It became less in Russia hearts.

And what heavy burden,
For the people, to whom they were
Simple Vitya and simple Gosha,
And simple Slava – during the recent days.[2]

O, Matrosovs[3] of the cosmodromes!
You left to us your regulations:
Even in space – by vein without having trembled,
To die at the work sites.

As much there are still difficulties
In the sky to be yielded!
And thus far humanity exists
The flame of future spaceships
Will be the eternal fire in your honour

You are as immortal as the cry:
"We have ignition!"
And it's not true that contact has been lost:
Between you and our native land
There is two-way link for ever.

---

[2] The poet referred to Viktor Patsayev as Vitya, Georgiy Dobrovolskiy as Gosha and Vladislav Volkov as Slava.
[3] Aleksandr Matrosov was made a Hero of the Soviet Union during World War II for sacrificing himself in an assault on an enemy bunker, and in so doing preserving the lives of his colleagues.

WESTERN SPECULATIONS

As soon as TASS made the announcement that the Soyuz 11 crew had been found dead in their couches, people all around the world began to consider whether their deaths were due to a technical fault or were the result of a fundamental limitation of the human body.

One of the prevailing theories was that man might not be able to survive for long periods in weightlessness. For several years there had been a serious debate among scientists about the effects of long-term exposure to weightlessness. In 1965 one of NASA's Gemini missions had spent 14 days in orbit in order to demonstrate that it was possible to remain in space for the length of time required to fly a lunar landing mission. However, there were indications that the heart grew lazy when exposed to weightlessness. In July 1969 the monkey Bonny died of heart failure after the 9-day flight of NASA's Biosatellite 3. After the 18-day flight of Andriyan Nikolayev and Vitaliy Sevastyanov on Soyuz 9 in 1970 the Soviets had discovered the debilitating effects of weightlessness: the loss of body fluids, the loss of calcium from the bones and the loss of muscle tone, including the heart. It had taken more than a week for them to readapt to gravity. Perhaps, it was suggested, the Soyuz 11 mission, having lasted six days longer than the previous record, had exceeded man's limits in space. Medical experts admitted that weightlessness could have played a part in the deaths, but were sceptical that the hearts of three men having different physiologies could have failed simultaneously.

According to one source, the crew of Soyuz 11 complained to the TsUP that they were having breathing difficulties soon after undocking from Salyut, but were told that it was normal.[4]

Western experts in space medicine did not think that the deaths of the cosmonauts resulted from the time they spent in weightlessness. Dr. Charles A. Berry, the chief physician at NASA's Manned Spacecraft Center in Houston, Texas, said: "There is no evidence whatsoever from either our experience or that of the Russians in space, or from ground-based experiments, to suggest that weightlessness could have been responsible." He thought that the accident may have been caused by the release of a toxic substance. Dr. Walton Jones, Deputy Director of Life Sciences at the NASA Office of Manned Space Flight, said that since the three men were found strapped in their couches, they likely died as a result of sudden decompression, such as would have occurred if a valve had leaked or if the cabin shell had ruptured or was struck and punctured by a meteoroid.

Within hours of the news of the loss of the crew, Kenneth Gatland of the British Interplanetary Society dismissed the effects of returning to Earth after such a long flight as the cause of death. There must have been a mechanical failure. But it was

---

[4] This was reported by *The Sunday Times*, but there is no direct evidence for this in the radio communications following undocking. However, it is not inconceivable that the cosmonauts had problems with breathing after a long day of transferring the final materials to the Soyuz and the stress resulting from the difficulty encountered in closing the hatch.

possible that after 24 days in space the cosmonauts were so tired that they had failed to verify all of the spacecraft's systems, or when an emergency had developed they had been unable to react sufficiently rapidly.

NASA was relieved when the official report ruled out weightlessness and physical deconditioning as causes for the accident. The American space specialists felt sure that the Soyuz must have suffered a mechanical or structural failure. Because the crew were not in protective pressure suits, they could have died from any number of causes: excessive heat, carbon dioxide fumes from a small fire, a nitrogen leak from the spacecraft's air-supply system, or a rapid drop in cabin pressure. Such theories were supported by unconfirmed reports that all radio transmissions – telemetry as well as voice – had ceased at the conclusion of the braking manoeuvre. In fact, most speculation centred on a failure in the oxygen supply. This was based largely on the rumour in Moscow that the cosmonauts had been found with serene expressions on the faces – such composure is characteristic of hypoxia, a starvation of oxygen that can produce a rapid and relatively painless death.

On learning of the difficulty in closing the hatch prior to undocking from Salyut, Western analysts theorised that if the hatch was insecure the mechanical stresses of re-entry could have made a minor leak into a disastrous one. But in September 1971 cosmonaut Dr. Boris Yegorov said that the disaster struck when the air leaked from the cabin during a period of several seconds as the orbital module was released. He insisted that the hatch was properly sealed, and said that suspicion had fallen on one of the valves used to equalise the pressures across the hatch.

The authorities had deemed the post-mortems sufficient to determine the cause of death, and had proceeded with the State funeral, but were waiting until they fully understood what had gone wrong before concluding the technical investigation.

Specific references

1. 'They Made Accomplishment'. Politika, Belgrade, 2 July 1971 (in Serbian).
2. 'Breathless clue to Soyuz space deaths', *The Sunday Times*, 4 July 1971.
3. 'Moscow to go ahead with plans for manned space stations despite Soyuz disaster'. *The Times*, 5 July 1971.
4. Stafford, Thomas P. with Cassutt, Michael, *We Have Capture – Tom Stafford and the Space Race*. Smithsonian Institution Press, 2002, pp. 154–156.
5. Kamanin, N.P., *Hidden Space, Book 4*. Novosti kosmonavtiki, 2001, pp. 333–338 (in Russian).

13

Thirteen seconds to eternity

COMMISSION

A special 12-member State Commission was formed to determine the specific cause of the Soyuz 11 tragedy. The chairman was Academician Mstislav Keldysh, and his deputy was Georgiy Babakin, who was the Chief Designer of the Lavochkin Design Bureau which developed lunar and interplanetary probes. The membership included Sergey Afanasyev, head of the Ministry of General Machine Building, and General Designer Valentin Glushko. Although Glushko developed the engines for Korolev's rockets in the 1950s, his relationship with Mishin was strained. The Commission set up ten subcommittees to investigate every aspect of a Soyuz flight, including launch, orbital operations, mission control, working with the Salyut station, undocking, the braking manoeuvre, re-entry and landing; and then to recommend ways in which to improve the design and operation of the spacecraft. Six of the subcommittees were led by the Air Force representatives, who included cosmonauts Shatalov, Nikolayev and Beregovoy. Interestingly, although General Kamanin was replaced in his post by Shatalov, he led the subcommittee that analysed conditions on Salyut and drew up recommendations for its future use. This would prove to be the final assignment of his 11-year career in charge of the Air Force's manned space programme.

The State Commission held its first meeting on 3 July, the day after the funeral, at which time it planned the investigation and specified the subcommittees. It had two weeks in which to undertake its investigation and submit its report. For its first operative meeting on 7 July, Keldysh invited the attendance of the most important TsKBEM people involved in the DOS programme – Mishin, Bushuyev, Chertok, Tregub, Shabarov, Semyonov and Feoktistov.

DECOMPRESSION

The first to present was Vasiliy Mishin, who described how the Soyuz 11 spacecraft differed from its predecessors. He pointed out that a total of 19 spacecraft had been launched since November 1966, with Soyuz 10 and Soyuz 11 being the 7K-T crew ferry. The main difference between the two recent ships was the modification to the docking system following its failure on the Soyuz 10 mission. According to Mishin, Soyuz 11 suffered no major problems until the separation of its modules. It is not clear whether he told the Commission of the difficulty in closing the hatch prior to undocking. Based on data recorded by the onboard memory device, Mir, the module separation occurred at an altitude of about 150 km (some sources say 168 km) and lasted just 0.06 seconds. The pressure in the descent module began to fall rapidly at that moment. At 1.47.26.5 a.m., two seconds prior to jettisoning the orbital module, the pressure in the descent module was 915 mm of mercury, which was normal. But some 115 seconds later the pressure had dropped to 50 mm, and was still falling. In effect, there was no longer any air in the cabin! In fact, the book relating the history of RKK Energiya (as the TsKBEM later became) states that the pressure fell even more rapidly than this, reaching near-zero in only 30–45 seconds.

Decompression could result from two causes: (1) the premature opening of one of two valves located at the top of the descent module, or (2) leakage from the hatch. Mishin presented diagrams featuring curves corresponding to these two modes of decompression. The curve calculated for a loss of pressure due to the valve opening exactly matched the actual loss of pressure recorded by the 'black box'. In addition, the force resulting from the air venting from this valve upset the stabilisation of the module, which prompted the automated control system to fire six 10-kg thrusters to compensate. The thruster firings calculated on the assumption that the air was being vented matched those recorded by the 'black box'. The maximum deceleration load of 3.3 g was recorded when the descent module reached an altitude of about 40 km, where the atmosphere began to thicken. At this point, air began to enter through the inadvertently opened valve. The second valve was automatically opened as planned, at an altitude of about 5 km. Although the cabin rapidly filled with fresh air, it was too late for the cosmonauts.

The conclusion was inescapable: one of the two valves had opened prematurely as the orbital module was jettisoned. The possibility of an incorrect command could be discarded because both valves were on the same electric circuit. Based on the 2-cm size of the valve's tube, the internal volume of the descent module, and the fact that the air would have passed through the valve at the speed of sound, the time for the pressure to diminish to near-zero was calculated at 50–60 seconds. If Dobrovolskiy, Volkov and Patsayev had been wearing pressure suits they would not have been in danger, but the Soyuz was a 'shirt-sleeve environment' and so they became the first men to die *in* space.

For the State Commission, two facts relating to the tragedy of Soyuz 11 crew were of crucial importance: spacesuits and valves. The decision to send cosmonauts into space without pressure suits had been taken years earlier. To create a 'spectacular' for Khrushchov, in early 1964 Korolev had ordered Feoktistov to adapt the Vostok spacecraft to accommodate three men, and in order to create the impression that this was an entirely new vehicle it was to be named Voskhod. As there was insufficient room for three men dressed in the pressure suits worn by the Vostok cosmonauts, it was decided that the crew should wear casual clothing. During a meeting on this issue Korolev said that working in the spacecraft in a pressure suit was as uncomfortable as working inside a submarine wearing a wet suit. Furthermore, to fit three couches into the capsule, it was necessary to discard the ejector seats, so the Voskhod crews were the first to be launched with no means of escape if their rocket were to have a malfunction during the first 27 second of its flight. Feoktistov was initially doubtful, but led the modification when Korolev promised that one of the designers could be a member of the first Voskhod crew. Because the descent module of the three-seat Soyuz was not much larger than the old spherical capsule, it was likewise designed for use without pressure suits.[1] In March 1964 Korolev advised Khrushchov of the possibility of sending a three-man crew into space. The American Apollo that was to be capable of carrying three astronauts was not expected to start flying until late 1966, so Khrushchov eagerly accepted Korolev's proposal; he was unperturbed that the cosmonauts would fly without pressure suits – for him the most important thing was once again to beat the Americans.

During Korolev's lifetime, only Kamanin had sharply objected to this idea. In fact, he had attempted to force a return to the use of pressure suits. On 5 and 7 July 1971 he made the following entries in his diary expressing his disappointment:

> Cosmonauts and the Air Force specialists insisted many times both verbally and in writing to the Central Committee of the Communist Party on the need to have on the ship pressure suits and equipment to pump air. But they were always refused – over a period of seven years! Responding to our requests, Mishin several times said that we were overcautious, that the decompression of the Soyuz spacecraft is completely excluded, meaning that "it is possible to fly [on it] in shorts".
> . . .
> The crews of our ships have flown without pressure suits for seven years. Cosmonauts have written to Khrushchov, Brezhnyev, Ustinov and Smirnov about the danger of such flights. Kutakhov sent a letter to Mishin concerning the fact that cosmonauts "fly in shorts", with a request to have pressure suits on board. But all our requests were refused – first by Korolev and in recent years by Mishin, who said that hundreds of unmanned satellites and piloted spacecraft have flown in space without a single case of decompression.

In the early phase of the Soyuz programme Mishin's responsibilities were related to rocketry; he had very little involvement in the design of manned space vehicles. When he succeeded Korolev as the Chief Designer in 1966 the development of the Soyuz was nearing completion. It would have been possible to modify it to accommodate a crew wearing pressure suits, but only by eliminating one of the couches.[2] Korolev's fundamental error, with the active

---

[1] In all other respects, of course, the Soyuz was more sophisticated than the Voskhod, particularly in having an escape system in case of a malfunction in the launch vehicle.

[2] Three pressure suits would have weighed a total of about 80 kg, and there would have to have been additional apparatus to support them independently of the cabin environment. The Soyuz spacecraft simply was not designed for such a configuration.

support of Feoktistov, was to have designed the spacecraft for use without pressure suits. As Feoktistov said 24 years after the loss of the Soyuz 11 cosmonauts, "the feeling of guilt persists". The second major error was the decision not to install the tanks which would have supplied additional air to the crew in the event of a decompression. This was accepted by Mishin despite the protests of General Kamanin and the specialists at the TsPK. Interestingly, no one at the OKB-1/TsKBEM had the courage early in the design to seriously analyse the risk of flying without pressure suits and then challenge Korolev and Feoktistov.

Later, in one of his interviews, Mishin defended Korolev's decision by saying that during over 1,000 tests of the descent module there had been no problems relating to decompression. Noting that for decades hermetically sealed aircraft have flown at altitudes of 10 km or greater carrying crew and passengers wearing casual clothes rather than pressure suits, Mishin said: "I think Korolev's decision was correct, and that after this it was necessary to focus attention not on individual protection, but on the protection of the entire module – on group protection. Our idea was to develop such a robust hermetic unit that we would not need a backup for each element."

While the descent module was at the landing site, it was established to be pressure tight. On its arrival in Moscow it was examined by experts from the TsKBEM. The hypothesis that a valve had been inadvertently opened when the orbital module was jettisoned looked good on paper, but despite being subjected to powerful shocks and vibrations the valves remained shut. The fortnight deadline allowed by the Kremlin for the investigation expired without such tests validating the hypothesis that on this occasion there had been an unexpectedly severe shock associated with the release of the orbital module. Later, Academician Keldysh pointed out that since the tests had been done in normal atmospheric conditions the forces would have been diffused by the air, and he suggested that the separation of the modules should be simulated in vacuum in an altitude chamber. Two tests were made in the TsPK, but in both cases the valves remained shut. Undeterred by this 'proof' of the design of the valves, the specialists devised tests involving incorrectly configured valves in an effort to gain insight into the issue. Tests that applied a variety of individual loads and modes of malfunction to the valve failed to open it. However, when these were all applied simultaneously, the valve opened. With this proof that it was possible for the valve to be shocked open, the premature opening of the valve during the separation of the modules of Soyuz 11 was officially accepted as the cause of the decompression.

On 10 July 1971, while the tests were underway, the State Commission released a 200-word statement. After pointing out that the flight of Soyuz 11 was normal until the onset of re-entry, it went on: "On the ship's descent trajectory, 30 minutes prior to landing, a rapid drop of pressure occurred in the descent module leading to the sudden deaths of the cosmonauts. This is verified by the medical and pathological-anatomical examinations. The drop in pressure was the result of a loss of the ship's hermetic seal. An inspection of the descent module showed there to be no failures in its structure. A technical analysis has determined several possible causes for the loss of the seal. The study of these continues." Incredibly, this is the only report ever to have been officially released describing the deaths of the Soyuz 11 crew!

The fact that the Commission's statement said that the cosmonauts died suddenly 30 minutes prior to landing owing to a pressure leak, whilst also saying there were no failures of the structure, led Western observers to conclude that the cosmonauts must have erred! In fact, two days after the tragedy some Western newspapers had reported an anonymous Soviet journalist who claimed the crew died because "they failed to seal the hatch of their spacecraft properly". At week's end, the *Evening News* in London reported that Russian scientists attending the funeral had blamed the cosmonauts. Victor Louis, the paper's Moscow correspondent, wrote: "human error and mechanical failure between them caused creeping depressurisation in the spacemen's 9-foot cabin and deprived the cosmonauts of life-supporting oxygen during the final phase of their journey". During the turbulent re-entry, Louis said, the spacecraft's hatch had opened sufficiently to allow the air to escape into space. Although there was some basis for this story – the difficulty in sealing the hatch just prior to undocking – the State Commission had ruled out the hatch seal as the cause of the decompression.

The Commission completed its investigation in early August and recommended a number of improvements intended to preclude a repeat of the Soyuz 11 tragedy. At the final meeting, Academician Keldysh pointed out that the "opening of the valve was due to a shock wave propagating across the metal structure of the spacecraft", and after noting that "to be simulated it is necessary to perform tens or hundreds of experiments in the altitude chamber" he suggested that if the steps proposed by the Commission were adopted then to continue "expensive and complicated tests" in an altitude chamber would "not make sense".

Interestingly, three of the most important documents about the Soyuz 11 tragedy were not made public, and presumably remain in the archives of either the Kremlin or the TsKBEM. These are:

- The final report of the State Commission, including the individual reports of its subcommissions.
- The data recorded by the 'black box' in the descent module prior to, during, and after the separation of the modules.
- The full reports of the autopsies by the Burdenko Military Hospital – even the Ministry of Heath's Institute for Biomedical Problems, which is the leading space medicine institution in Russia, does not have copies of the autopsy results!

As in the case of Komarov's death, the Kremlin hid the truth about the Soyuz 11 tragedy from the Soviet people. The fact that Dobrovolskiy, Volkov and Patsayev died as a result of a valve inadvertently opening was revealed by the *Washington Post* on 29 October 1973 – more than two years after the fact! In planning the joint mission during which an Apollo was to dock with a Soyuz in the summer of 1975, the NASA officials said during a visit to Moscow that they had a need to know what had gone wrong with Soyuz 11. The *Washington Post* reported that a vent valve was accidentally forced open, and that the air in the descent module leaked to space in a matter of seconds. The valve had opened just after the orbital module was jettisoned. This procedure involved the firing of explosive bolts, and it was reported that the shock, which was greater than that expected, had been sufficient to cause the valve to open. Two of the cosmonauts had tried to unstrap from their couches in order to close the valve but had not been able to act fast enough. In ten seconds the cabin pressure was so low that it could no longer support human life. After a further 45 seconds there was no air left at all. Following a period of unconsciousness, the crew died from pulmonary embolisms. The tissue damage to their bodies was due to the boiling of their blood during the 11.5-minute interval that they were exposed to vacuum – a symptom that could at first have been misinterpreted as being indicative of an instantaneous and catastrophic decompression.

THE VALVE

Let us consider the function of the valve which was the technical cause of the loss of the Soyuz 11 crew. The limited capacity of the launch vehicle obliged Feoktistov and his design team to make the Soyuz descent module a very small vehicle – it is so cramped that it is right on the limit for accommodating the human body. In fact, the bell-shaped module stands 2.16 metres tall, has a maximum diameter at its base of 2.2 metres and weighs only 2.8 tonnes. Yet it had to contain couches for three cosmonauts and all the necessary life-support equipment, together with the systems to operate the spacecraft in space and two large parachutes for landing. The 'free volume' of the cabin is a mere 2.5 cubic metres, which is less room per cosmonaut than the Vostok capsule! The air in such a cramped module can support the lives of three men for only a short time – but this is viable because it operates autonomously only for the 30 minutes from the separation of the orbital and propulsion modules through re-entry and landing. Nevertheless, once the main parachute deployed at an altitude of approximately 5 km, two valves were to be opened to allow fresh air to enter the cabin; both to equalise the internal and external pressures and to eliminate the risk of the cosmonauts asphyxiating in the event of their having to remain inside for some time after landing, as might occur if the hatch were unable to be opened as a result of a technical problem or if the module were to land in water and the hatch was partially submerged.

The fact that both valves are closed during the majority of the mission and then opened only a few minutes prior to landing confused the State Commission. Surely the recovery team would open the hatch promptly, or if the module landed off target the cosmonauts would open it themselves! Given that the premature opening of one of the valves caused the deaths of three cosmonauts, what where the valves actually for? Was their inclusion a terrible error by the designers? The explanation from the TsKBEM of the risk of asphyxiation if for some reason the hatch was unable to be opened promptly was inconclusive. An additional confusion concerned the fact that each valve had two shutters. In fact, this aspect of the design would prove to be one of the most important factors in the Soyuz 11 tragedy.

To understand what happened, we must examine the valve's structure. The design was straightforward, involving a cylinder of cork with a rubber ring and a piston rod supported by a ball-lock shutter that was automatically controlled. The crew had no control over the automatic shutter, which would be opened by a pair of pyrotechnic charges after the deployment of the main parachute. Next to the automatic shutter was one that the cosmonauts could open manually by a small rotating knob. So long as at least one shutter remained closed, the valve ought to be shut. The valves were placed below the ring of the hatch: the No. 1 valve above Dobrovolskiy's couch and the No. 2 valve above Patsayev's couch, on opposite sides of the hatch so that if the module were to land on water there would be no chance of both of the valves being submerged. In the event of a splashdown, the manual shutters would be operated as required to prevent water ingress. This was the only circumstance in which the crew were to operate the manual shutters.

Why did the automatic valve open at an altitude of approximately 150 km, rather than at 5 km? The orbital and descent modules were connected by a dozen bolts in the ring that housed the hatch. During the assembly of the spacecraft, the bolts had been fastened using a special tool, then the joint was checked in an altitude chamber to ensure a hermetic seal. The combined force of all the bolts was about 100 tonnes. To separate the modules in space, the bolts had to be severed simultaneously. Hence each bolt incorporated a small explosive charge and an electric circuit. According to the programme, a timer would cause electricity to be supplied to the bolts in order to detonate the explosive charges and sever the bolts, applying a force of 100 tonnes for

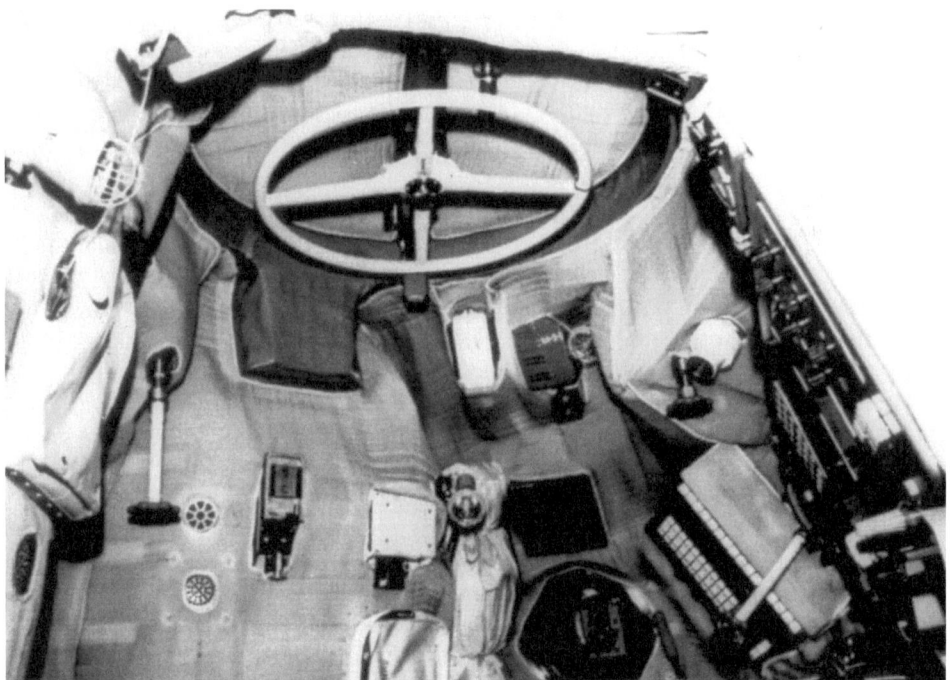

At the top of the Soyuz cabin is the hatch, with one of the ventilation valves visible under its ring on the right, next to a black box. (From the book *Soyuz – A Universal Spacecraft*, courtesy Rex Hall)

a duration of one microsecond, in the process sending a shock wave across the metallic surface of the craft. The valves were located close alongside the connecting ring, and so would have been particularly sensitive to the propagation of this shock. In the case of Soyuz 11 this caused an automatic valve to pop open. The fact that particles of gunpowder were found inside one valve was conclusive proof that it had opened at the moment of separation.

THE AGONY

How did the cosmonauts react? It is possible to make inferences from the analyses performed by the medics, the state of the cabin, and the data recorded by the 'black box'. During the descent, each cosmonaut wore a medical belt with various sensors and the data on their vital functions was recorded. Prior to their return, the general physical state of each man was good. Dobrovolskiy's pulse in a normal, unstressed state was 78–85 beats per minute. Volkov, being more dynamic and emotional, was usually higher, and at the time of undocking from Salyut his pulse increased to 120, perhaps reflecting his concern about the hatch seal. Patsayev's pulse was between 92 and 106.

During the first second after the separation of the spacecraft's modules the pulses of all three men dramatically increased. Dobrovolskiy rose to 114. Volkov shot up to 180! Four seconds after the onset of depressurisation Dobrovolskiy's respiration rate was 48 breaths per minute; the normal rate is 16. Such rates are characteristic of a sudden oxygen starvation. The rapid increases in pulse and respiration indicate that the crewmembers were immediately aware of what was occurring. In addition to hearing the air leaking out and feeling the pressure fall, they would have heard a loud siren and seen the value of the cabin pressure decline on the indicator set in the lower left corner of the main instrument panel. There would also have been physical indications, including a rapid fall in temperature and air fogging as the water vapour condensed. They would have suffered the effects of decompression – an immediate strong pain in the head, chest and abdomen, followed by burst eardrums and blood streaming from the nose and ears. Their heart rates rose during the first 20 seconds, but by 60 seconds had reduced to just 40 per cent of the baseline.

Death was not instantaneous. Due to out-gassing of oxygen from the venous blood supply to the lungs, the men would have remained consciousness for 50–60 seconds. However, they could have moved about and tried to remedy their plight only during the first 13 seconds; this being the 'time of useful consciousness', corresponding to the time that it took for the oxygen-deprived blood to pass from the lungs to the brain. Because the valves were situated above their couches, Dobrovolskiy and Patsayev attempted to take action. Being in the centre, nearest the hatch, Dobrovolskiy was in the best position to act. However, the cosmonauts did not know the actual source of the leak. Recalling the difficulty that they had faced in sealing the hatch, their initial diagnosis must have been that the air was leaking through the hatch. Dobrovolskiy unbuckled and pulled himself to the hatch. However, it was properly closed. When Volkov and Patsayev switched off the radio equipment in order to listen to the hiss in an effort to identify the source of the leak, this was realised to be one of the two valves. But which one? Valve No. 2, above Patsayev, was marked as 'open', so he went to try to close it. But it was No. 1 which was open. It is difficult to know who did so, but either Patsayev or Dobrovolskiy began to close the hand-operated shutter of valve No. 1. However, in normal circumstances it required at least 35 seconds to close the valve by hand, and by the time they passed out it was only partially cycled. Volkov was too far away from the valves to assist, so he remained strapped into his couch. By virtue of being more active, Dobrovolskiy and Patsayev would probably have lost consciousness before Volkov, for whom the frustration of being unable to assist must have been intense.

They died rapidly. The initial paralysis due to oxygen starvation would have been followed by general convulsions. During this time, water vapour rapidly formed in the venous blood, and in soft tissue. Blood and other bodily fluids boiled and turned to vapour, causing the body to swell to perhaps twice its normal volume. The heart rate initially soared, but then diminished to an unsustainable rate. The arterial blood pressure dropped to zero after about 60 seconds, but the venous pressure rose due to gas and vapour distending the venous system. Within a minute, the venous pressure exceeded the arterial pressure. In effect, there was no circulation of blood. After the initial rush of gas from the lungs during decompression, gas and vapour continued to flow out through the airways, and the sustained evaporation of water chilled the mouth and nose to almost freezing temperatures. The remainder of the body would have cooled more slowly. The first fatal damage occurred in the cosmonauts' lungs, as the most vulnerable part of the body in such circumstances. They naturally tried to hold their breath, but as the cabin pressure declined the lungs and thorax became over-extended, tearing and rupturing the lung tissue and capillaries. The trapped air was forced directly into the blood, following the ruptured blood vessels and creating massive air bubbles in the vital organs, including the heart and brain. Clinical death began after 90–100 seconds, simultaneously in all three men. By 110 seconds after the separation of the modules there were no heart or respiration rates recorded. Ten seconds later, life was extinct. The cabin remained in vacuum for 11.5 minutes, then began to fill with air from the upper atmosphere.

COULD THE COSMONAUTS HAVE SURVIVED?

In analysing the actions of the Soyuz 11 cosmonauts during the decompression to assess whether they might have saved themselves, there are two basic approaches. Mishin and the TsKBEM engineers concluded that the crew should have been able to halt the leak – but they had panicked and failed to identify the source of the leak in time. But General Kamanin and the military cosmonauts at the TsPK thought that the decompression occurred so rapidly that the crew had no real chance of manually closing the shutter on the leaking valve.

Of the official sources, Kamanin provided probably the most realistic description of the fateful events. As he wrote in his diary, following the braking manoeuvre the cosmonauts felt the onset of deceleration – which meant that the ship had begun its descent:

Aboard, everything is normal. However, the cosmonauts, remembering their recent difficulty with the transfer hatch, carefully monitored the pressure in the cabin. A bang is heard – there is the separation of the modules. But what is this? The pressure in the cabin begins rapidly to fall. . . . Decompression! After unfastening from his couch, Dobrovolskiy goes to inspect the hatch. It is airtight, but the pressure continues to fall. They can hear the whistle of air venting to space. Because of noise from the transmitters and receivers, they cannot trace the source of the whistle. Volkov and Patsayev unfasten their shoulder straps and switch off the radio apparatus. The source of the whistle is above the centre couch – where a vent valve is located. Dobrovolskiy and Patsayev attempt to close the valve, but because they are weakened they fall back to the seat. As he loses consciousness, Dobrovolskiy manages to fasten the waist lock of his entangled straps.

Mishin argued that once the crew had realised that one of the valves had opened prematurely, they should have blocked the flow by placing a thumb over the inlet.[3] Some sources have pointed to a bruise on Dobrovolskiy's thumb as evidence that he had indeed attempted to do this. Another source says the bruise was on Patsayev thumb, although this may actually have been a reference to his facial bruise. Mishin never accepted that the crew could not have saved themselves. To the end of his life he insisted that if only they had been better trained then they would have reacted properly, and therefore probably have survived: "How can you describe the deaths of great and brave people – a disaster which caused such pain to their relatives? It is more painful to know that it was avoidable. During separation, the explosive bolts generated a force that was too strong and the ball left its nest, opening the valve prematurely. The cosmonauts could hear the air whistling. Patsayev tried to close it using his thumb, but failed. There was a manually operated shutter with which it was possible fully to protect the cabin, but they either forgot it, or did not know, or missed it in their training." In another interview, Mishin again criticised the crew: "During the decompression, the air would have escaped to space at such a high speed that the men had to have heard the whistling – a signal of imminent disaster. It was necessary to unfasten the belts, to stand up and to shut the valve. They could block the valve even using a thumb! . . . But the cosmonauts were disoriented. . . . Perhaps they were lost. . . . Patsayev seems to have realised what was the matter. He unfastened from his couch, but did not have time to stand up."

While Mishin was trying to blame the crew and to justify the spacecraft's design, representatives of the TsPK thought differently. In contrast to Mishin, Kamanin was confident that the crew fought to save their lives right to the end: "Is it possible to accuse them of not knowing how 'to plug the hole in the ship by finger'? We cannot presume this to be feasible at all, as no one has yet tried to do it. Indeed, outside the ship is the deep cold of the vacuum of space, which causes the instantaneous boiling

---

[3] In making this remark, Mishin gave the impression that he expected that a cosmonaut would hold his finger in place to stem the air leak right through the re-entry process, until the capsule was in the atmosphere. However, the real value in interrupting the leak in this manner would have been to buy the time required to close the manual shutter on the valve. Yet there was no tank to replenish the lost air.

of the blood. I think even in normal conditions it would be hard to hold a finger in open space for a period as brief as several seconds. In addition, the crew had first to locate the source of the decompression and then, 'after plugging the hole by thumb', to retain the hermetic seal of the cabin for 15–17 minutes during which they would be subjected to the increasing deceleration loads of the descent."[4]

Commenting on Mishin's claim that a man could have blocked the valve using a finger, Dr. Yevgeniy Vorobyev pointed out that in such a rapid decompression the state of consciousness would have been diminished after 20 seconds. "To unbuckle, locate the hole under the cover and block this in 20 seconds would be unrealistic. It would have been necessary to train to do so. We tested the possibility of closing the valve manually, in the case of a splashdown. Even in a calm situation, this operation took 35–40 seconds. Thus, they had no chance of surviving."

General Shatalov openly condemned Mishin for his ongoing efforts to blame the crew. Cosmonaut Leonov tested manually closing the valve of the Soyuz simulator in the TsPK, taking 52 seconds, which was four times longer than the time available to the Soyuz 11 crew.

Although further explanations were given in Chertok's memoirs, colleagues of the Soyuz 11 crew – in particular Yeliseyev, Kubasov, Shatalov and Leonov, together with Viktor Patsayev's wife Vera – contributed the most to a full understanding the tragedy.

Aleksey Yeliseyev's insight into the valves leads to the inevitable conclusion that the tragedy ought never to have occurred!

As noted, each valve had both an automatic and a manual shutter. However, when the designers devised the valve no one considered the possibility that the automatic shutter might open spontaneously. In accordance with instructions, prior to launch both shutters (automatic and manual) on valve No. 1 were required to be closed – in the mode 'closed-closed'. On the other hand, on valve No. 2 the automatic shutter was to be closed and the manual shutter open – in mode 'closed-open'. What does this mean? During the descent, four pyrotechnic charges were to open the automatic shutters on both valves at an altitude of 5 km. However, because the manual shutter on valve No. 1 was closed, air would flow into the cabin only through valve No. 2, whose manual shutter was already open. As noted, the reason for there being a pair of valves was to ensure that in the event of a splashdown in which water leaked into the module through valve No. 2, there was another valve on the opposite side which would allow in air – the research cosmonaut seated near valve No. 2 would close its manual shutter to halt water penetration while the commander opened the manually operated shutter on the valve positioned directly above his couch. However, during the preparation of the ship there was a mysterious change to the procedure! Instead of being 'closed-closed', valve No. 1 was actually set 'closed-open'; and instead of 'closed-open', valve No. 2 was set 'closed-closed'. As the valves were identical, the technicians did not pay special attention to this change.

---

[4] Kamanin has interpreted Mishin's remark about a cosmonaut stemming the air leak by holding his thumb over the hole literally, and is criticising the expectation that this could have been sustained as the deceleration loads increased and forced the crewman back into his couch. In fact, if all that was intended was to buy time to close the manual shutter in the valve, then this criticism of the idea does not apply.

Top: Cosmonaut Lazaryev works on the hatch inside the Soyuz simulator, with one of two valves installed in the vicinity. Three bottom photos show the opening of one of the valves (left), the control for the manually operated shutter and the ventilation switches.

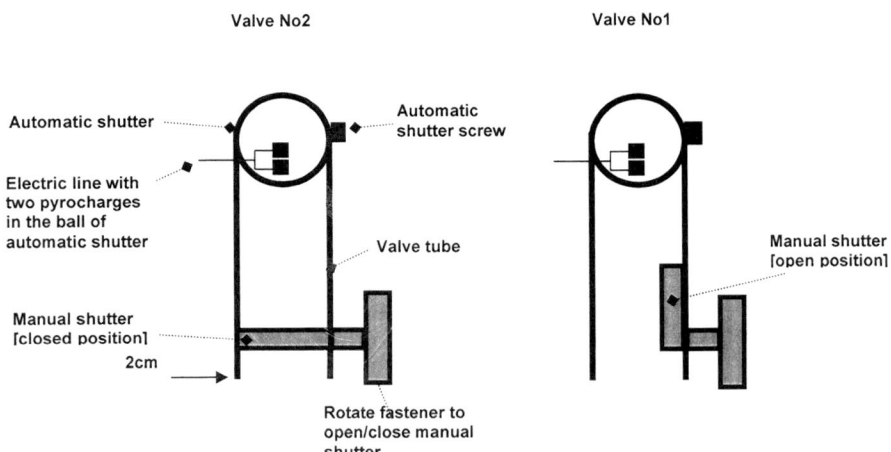

A simplified depiction of the operational structure of the two ventilation valves.

As the explosive bolts were fired to separate the orbital module, the shock caused the automatic shutter on valve No. 1 to open. This valve was positioned very near to two of the bolts, and thus was exposed to the greatest stress by the explosive action. Since the manually operated shutter had been left open, air was able to leak to space. A detailed analysis of the telemetry recorded by the 'black box' established that the automatic control system had fired the attitude control thrusters to counter the force of the air venting at speed through this valve. After inspecting the seal of the hatch, the cosmonauts quickly realised that the automatic shutter had inadvertently opened in one of the valves. Knowing that both shutters on valve No. 1 were *supposed* to be closed, they directed their attention to valve No. 2, which they believed had been set to 'closed-open' and was 'open-open' as a result of the shock of firing the bolts, but in fact it was still hermetically closed. Patsayev's effort to close the manual shutter of valve No. 2 was foiled by the fact that it had already been closed prior to launch! Realising that valve No. 2 was closed, Patsayev or Dobrovolskiy set about closing the manual shutter of valve No. 1, but managed only to partially do so before losing consciousness.

According to Yeliseyev the cosmonauts forgot, or in panic missed the fact that the order of the valves had been changed! He said: "If they had just remembered this! If even they did not recall, but they had begun to close both valves just in case, they would have saved themselves." The revision to the manually operated shutters was also noticed by Kubasov, who, in addition, noted another important detail: "At the cosmodrome, according to instructions, the manually operated shutter on one of the two valves is open and on the other is closed. This is specified in both the onboard documentation and the documentation of the manufacturer. But on Soyuz 11, ... according to the onboard documentation the valve that prematurely opened ought to have had its manually operated shutter closed, and in the documentation of the manufacturer it should have been open." Thus we see that valve No. 1 had state 'closed-closed' in the onboard documentation, and the crew did not simply forget or in a blind panic miss the order of the open/closed shutters. They firstly tried to close valve No. 2 because in their documentation its manually operated shutter was specified as being open, but in reality it was closed! As in the case of the Soyuz 1 tragedy, the technicians who prepared the spacecraft had not followed the rules!

Vladimir Shatalov, who was member of the State Commission which investigated the Soyuz 11 tragedy, reported some details of his inspections related to the lack of technical discipline in the installation of the valves:

> The most likely cause was a design fault or omission during the installation of the valves during the assembly of the spacecraft. Both valves had to be torqued to the certain level by using a special tool, even though access to the valves was problematic. ...
>
> During an inspection, it was found that for both valves the screw had not been sufficiently tightened, and the ball was free to jiggle about. When they examined the valves on already flown craft, including my Soyuz 10, it was noticed that the screws were torqued differently. The required force was 50 kg, but some of the descent modules had valves torqued at 30 kg, some at only 20 kg, and one had valves whose screws were barely tightened! There were no spacecraft already flown in space with valves torqued to the proper degree. I could not believe this. Well, it was an accident waiting to happen!

In the book *Two Sides of the Moon* published in 2004, Aleksey Leonov states he was in the communication centre in Kaliningrad for the undocking. As the crew worked through the checklist, he advised them to close the shutters of both valves, but to remember to reopen one once the parachute had deployed.

> "Make a note of it in your logbook," I instructed them.
>
> Although this deviated from the flight regulations, I had trained for a long time for the mission they were flying, and in my opinion this was the safest procedure. According to the flight programme the vents were [to start closed] and then open automatically once the parachute had deployed after re-entry. But I believed there was a danger, if this automatic procedure was followed, that the vents might open prematurely at too high an altitude and the spacecraft [would] depressurise.
>
> It seems the crew did not follow my advice. Unfortunately, my intuition proved right. ...
>
> The loss of the Soyuz 11 cosmonauts was a terrible blow to the morale of the whole corps. Everyone understood that we were in the business of testing spacecraft, and the deaths of these three men undoubtedly saved the lives of later crews, because of the substantial modifications made, but their loss was a tragedy. Not only was I deeply saddened by what had happened, but I was frustrated, too. Had I been allowed to fly in their place I am sure my crew would have survived.

Leonov also wrote that he never told anyone that the crew had failed to follow his recommendation. Many years later, Vera Patsayeva, who worked in the TsNIIMash and had access to the radio exchanges, "recognised the crew's tragic mistake of not following my advice and made that fact public". He tried to avoid the children of the lost crew: "I could not bear to look into their eyes. Even though it was not my fault, I blamed myself for what had happened. It was not until much later that the children learnt how desperately I had tried to avert the tragedy."[5]

Leonov also noted that the cardiogram data showed that Volkov, who remained in his couch, died 80 seconds after the decompression, Patsayev after 100 seconds and Dobrovolskiy after 2 minutes. Leonov's claim that if he had been in command then his crew would not have succumbed to such a failure was contradicted by Kubasov in an interview with *Novosti kosmonavtiki*, who, after analysing what they would have done, had concluded that death was inevitable.

In contrast to Mishin, who insisted that the crew had been at fault, the strongest criticism of the cosmonauts' action ever to be made by any representative of the Air Force was Leonov's claim that they had not accepted his advice to close both valves and reopen one after the parachute had deployed. On the other hand, this advice was contrary to the rules. It would have protected against a valve opening prematurely, but to have required that a valve be opened manually would have placed the crew at risk of asphyxiation in the event of stronger than expected dynamic loads during the re-entry rendering them unconscious – it was, after all, to preclude this outcome that the valves were designed to work automatically. But it indicates that Leonov's crew had trained to perform re-entry differently to their backups. Furthermore, in training Leonov seems not to have described this alternative procedure to Dobrovolskiy.

FROM VERA PATSAYEVA'S NOTES

Until her death in 2002, Vera Patsayeva collected information on the worst tragedy in the Soviet manned space programme – which claimed the life of her husband. An expert in remotely sensing the Earth from space, she worked at the TsNIIMash, which was located alongside the TsKBEM. She was close to many designers and specialists from the TsKBEM, including Yeliseyev and Raushenbakh, and had access to secret information on the mission. Courtesy of her daughter Svetlana, we can now publish for the first time a chapter from the notes of Vera Patsayeva entitled 'Was there a chance for survival?'

> I recall that several days before the end of the flight of the Soyuz 11 crew, I queried well-known cosmonaut K.P. Feoktistov: is it possible that there will be trouble during the landing? I always thought that on a space mission the most dangerous operations are at its start and its end. If anything happens to the hatch in orbit decompression will be instantaneous, and death inevitable for the crew in the cabin without pressures suits.

---

[5] When asked about this by the author, the cosmonauts' children Marina Dobrovolskiy and Svetlana and Dmitriy Patsayev could not confirm Leonov's remark. Also, his remark about Vera Patsayeva is not recorded in her meticulous diary.

Konstantin Petrovich Feoktistov replied that the cabin is reliably protected from decompression. If the hatch is defective, the automated systems would not permit the ship to begin de-orbit. The ventilation system of the cabin can be opened to the environment automatically only during landing, when the external atmospheric pressure reaches a specific value. The cosmonauts can also open and close the shutters of the valves manually, but in space they are automatically closed. So he said there was no reason to worry. The Soyuz spacecraft had been repeatedly tested in landings and had proved itself to be reliable. Furthermore, he said that the reliability of the descent module made it unnecessary for a crew to wear pressure suits.

For a long time after the loss of my husband, I was unable to ask about the causes of the tragedy. I thought the official version of a random loss of cabin pressure explained everything.

But 15 years later I read an article by Vasiliy Pavlovich Mishin in which he said the Soyuz 11 crew had missed a chance to save themselves. They did not close the shutter of the ventilation system in time. Apparently they were unaware of how to act in this emergency. All they needed to have done was to close by hand the ventilation duct through which air was leaking to space. He laid the entire blame for the tragic outcome on the crew. In his opinion, the designers of the spacecraft and the mission planners were not at fault. The crew was lost due to their ignorance of a vitally important system of the ship, and because they were confused.

For me, it was very painful and offensive to read what the Chief Designer wrote about the tragedy. It is painful because he says that the cosmonauts had a chance to save themselves but failed to take it. And it is offensive to the cosmonauts, who are spoken of as if they are guilty for their own loss, and who have no opportunity to defend themselves.

So did they not know how to act? Or did they know, attempted to act, and were unsuccessful? For sure they realised what was happening, because they attempted to unfasten their seat belts in order to reach the source of the air leak. How much time did they have to resolve this? To achieve this was not an easy task in the active phase of the descent, when the dynamic loads pressed them into their seats. Even though the valve that was leaking air to space could be closed by hand, without pressure suits the time available to do so was not great. True, the Chief Designer asserts that it was sufficiently simply to raise a hand. ... He adds that Viktor Patsayev perished attempting to turn the valve manually, but there was insufficient time for this. However, this proves that the cosmonauts knew the cause of the emergency.

What was the chance of the cosmonauts saving themselves? ... I decided to find out. Firstly I wanted to understand in more detail what happened in the cabin on the night of 29/30 June 1971 when the de-orbit procedure began. And, in particular, I wanted to know whether the Chief Designer was correct in asserting that the tragedy could have been averted if the cosmonauts were better prepared or had not become confused. ... On the other hand, was it an inevitable accident? A random event that can strike anyone, irrespective of his experience or preparedness? Or was it a question of fate? Perhaps there was some sort of a prior warning which the cosmonauts did not know how to recognise in preparing for the flight? Or was it the case that there were flaws in the making of the spacecraft of which the Chief Designer did not wish to talk? The cosmonauts lacked the suits that could have protected them in the event of the cabin losing pressure. Was this not an error in the design of the spacecraft?

What was the reason for the crew's loss – their own errors, or the errors of the designers? What was to blame – confusion during the emergency, or the malfunction of a system that should not have failed in that manner?

So I began to question the specialists. I started with the designers because it was the Chief Designer who had put the blame on the cosmonauts. It was very complex to reach Konstantin Feoktistov by telephone and set a meeting to discuss the reason for the tragedy, but he readily accepted. Contrary to my expectation, he was very affable and agreed to answer my questions. Soon I felt free and comfortable with him. After asking permission to switch on my tape recorder, I asked what I considered to be the most important question: During the emergency, was there any realistic possibility of the cosmonauts avoiding catastrophe.

"They must, and they could! They had 30 seconds for that. It is sufficient to unfasten the seat belts and reach up to the valve of the ventilation system in order to close it by hand."

He explained to me how the valve was designed, and why its operation on this occasion caused the decompression of the descent module. It was to open automatically in the Earth's atmosphere, but the unforeseen had occurred. It was 'unseated' at the time that the descent module separated. Apparently, there was a manufacturing defect.[6]

Feoktistov does not accept even the slightest possibility that the designers were responsible for the loss of the crew. The decision not to use pressure suits was correct, he thinks, since it facilitated three couches rather than two. Regarding safety, it had to be ensured by the control and manual blocking of the automation using the flight engineer's command panel. Moreover, from the time of Voskhod, when cosmonauts first flew in space without pressure suits, the reliability of the descent module had proven itself by many flights.

Feoktistov also explained: "I place the moral risk above the physical risk. A cosmonaut always risks his life on a mission, because it is not possible to predict all possibilities."

Returning to my question, he repeats: "They wasted time! It was necessary to act immediately. The flight engineer was obliged to know how to act. He had 30 seconds available."

So the cosmonauts had 30 seconds, and a designer said that was sufficient to

---

[6] Here Feoktistov told Vera Patsayeva of the defect noted by Shatalov. The automatic shutter took the form of a ball fixed in its 'nest' by a screw, but the screw on valve No. 1 was not fastened properly and the shock of the pyrotechnics unseated the ball from the nest.

access the valve's manually operated shutter and close this to halt the leak of air from the cabin.

The fact is that the ventilation system has two openings on opposite sides of the descent module. These are pyrotechnic valves.[7] They remain closed in space, and open automatically during the descent through the atmosphere. In addition to the pyrotechnic valves, each opening has a manual shutoff and a fan. One fan directs air into the cabin and the second draws it off. In orbit, when the modules of the ship are connected, both openings are protected by the frame of the orbital module. The operation of either of the pyrotechnic valves opens a passage to space which is about two centimetres in diameter. After midnight on 30 June, when the orbital module was jettisoned, one of the valves inadvertently opened.

The Chief Designer says that the cosmonauts heard the air whistling, and that Patsayev unfastened and reached up to halt the leak, but did not succeed. On the other hand, he said that the tragedy could have been avoided if only the crew had recalled in time the existence of the manually operated shutter. As noted, prior to initiating the descent the crew was to have confirmed that the manually operated shutters in the valves were set according to the flight instruction. Mishin states that it is unknown whether the cosmonauts did not know that they were to do this, or whether they simply forgot to do it.[8]

Did the cosmonauts forget, or did they not know? In any case, there is the flight instruction in which all of the actions of the crew prior to de-orbit are specified in detail. In addition, there is the Control Group in mission control. It would have ensured that the crew followed the flight instruction without missing out any steps.

In the log book of Soyuz 11, which is stored in NPO Energiya, is a page that lists the actions prior to the descent. In the instruction, it states that both at the cosmodrome prior to launch and before de-orbit the crew must verify the settings of the manually operated shutters in the valves of the ventilation system.

The records of the radio communications between the cosmonauts and the operators of the ground-based services at the cosmodrome confirm that they made this test prior to launch.

Arkadiy Ilyich Ostashev, the tester at the ground-based complex, said that in checking the Soyuz 11 spacecraft at the TsKBEM and in preparations at Baykonur a discrepancy was noted between the onboard documentation and that of the manufacturer about the ventilation valves. According to Ostashev, the operator instructed that the settings of the manual shutters of the valves be altered. As a result, the inscription 'Closed' meant that the shutter was open!

---

[7] As explained earlier, there was a small pyrotechnic charge in each valve to release the ball from its nest. Both valves were on the same electric circuit.

[8] As may be inferred from Mishin's remarks, he made contradictory accounts in interviews given many years later. The fact that Leonov says he discussed the valves with the crew proves that they were aware of their settings, because they decided to use the settings which were specified in their onboard instruction. The thrust of Mishin's argument was that he wished to place the blame on the crew's training (which was the responsibility of Kamanin) rather than on the design of the craft by his bureau.

When the automatic shutter on a valve became unseated, a decompression was therefore inevitable.

In an attempt to explain why and how this error occurred in the onboard documentation, I turned to Viktor Petrovich Varshavskiy, who was our great authority for onboard documentation. During the flight of the Salyut station, he was the leader of the group that made the instructions for the cosmonauts, and oversaw their operation in orbit.

"When the first orbital station was launched, the mission control centre was not as it is today. There was no computer to process flight information. Communication with the cosmonauts was undertaken from Yevpatoriya. We were on duty 24 hours per day. Revisions into the instruction and the flight programme were made daily. Often the cosmonauts grew angry because we made so many changes and because sometimes these were inconsistent with the state of apparatus. In terms of organisation, at that time we weren't ready to the operate such a station. The first design imperfections were revealed by Soyuz 10, when a problem in its docking mechanism prevented the docking with the station. That crew was obliged to return to Earth. We were allowed just a month to make the modifications to the docking system. Soyuz 11 was urgently transported to the cosmodrome to meet a schedule for launching to the station. There was very little time to draw up the new instruction and to compare it with the manufacturer's documentation for that particular vehicle. That is how the divergence arose. And unfortunately when the cosmonauts began their preparations for the de-orbit manoeuvre, our 'controllers' forgot to remind them of the corrections to the onboard instruction."

At the moment of separating the modules after the de-orbit manoeuvre, the shock from the explosive bolts unseated the ball of the automated part of the valve above Dobrovolskiy's couch, allowing air to escape. In addition to the whistle of the air, the signal horn warned of a decompression. Immediately, a thick fog was formed in the cabin. The crew had only seconds to analyse the situation, and act. It was necessary to unfasten the seatbelts, stand up and stop the leak. Unbeknownst to the crew, however, in the case of the valve above Patsayev 'Open' meant closed, and for the valve above Dobrovolskiy 'Closed' meant the automated shutter was open. They had to act on the indications. However, the barographic data shows that the pressure fell so rapidly that after 13 seconds they were rendered ineffective.

Thirteen seconds to eternity. Was this sufficient? If the situation is regular (as cosmonauts say) then it is a lot. But it is far less if the situation is out of control – as it was to men exposed to the vacuum of space without pressure suits.

Gennadiy Fyodorovich Isayev, who for many years observed and analysed the actions of cosmonauts in space, answered my question very emotionally.

"It is not possible to raise the question in that way! What is enough time, and what is insufficient time? There were only a few seconds available! This was, as we say, a non-standard situation. There was no standard solution. A cosmonaut performs his work in accordance with instruction, and such work is called regular. He cannot train for uncertainty. In a non-standard situation time is required to evaluate the situation and devise a solution. This time can be completely different to the one calculated by training instructors on Earth. The tragedy of Soyuz 11 clearly shows that there were flaws in the project. The cosmonauts could not control the automation in the most important system protecting their lives, so they could not immediately counter the malfunction. Starting with the first Voskhod flight, the spacecraft designers deprived the cosmonauts of recovery facilities in the case of a cabin decompression. It is the inevitable result of the generally inadequate relationship of society to the man. It is the philosophy of the totalitarian system in which we lived."

In 1971 Skella Aleksandrovna Bugrova worked for the Control Group in Yevpatoriya. She recalls that the 'sliding' circadian rhythm imposed on the first Salyut crew had a serious effect on their health and relationships with the people on Earth. With added fatigue and the influence of weightlessness, they became spiritually and physically overloaded. "And we on Earth could not manage to analyse the flight information. As a result, we were slow to respond to the questions and observations of the crew, which irritated them. In our memories, even today, are the last conversations with them before the descent. If we had not rushed the crew to prepare for the descent, if only the flight director had had the courage to postpone the undocking from Salyut in order to fully investigate the absence of the signal from the hatch, and then reviewed once again the status of the life support system, then maybe the tragedy would have been avoided."

PEOPLE AND OMISSIONS

A characteristic of the development and operation of the first Salyut space station was continuous work under time pressure. It is true that the TsKBEM's designers, disappointed by the failure of the lunar programmes, worked with great enthusiasm because DOS was new and of major significance to the prestige of the Soviet space programme. However, the deadlines were simply unrealistic. Although the Kremlin told the team not to hurry, they were well aware that Moscow wished the station to be launched and occupied as soon as possible.

The list of achievements during only 16 months is almost unbelievable:

- The station was designed, constructed, tested, modified and launched into space.
- All of the station's equipment, including the sophisticated apparatus for the scientific programme, was developed, tested and installed.
- The entire flight programme was prepared, including the extremely complex mission control and data processing.
- Two Soyuz spacecraft were fitted with the new docking system for internal transfer to the station, and after the Soyuz 10 failure this was modified for Soyuz 11.
- One Proton and two Soyuz launch vehicles were constructed.

- Three crews (a total of nine cosmonauts) were trained in how to operate the station, and underwent numerous reassignments.

But owing to the unrelenting pressure of time, the designers and managers of the Soyuz 11 mission made numerous omissions and errors, repeating the same basic mistakes which led to the loss of Vladimir Komarov on Soyuz 1 in 1967. After only limited testing, and without attaining the standards required for a successful mission, they launched an inadequately prepared spacecraft with a crew who had expected to train for longer before flying, supported by a control team which was technically and operationally ill-prepared for such a complex mission. In assessing the tragic loss of the Soyuz 11 crew, it is necessary to recognise that it resulted both from technical factors such as the design of the spacecraft and its operating regime, and from the omissions and errors of people right across the programme – politicians, generals, managers, designers, controllers and cosmonauts.

Before closing this gloomy chapter, it is worth summarising all the factors that are known to have contributed in some way to the tragedy.

Ventilation valves:
- **Valve screw** – The screws on the ventilation valves of the descent module had been insufficiently torqued. The automatic shutter used a ball which was held in its nest by the screw. But the screw on No. 1 valve was not fastened properly, and when the pyrotechnics fired to jettison the orbital module the ball was unseated from its nest. Shatalov, who wrote of the discovery of this problem, did not specify how the screws were torqued on the valves of the Soyuz 11 spacecraft. But it is reasonable to infer from the fact that the ball was unseated on valve No. 1 that the screw was insufficiently torqued. The fact that the screw on the automatic shutter of valve No. 2 remained in its nest implies that this one was torqued to a higher force. It can therefore be concluded that if the screw on valve No. 1 had been somewhat tighter then the tragedy would not have occurred.[9]
- **Positions of the manually operated shutters** – During the preparation of the spacecraft at the cosmodrome the technicians changed the positions of the manually operated shutters on both valves, making them inconsistent with the onboard documentation. Second only to the loose screw, this is another key factor relating to the valves of Soyuz 11. It is also ironic that on the first occasion that an automatic shutter failed, the actual settings of the manually operated shutters were inconsistent with the onboard documentation. When the configuration of the valves was changed at Baykonur the technicians, as Yeliseyev has written, "did not pay special attention to this change" because the valves were identical. In fact, they seriously erred in setting the valves in accordance with the manufacturer's specifications. The manufacturer had no

---

[9] In view of the poor workmanship and the fact that there were no post-flight checks until after the Soyuz 11 accident, a decompression at this phase of the mission was an accident waiting to happen, and if it had not occurred on Soyuz 11 it may well have done so on a later mission.

knowledge of the onboard documentation. The job of the technicians was to set the hardware to the configuration required by the crew, who expected the valves to be set differently, and trained to operate them in case of landing on water. In assembling and testing the spacecraft at the cosmodrome it ought to have been understood that irrespective of how the valves were supplied, once installed in the spacecraft they had to be set according to the onboard documentation!

- **Manually closing the valves** – The construction of the valves required direct manual operation, which in turn meant that the cosmonaut had to unbuckle from his couch and stand up; the valves should have been able to be closed via the command panel.
- **Closing time** – The closing of the manually operated shutter was too time consuming. In ideal conditions it took at least 35 seconds, which, in view of the dire consequences of the failure of the automated part of the system, was much too long.
- **The location of the valves** – Valve No. 1's position above the centre couch was very close to two explosive bolts, making it susceptible to damage from the shock that was propagated through the structure by the jettisoning of the orbital module.
- **Valve malfunction** – It cannot be proved from the technical documentation, but it is possible that errors were made in manufacturing the valves.

Soyuz life support:
- **Risk assessment** – Having decided that decompression was impossible, the designers of the spacecraft did not provide an efficient means of protecting against it. The TsKBEM neglected to conduct a full risk assessment of all the factors which could lead to the loss of the crew as a result of not wearing pressure suits. This was done only after the Soyuz 11 tragedy.
- **Automation** – As on all previous Soviet spacecraft, Soyuz was designed to have the maximum of automation. For example, all landing operations were fully automated and did not require the involvement of the crew. On the one hand this enabled the craft to be flown unmanned; on the other, this made it difficult for a crew to intervene – as was demonstrated when Soyuz 10 was unable to dock with Salyut, and the resulting decision to modify that system for Soyuz 11 to give that crew a degree of control over the docking process. However, as Soyuz 11 demonstrated, the great mistake was to minimise the role of the crew in operating the most important of the vehicle's systems – the life support system.
- **Openings on the descent module** – The descent module had three openings of critical importance for its hermetic seal. Two valves with tubes, each with a diameter of about 2 cm, and the hatch at the top of the module which had a diameter of 60 cm.
- **Pressure suits** – There were no pressure suits to protect the cosmonauts in the event of a decompression.
- **Oxygen masks** – The crew were not even given simple oxygen masks of the

type that deploy automatically if the pressure drops in a commercial airliner. The spacecraft had a loss-of-pressure alarm. If this had deployed masks, the cosmonauts may have been able to function for several minutes after a rapid decompression, which would have been more that sufficient time to shut the leaking valve.

- **Air decompression tanks** – Due to its severely limited volume, the descent module did not have its own air tanks, and could not have replenished the cabin in the event of decompression. But on the other hand, it was believed that the possibility of decompression had been designed out.

- **Inspection of the valves on the previous Soyuz descent modules** – It was not the practice of the specialists at the TsKBEM to inspect the state of the valves on a descent module after its mission. If they had done so they would have noted the varying degrees to which the screws of the valves were being torqued during assembly.

- **Explosive bolts** – The explosive bolts were installed on the connecting ring that incorporated the hatch, which was another part of the descent module which was of critical importance for the crew safety. Westerners speculated that the twelve bolts were supposed to have fired in sequence, but for some reason went off simultaneously, thereby generating an intense shock which forced open the valve. However, the bolts were on the same electric circuit and were meant to fire simultaneously. The shock from their detonation was the same as on previous missions, but the valve was not. Based on all of the available sources related to the technical factors which caused the premature opening of the valve, it is possible to conclude that the main technical cause was its inappropriate assembly, possibly aggravated by a manufacturing fault.

Mission Control:
- **Organisation** – The organisation of the Soyuz 11 mission was one of the weakest links in the chain of factors leading to the tragedy. The spacecraft was modified and tested much too hastily. The programme for the mission and the organisation of the crew's activities were also developed in a hurry, and without full consideration of the implications of a prolonged exposure to weightlessness. For Mishin, who was antagonistic to the DOS programme, the main event in June 1971 was the third launch of the N1 lunar rocket. It was to oversee this that immediately after the docking of Soyuz 11 with the Salyut station he reassigned the flight directors. Kamanin and Chertok had only praise for Yeliseyev, but his nomination as the new flight director for the Soyuz 11 mission whilst it was underway was strange. In addition, the difficulty in coordinating the tracking ships in the final phase of the flight is further evidence that there were gaps in the organisation.

- **The technical documentation** – After the return of Soyuz 10, the revisions to the docking system were made within a month. Due to the tight schedule, Soyuz 11 was launched with onboard documentation and instruction which was inconsistent with the true situation – in particular, the manual shutters of the ventilation valves: No. 1 was 'closed-open' instead of 'closed-closed', and No. 2 was 'closed-closed' instead of 'closed-open'. Consequently, when the cosmonauts realised that a valve was leaking and went by the onboard documentation they wasted valuable time trying to close a valve which was already closed, while one that they thought was closed was actually leaking.

- **Carelessness** – The controllers at the TsUP knew the valves in the descent module were not as specified by the onboard documentation, but appear to have forgotten to inform the cosmonauts during the preparations to return to Earth.
- **Inspections** – There were gaps in the organisation of the inspection of the spacecraft's systems before undocking from Salyut. This was mainly left to the crew, who were exhausted after the longest space mission in history. In addition, the coordination between the TsUP and spacecraft was aggravated by the briefness of the periods of radio communications.
- **Hastiness** – When a problem developed with the hatch during preparations to undock from Salyut, the TsUP failed to halt the proceedings. Instead of pausing to investigate the problem, the controllers improvised to circumvent the issue with a strip of insulating tape! The problem with the hatch was the warning bell that no one heard. The flight director should have intervened to review with his controllers the status of the life support system, and had the crew repeat the setup of the vital life support elements of the descent module – the ventilation valves as well as the hatch. With the spacecraft docked at the station, time was on their side.

Training:
- **Crew teamwork** – With a new commander assigned less than four months prior to the mission, Dobrovolskiy, Volkov and Patsayev were members of the third crew until the launch of Soyuz 10 in late April 1971. They had not trained as intensively as the first two crews. Yevgeniy Bashkin, who was an instructor, says that this crew was not given the same attention as the others, since no one expected them to fly to DOS-1. Cosmonaut Gorbatko even said that the majority of the TsPK staff did not recall seeing Patsayev during all his months of training! In addition, after the failure of Soyuz 10 to dock, the crew which expected to fly Soyuz 11 – Leonov, Kubasov and Kolodin – lost about one month of their training time when the launch was advanced from the middle of July to early June. They had only one month to train with the revised docking procedure, and the TsPK staff concentrated on this activity. Dobrovolskiy, Volkov and Patsayev, who became backups after the failure of Soyuz 10, trained in the shadow of the prime crew. But a few days prior to the launch they found themselves assigned the flight.
- **Decompression training** – In the training programme there was no basic cosmonaut decompression training.
- **Differences in training procedures** – The prime and backup crews for the Soyuz 11 mission trained using different re-entry procedures! Contrary to the training regulations, Leonov trained with the manually operated shutters of

both ventilation valves closed during the descent. Dobrovolskiy trained according the rules, with one manually operated shutter open and the other closed.

Cosmonauts:

- **Wrong valve** – Owing to the error in the onboard documentation and the fact that the flight controllers neglected to warn them otherwise, the crew of Soyuz 11 undocked from Salyut believing that the manual shutters of the ventilation valves were set in the opposite sense to that which was the case, and when they realised that a valve was leaking they directed their attention to the wrong one.
- **Dobrovolskiy and Patsayev** – As the two valves were located above their seats, Dobrovolskiy and Patsayev were involved in the emergency action.
- **Volkov** – Feoktistov has argued that Volkov, being the flight engineer, was responsible for the onboard systems, and that he failed to conduct a proper inspection of the valves – he could have detected the difference between the onboard documentation and the actual settings of the valves. An important question is what the cosmonauts should have done if Volkov had noted that the valves were set differently? Should he have closed No. 1 and opened No. 2 according the onboard documentation? It would have been logical to tell the TsUP about the difference, and let the flight director decide on the action. Would the fight director order Volkov to leave the valves as they were, or to adjust them to match the onboard documentation? Here is one of the key points concerning the actions of the crew. During his checks, Volkov ought to have discovered that the valves were set differently to the specification in the onboard documentation, but he didn't. If he had, and had reset the state of the manually operated shutters to the onboard documentation, with No. 1 closed and No. 2 open, then when the automatic shutter on No. 1 became unseated there would not have been a decompression.
- **Leonov's advice** – Dobrovolskiy did not accept Leonov's intuitive advice in preparing for the descent. If the crew had disregarded what was specified in their instructions and had closed the manually operated shutters in both of the valves, when the shock wave from the explosive bolts firing opened the automated shutter of valve No. 1 this could not have led to a decompression. (It is ironic that if this had been done, the lax post-flight inspection routine would probably never have revealed just how close to disaster the Soyuz 11 crew had come!)
- **A finger on the valve** – Could the Soyuz 11 crew have done more to save their lives? Might Dobrovolskiy or Patsayev have been able to stem the air leak by placing a finger over the valve inlet whose aperture was no larger than a coin. Although Kamanin and the medics said no, Mishin persistently claimed that this could have been done! Could a cosmonaut survive with a part of his skin in direct contact with space? NASA has had one experience of a suit puncture. During a spacewalk on Shuttle mission STS-37 the palm restraint in an astronaut's glove came loose and migrated until it punched a 1/8-inch hole in

the pressure bladder between his thumb and forefinger. He did not realise that his suit had developed a puncture until after he was back inside the spacecraft and discovered a painful red mark on his hand. There had not been a decompression because when the metal bar holed the glove his hand spanned the opening, he bled into space, and the coagulating blood sealed the opening and served to 'glue' the bar in the hole, sealing it again.

- **Slower reaction** – The fatigue and disorientation of the Soyuz 11 crew after 24 days in space, together with the inadequate organisation of the flight, the tensions with the TsUP, the anxiety of the fire, and the difficulty closing the hatch all probably served to slow the reaction time of the cosmonauts during the rapid decompression.

Taken together, all of these factors led – directly and indirectly – to the deaths of the Soyuz 11 cosmonauts.

Although it is frowned upon to discuss the offenders, even although most of them are now dead, this remains a fundamental question. The people who knew that their actions or inactions contributed to the deaths of the cosmonauts had to live with this knowledge. Contrary to expectation, the Kremlin did not issue severe punishments, perhaps because the principal offenders included people from outside the TsKBEM who, against Kamanin's protests, supported the decision to eliminate pressure suits. The most senior manager punished was Pavel Tsybin, a Deputy Chief Designer who worked for Feoktistov on the development of the transport version of the Soyuz. As after the Soyuz 1 disaster, therefore, the principal blame was assigned to a man who had no *direct* responsibility for the root cause of the problem. The decision for the crew of a Soyuz ship to fly without pressure suits was made many years earlier. The motivation to have a crew of three was probably to match the Apollo spacecraft that was being developed by the Americans. The concept of the spacecraft was for three modules, with the descent module in the middle. The small size of the capsule required the cosmonauts to fly without pressure suits. Interestingly, when Dmitriy Kozlov, the Chief Designer of Branch No. 3 of OKB-1, revised the design of the spacecraft for use by the military, he reduced the crew to two cosmonauts wearing pressure suits. But Mishin, with the support of Afanasyev (and obviously Ustinov), stopped this project and thereby ended any chance of radically revising the basic concept of the Soyuz spacecraft. But this early work was not lost, and when Bushuyev recommended a detailed redesign of the spacecraft in the wake of the Soyuz 11 tragedy, some of Kozlov's arguments were reconsidered and accepted.

Although Soyuz was built as a cooperative project with numerous design bureaus and civilian and military structures, the TsKBEM was in charge. The TsKBEM was responsible for all technical aspects of the spacecraft, and also for the organisation and technical control of a flight. So people from the TsKBEM must be at the top of the list of offenders. First is the Chief Designer, Vasiliy Mishin. Then Konstantin Bushuyev, his deputy for manned spacecraft. Then the Soyuz design team headed by Konstantin Feoktistov. Then the managers led by Yevgeniy Shabarov, who were responsible for assembling and testing the Soyuz apparatus. They neglected to test the torque on the screws of the automated ventilation valves, and made changes to the positions of the manually operated shutters. Even if a valve had a technical malfunction, the TsKBEM was responsible for detecting this and, in coordination with the manufacturer, fixing it. On the list must also be the people who were responsible for post-flight assessment of the descent module's apparatus (which would have revealed the earlier problems with the screws in the ventilation valves); those who tested procedures and managed the technical documentation for the cosmonauts; and the mission controllers led by Yakov Tregub and his assistant Aleksey Yeliseyev.

Are there other offenders outside the TsKBEM? The decompression training that Mishin highlighted is an issue for discussion. Decompression was not included in the training, largely because the descent module was believed not to be susceptible to decompression. But the question was what three men squeezed in the cramped descent module *could* do in the way of training for such an event? What would be the standard procedure for a period as brief as 13 seconds? Were they supposed to stand up and (as Mishin said) block the valve using a finger? They clearly trained to cycle the manually operated shutters after landing in water. Perhaps they could have been trained to rapidly find the source of a decompression, and to work as a team in such a situation. But how could they have closed a shutter in a mere 13 seconds during an emergency which took at least 35 seconds to shut in ideal circumstances? There was no technical support (such as automatically deployed oxygen masks) to enable a crew to survive decompression long enough to close the valve, and even if they managed this there was no reserve air tank to replenish the cabin! Training was the responsibility of Generals Kamanin and Kuznetsov and, as the main critic of the decision to dispense with pressure suits, perhaps Kamanin should have been more insistent that measures be taken to enable a crew to survive a decompression.

But the most significant omission in the training of the cosmonauts was related to the safest management of the ventilation valves. If both valves had been operated in the 'closed-closed' mode, the Soyuz 11 crew would not have died! Contrary to the specification in the training and flight instruction, Leonov's crew were trained with this mode. The fact that Dobrovolskiy's crew were trained to operate differently is not their omission, it is that of the trainers at the TsPK.

In addition to all these technical and managerial problems related to the Soyuz 11 mission, there is still the question of whether the cosmonauts could have done more to survive. Their actions can be analysed in relation to operations before and during re-entry/decompression. What could they have done before re-entry? While docked with Salyut, the crew performed a detailed inspection of the spacecraft's systems by referring to the onboard technical documentation. Here is the first point where the coordination between the TsUP and the crew faltered. The controllers failed to point out that the onboard documentation was incorrect. The crew inspected the valves, but failed to realise the differences between their true status and that in the onboard documentation. As noted, if they had reset the valves to match their documentation, there would have been no decompression. During the checking procedure prior to undocking from the station, Leonov advised that they close both valves and reopen one once the parachute had deployed, but the crew followed their flight instruction. From their point of view, this was reasonable – it was the way that they had trained.

During the decompression the cosmonauts attempted such actions as they could in an effort to halt the leak, but (in view of all the technical limitations in the descent module listed above) they did not have a realistic chance. First, they lost valuable seconds inspecting the hatch seal. Upon realising that the air was leaking from one of the ventilation valves, they attempted to close the manually operated shutter of valve No. 2, which they believed was open but in reality was closed. In desperation, they turned to valve No. 1, which was supposed to be closed, and found that it was actually open! Yeliseyev said they should have worked as a team, and closed both of the valves simultaneously. But even if they had not wasted time on the hatch, and Dobrovolskiy and Patsayev had each moved to close one manually operated shutter, the 13 seconds available before they were rendered ineffective was insufficient to have completed the task – without prior planning and some technical assistance, they had stood no chance of saving themselves.

Specific references
 1. Kamanin, N.P., *Hidden Space, Book 4*. Novosti kosmonavtiki, 2001, pp. 338–340 (in Russian).
 2. Chertok, B.Y., *Rockets and People – The Moon Race, Book 4*. Mashinostrenie, Moscow, 2002, pp. 341–348 (in Russian).
 3. Salahutdinov, G.M., 'Once More about Space'. Aganyok, No. 34, 1990 (Interview with Vasiliy Mishin).
 4. Tarasov, A., 'Missions in dreams and Reality'. *Pravda*, 20 October 1989 (Interview with Vasiliy Mishin).
 5. Yeliseyev, A.S., *Life – A Drop in the Sea*. ID Aviatsiya and kosmonavtika, Moscow, 1998, p. 82 (in Russian).
 6. Novosti kosmonavtiki (in Russian)
    No. 4, 2002 (Interview with Vladimir Shatalov)
    No. 3, 2005 (Interview with Valeriy Kubasov)
 7. Scott, David and Leonov, Alexei, *Two Sides of the Moon – Our Story of the Cold War Space Race*. Simon & Schuster, 2004, pp. 263–265.
 8. Afanasyev, I.B., Baturin, Y.M. and Belozerskiy, A.G., *The World Manned Cosmonautics*. RTSoft, Moscow, 2005, p. 230 (in Russian).
 9. 'Cosmonauts died because valve was forced open', *The Washington Post*, 29 October 1973.
10. Email from Svetlana Patsayeva with materials from Vera Patsayeva, 1 August 2007.
11. Email from David M. Harland, 13 November 2006, 'Soyuz-11 and the silence of the cosmonauts'.
12. *Raketno-Kosmicheskaya Korporatsiya ENERGIYA imeni S. P. Koroleva (RKK Energiya: The aerospace corporation named after S.P. Korolev) 1946–1996* (in Russian).

14

The fall of the Chief Designer

SALYUT'S LAST DAYS

The tragedy that befell the Soyuz 11 crew had not only dramatic effects on the plans for further use of the world's first space station, but also the entire Soviet manned space programme. On 9 July 1971, while the investigation of the accident was underway, the State Commission decided to halt preparations for the next flight to Salyut. This was despite Leonov's assurance that his crew was ready for a 1-month mission. But after such a terrible tragedy, no one wished to take the risk. Salyut was in very good condition, continuing to orbit in its automated regime. It executed two manoeuvres: on 19 August raising its orbit to 290×308 km and on 25 September lowering it to 224×262 km. The controllers at the TsUP in Yevpatoriya continued to monitor its systems. However, when it became clear that there was no prospect of revisiting the station, it was de-orbited on 11 October 1971 by lowering its orbit so that it would enter the atmosphere over the South Pacific, where it burned up. It had been in orbit for 175 days.

LOST AT LAUNCH

Meanwhile, the TsKBEM engineers were hard at work developing modifications to the Soyuz to eliminate the weaknesses revealed by the investigation conducted by the State Commission of Academician Mstislav Keldysh. At the recommendation of Konstantin Bushuyev, the issue of pressure suits was reconsidered, and it was duly decided that henceforth cosmonauts should wear them for launch and the return to Earth, even though this would mean reducing the number of couches to two in order to accommodate the system that would automatically pump air into the cabin in the event of a decompression. In fact, this oxygen supply system was designed in such a way that the crew would be able to survive decompression even if they were not protected. Gay Severin quickly adapted the Sokol ('Falcon') stratospheric pressure suit, designating the cosmonaut version the Sokol-K.[1] As regards the problematic valve, this was modified in such a manner that a premature opening would cause it to reclose automatically. Once the list of revisions was agreed, it was found that the Soyuz would be overweight for its launch vehicle – at 6.8 tonnes it would be some 100 kg heavier than its predecessor. Something had to go. Because the spacecraft was to be used to ferry crews to and from space stations, and hence would require an endurance in independent flight of only two or three days, it was decided to discard the heavy solar panels in favour of chemical storage batteries. The revisions were completed within a year. One curious fact is that although the new model was significantly different from its predecessor, the 7K-T designation was not extended by an 'M' to indicate that it was a modified version.

The spacecraft which would have flown as Soyuz 12 to deliver Leonov's crew to Salyut was launched unmanned as Cosmos 496 on 26 June 1972, and placed into an orbit with the parameters 195 × 342 km to test the modifications, returning six days later having suffered no problems.[2] This success prompted the TsKBEM managers

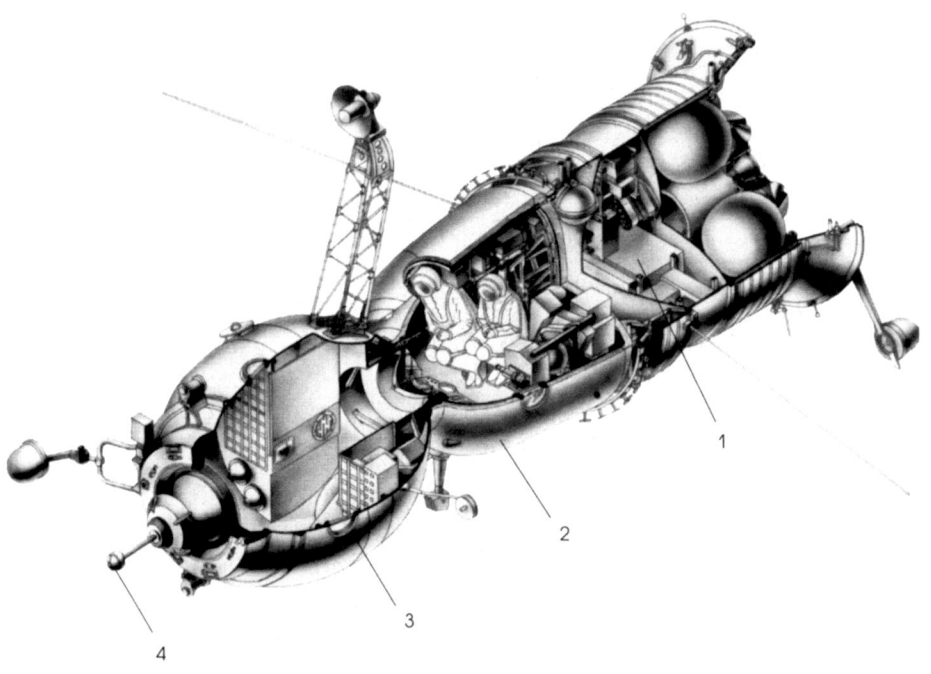

The revised Soyuz spacecraft: 1, the propulsion module without solar panels; 2, the descent module for two cosmonauts wearing pressure suits; 3, the orbital module; and 4, the active docking mechanism.

[1] In Sokol-K, the 'K' was for 'космос', the Russian word for 'space'.
[2] This test could last six days because the unmanned spacecraft placed a lower load on its batteries.

to push ahead with the manned programme by preparing the DOS-2 space station, which was identical to Salyut in terms of construction and apparatus for the reason that it had been the backup vehicle to its predecessor.[3]

In October 1971 four teams of two cosmonauts had been formed, two of which were to fly to DOS-2:

- Aleksey Leonov and Valeriy Kubasov
- Vasiliy Lazaryev and Oleg Makarov
- Aleksey Gubaryev and Georgiy Grechko
- Pyotr Klimuk and Vitaliy Sevastyanov.

Initially, Rukavishnikov was considered for the first crew, but when it was certain that Kubasov did not have tuberculosis Leonov succeeded in having him appointed

The 'first crew' for the DOS-2 station: Leonov (left) and Kubasov wearing the new Sokol-K pressure suit.

[3] DOS-2 was DOS-7K No. 2, 17K No. 122.

as his engineer. Gubaryev and Sevastyanov were carried over from DOS-1. But the fact that the new crews had an Air Force cosmonaut as commander and a TsKBEM cosmonaut as flight engineer meant that the military cosmonauts who had trained to serve as engineers for DOS-1 were dropped, which was bad news for Pyotr Kolodin and Anatoliy Voronov.[4] Lazaryev, Makarov, Grechko and Klimuk were transferred from the Contact programme, which had been terminated some time earlier, to train for DOS crews.

DOS-2 in the Assembly-Test Building at Baykonur. The insert shows the station in its shroud, ready for mating with its Proton launch vehicle. The name 'Salyut 2' is written on the side of the station's main compartment.

---

[4] Although recruited as military cosmonauts, the fact that Kolodin and Voronov were not military pilots meant that they were unlikely to be assigned as spacecraft commanders.

During the first half of 1972 two of the new spacecraft were built and sent to Baykonur along with DOS-2. All the necessary preparations were concluded by the end of July. With the four crews in attendance, a Proton lifted off at 6.21 a.m. on 29 July 1972 with DOS-2, but 182 seconds later an engine on the second stage failed and the vehicle fell to Earth. If the station had made it into orbit it would have been named Salyut 2, which was the name written on its side. However, since the flight failed at such an early point, the launch was never declared, with the result that for many years the existence of this station was kept secret.

After the loss of DOS-2 the first three crews were reassigned to an autonomous mission to be flown in August-September 1972 using one of the Soyuz spacecraft, but when this was cancelled all four crews from DOS-2 began to train to operate the DOS-3 space station, the construction of which was underway. However, there was a parallel development in progress.

THE ALMAZ-1 DRAMA

The loss of DOS-2 was a blow to the Kremlin, which wished to have another Soviet station in orbit before the Americans launched their Skylab in May 1973. But there was still hope, because there was another project – the military Almaz (OPS). Could this be prepared in time? If so, then by naming it Salyut 2 the impression could be given that this was an improved form of the DOS design, and thus hide its military role. Vladimir Chelomey objected to having his station bear the name of Salyut, but accepted that it was important that the world did not realise that there was a military space station programme.[5] When ordered to proceed, the engineers at the TsKBM and the Khrunichev factory worked around the clock to prepare the first Almaz (No. 101-1). The sudden sense of urgency came as a welcome relief to all concerned.[6]

In September 1972, when OPS-1 was undergoing its final checks at the TsKBM, four two-man crews were selected from the 28 military cosmonauts assigned to this programme:

- Pavel Popovich and Yuriy Artyukhin
- Boris Volynov and Vitaliy Zholobov
- Gennadiy Sarafanov and Lev Dyomin
- Vyacheslav Zudov and Valeriy Rozhdestvenskiy

In contrast to the DOS crews, which combined military and civilian cosmonauts, only Air Force officers were to fly to OPS-1.

---

[5] The irony, of course, was that Salyut was a civilian development of Almaz, and as Ustinov had realised early on, launching a scientific station first would serve as a *maskirovka* to hide the real project.

[6] Rodion Malinovskiy and Andrey Grechko (Ministers of Defence from 1957 to 1967 and 1967 to 1976 respectively) and Marshals Konstantin Vershinin and Pavel Kutakhov (Commanders in Chief of the Air Force from 1957 to 1969 and 1969 to 1984) had persistently urged that that the construction of the first Almaz station be accelerated.

After the station was delivered to the cosmodrome in January 1973 the TsKBM's engineers braved the extremely cold weather to make the final checks of its systems. Meanwhile, Chelomey attended the final crew training at the TsPK in February, and then all four crews flew to Baykonur for the launch.

Ten days after the station was launched, Popovich and Artyukhin were to follow in Soyuz 12. The docking was scheduled for the next day, on the station's 160th orbit. But with the Proton standing on the pad and loaded with propellant, it was announced that owing to a technical problem the launch of Soyuz 12 would not be possible until the start of the second week of May. It was decided to go ahead and launch the station on schedule, and OPS-1 lifted off on 3 April 1973 after three months of preparation. Eight and a half years had elapsed since the decision to start the Almaz programme! In announcing the launch, TASS named the station Salyut 2 and said that it was to continue its predecessor's programme of scientific research. Chelomey cleverly ordered that the name Salyut 2 be written on the ring on the third stage of the Proton rocket that supported the station in order that when the station shed this ring upon entering orbit it would fly on bearing only 'CCCP' in red on its side. The Kremlin was delighted to have successfully launched a station ahead of the Americans.

In the first phase of the flight the TsUP controllers at Yevpatoriya checked all the onboard systems, confirming that the solar panels and antennas were deployed and that the interior environment was normal. After two manoeuvres, the initial obit of 215×260 km was increased to 261×296 km. All was well when the station left the communication zone on 14 April, but when it re-entered the zone on its 193rd orbit on 15 April it was found that the main telemetry system was inoperative. When the backup system was turned on this indicated that the internal pressure had fallen and that the force of the venting air had disturbed the station in space. As the controllers watched, the station's systems failed one by one, and soon it was dead.

It was initially supposed that the air leak was caused by a problem with the supply system, which was in the propulsion compartment. This was accepted by the State Commission. The experts at Chelomey's TsKBM and the Khrunichev factory that manufactured the station initiated a detailed analysis of all the data which had been received from the station. While this investigation was underway, on 30 April the American magazine *Aviation Week & Space Technology* said that the station broke up on 14 April, and many of the fragments had since burned up in the atmosphere. At that time, preparations for a joint mission of an Apollo and a Soyuz in 1975 were at an advanced stage and Konstantin Bushuyev, the TsKBEM's technical director for this project, returned from America with tracking data for the third stage of the Proton launch vehicle and other objects which had entered orbit with the station. In the American catalogue the station was object 1973-017A. Of the 24 other objects listed, 17 had re-entered the atmosphere prior to 14 April. What of those remaining? And were there any objects that had not been detected by the American radars? The TsKBM engineers had expected only the third stage and the joint ring to reach orbit along with the station. The ring was jettisoned 774.5 seconds into the flight, and its departure was observed by a TV camera on the station. It separated cleanly, and did not break up. The fact that the station functioned perfectly until 14 April meant that it could not have been the source of so many fragments. The analysts examined the third stage. This was jettisoned 584.4 seconds into the flight and separated in such a manner that after the first orbit it ought to have been 110 km from the station, and then re-entered the atmosphere six days later. But according to the Americans it was gone after three days! Might it have exploded? Might some of the debris from this explosion have hit the station? At engine shutdown, the third stage should have held about 290 kg of propellant, and this could have caused an explosion. If a fragment of the stage had hit the station, it would have done so at a speed of about 300 m/s. Based on a model of such an explosion, a ballistic analysis verified that 21 of the objects that were tracked by the Americans could have been pieces of the third stage. It was also found that the orbits of five of these pieces intersected that of the station. In view of this analysis, the State Commission revised its conclusion and accepted that OPS-1 was crippled by being struck by a piece of debris. The fact that the station had operated perfectly prior to this suggested that its design was sound.

In hindsight, Mishin's procrastination in preparing the Soyuz which was to deliver Popovich and Artyukhin to the station precluded yet another tragedy. On the original plan, the Soyuz would have docked during the station's 160th orbit. The station was crippled between its 177th and 190th orbits, while out of the communication zone. Popovich and Artyukhin would have been on the station, and quite possibly asleep. It is evident that the station lost its integrity so rapidly that it is doubtful they would have been able to escape to the Soyuz (presuming that this was undamaged) and undock as the station broke up!

Another irony is that even although the Soviets referred to the first Almaz station as Salyut 2 and gave the impression that it was to continue the scientific work of its predecessor, Western analysts soon found that the OPS transmitted at 19.944 MHz, which was a frequency commonly used by Soviet military reconnaissance satellites. Because the name Almaz was a secret, the OPS stations became known in the West as 'military Salyuts' – which is precisely what the Kremlin had hoped to avoid!

Left: Almaz-1 with a large solar panel on each side of the passive docking cone at the rear. On the right, Almaz-1 is mated with its Proton launch vehicle. Notice the sign Salyut 2 on upper stage's support ring.

DOS-3: AN IMPROVED STATION

In sounds strange, but between December 1972 and April 1973 two stations were simultaneously in preparation at Baykonur, which was fairly buzzing with activity. One was OPS-1 for the military and the other was DOS-3 for the Soviet Academy of Sciences. The relationship between the TsKBM and the TsKBEM was strained by competition for access to the altitude chamber and other service/test facilities. Mishin's engineers had also to prepare the Soyuz that was to deliver the first Almaz crew. And Chelomey's people were also preparing two Protons: one for OPS-1, the other for DOS-3. All this activity followed the fiasco of the fourth N1 lunar rocket on 22 November 1972, which exploded after 107 seconds, a few seconds before the first stage was to have shut down and been jettisoned. The Kremlin finally accepted what had long been evident to many at the TsKBEM – the N1, and indeed the entire N1-L3 programme, was so complex that to perfect it would take much more money, resources and time than anybody had ever expected. Following the final American manned lunar landing in December 1972 the Kremlin turned its back on the Moon, preferring instead to pursue manned stations in low Earth orbit. Having lost OPS-1 precisely one month before the Americans were due to launch Skylab, the Kremlin demanded that every effort be made to launch DOS-3 ahead of its rival.

Even as DOS-1 and its backup DOS-2 were being built, the TsKBEM's engineers were designing an improved station. Two identical vehicles were built: DOS-3 and DOS-4.[7] The testing of DOS-3 was completed at the TsKBEM in the second half of 1972, and it was delivered to Baykonur in December.

One of the limitations of the first two DOS stations was the power supply. DOS-1 had two pairs of solar arrays, one pair at the front and the other pair at the rear, and they were in a fixed alignment. To provide the maximum power output, the station had to maintain an orientation in which its arrays were illuminated by sunlight. But having to manoeuvre in such a way consumed precious fuel. And holding this solar-inertial orientation made it difficult to make astronomical or terrestrial observations. The improved DOS-3 design had three much larger solar arrays, all mounted on the narrower section of the main compartment, and which could rotate to face the Sun while the station was oriented optimally to perform specific observations. The total collection area was 60 square metres and the power was 4 kW, which was double that available to DOS-1. It is interesting that these solar arrays were borrowed from the TKS spacecraft that Chelomey had intended to use to supply his Almaz military station. But to compensate for the mass of these large solar arrays the fuel capacity of DOS-3 had to be reduced, which in turn required that the operating altitude had to be increased to about 350 km – recall that the higher a spacecraft's orbit, the less it is required to fire its engine in order to sustain that altitude. There were also some changes in the propulsion compartment. Other additions were the Delta navigation system and Kaskada (Cascade), which was a new and more economical system for controlling the station's orientation in space. For the first time, the water supply for

[7] DOS-3 was 17K No. 123 and DOS-4 was 17K No. 124.

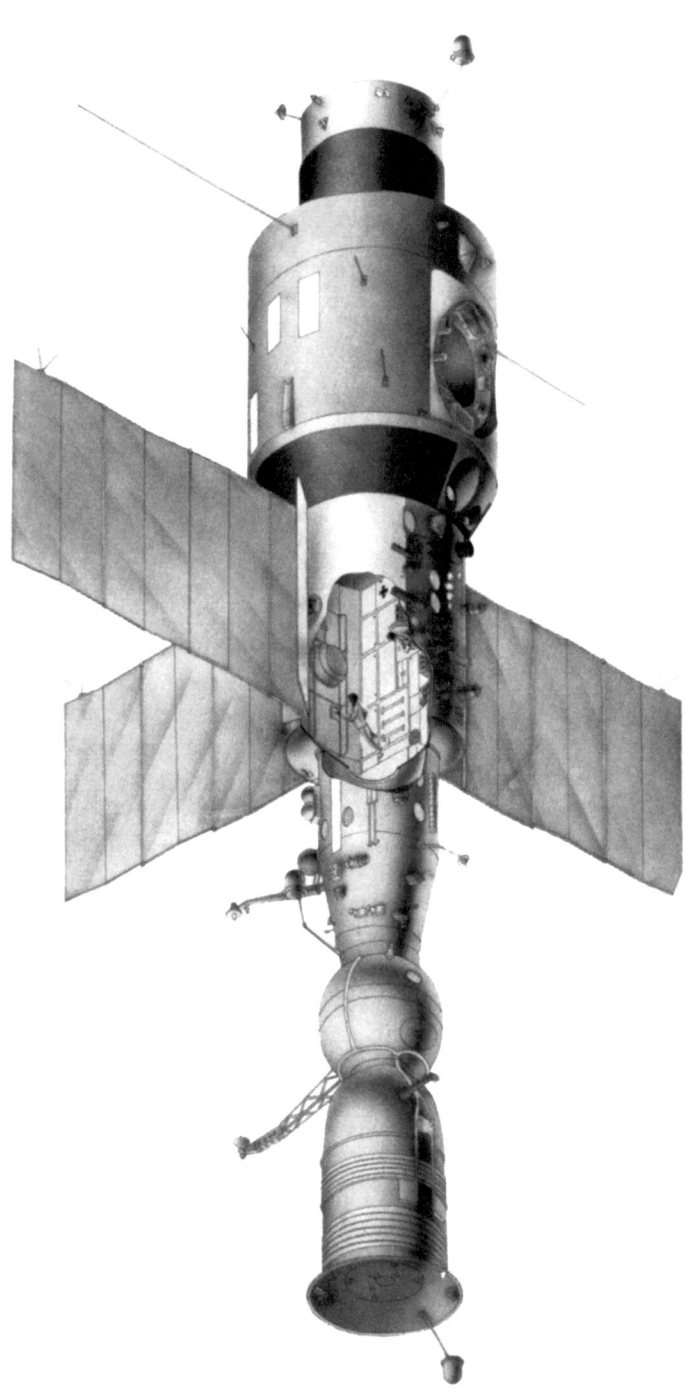

The improved DOS-3 space station.

the crew would by partly recycled using a condenser in the air conditioning system. And the scientific payload was increased to about 2 tonnes. The improved DOS had the capacity to support two men for 180 days, and the plan was to send three crews, each of which would spend two months on board. It was fully expected that DOS-3 would significantly upstage the American Skylab.

In the period October 1972 to April 1973 the crews who had trained for the lost DOS-2 switched their attention to DOS-3:

- Aleksey Leonov and Valeriy Kubasov
- Vasiliy Lazaryev and Oleg Makarov
- Aleksey Gubaryev and Georgiy Grechko
- Pyotr Klimuk and Vitaliy Sevastyanov

However, although DOS-3 was successfully launched on 11 May 1973, it became one of the rare spacecraft of the manned space programme over which control was almost immediately lost. This fiasco illustrated all the institutional deficiencies that had accumulated over the years, and it is therefore worth examining in detail.

The first in a series of details that brought about the demise of DOS-3 was related to the altitude of the initial orbit. Although its overall dimensions were the same as those of DOS-1, because DOS-3 carried more scientific equipment it was a little bit heavier. Whereas the Proton had been able to insert the lighter DOS-1 into an orbit of 200×222 km, the best that it would be able to achieve with DOS-3 would be 155×215 km. The station's first assignment would therefore be a series of manoeuvres designed to achieve its 350-km circular operating orbit. Before each engine firing, it would have to orientate itself appropriately. An ionic sensor was to be used to sense the orientation of the station relative to the ionosphere through which it was passing. DOS-3 was the first station to be provided with this sensor. An analysis showed that the station would have at most four days to escape from its initial orbit, as after this the orbit would have decayed to such an extent that the engine would not be able to achieve the desired operating orbit. It was therefore vital to raise the orbit as soon as possible.

A second factor which contributed to the loss of DOS-3 was the use of the ionic sensor. Because such a sensor detects not only ions but also glowing particles from the attitude control thrusters, it is an unreliable means of finding the orientation of a spacecraft that is manoeuvring. In fact, before the activation of the ionic orientation procedure, the station's thrusters must be fired to orient it to maximise the number of ions entering the sensor's tube. Measuring their angle of entry provides a point of reference for controlling the station. Once the station has rotated to place the ionic flow at the desired angle, the KTDU-66 main propulsion system can be activated to manoeuvre towards a higher orbit. If the orientation is not performed accurately, a spacecraft can end up in the wrong orbit, possibly decreasing its altitude instead of increasing it – in the worst case diving back into the atmosphere! But the problem with the ionic orientation method is that the tube is exposed to particles in the efflux of the thrusters, which can confuse the analysis. A further complication is that the efficiency of ionic orientation varies with the strength of the Earth's magnetic field, and so with the station's geographical latitude. In given conditions, ionic orientation

Left: Kubasov and Leonov training for the first mission to DOS-3. Right: a model of DOS-3 at the TsPK, with the descent module of Soyuz 2 visible in the distance and cosmonaut Shatalov on the left.

The 'first crew' for DOS-3: Kubasov (left) and Leonov by the Soyuz simulator.

The 'second crew' for DOS-3: Lazaryev (left) and Makarov.

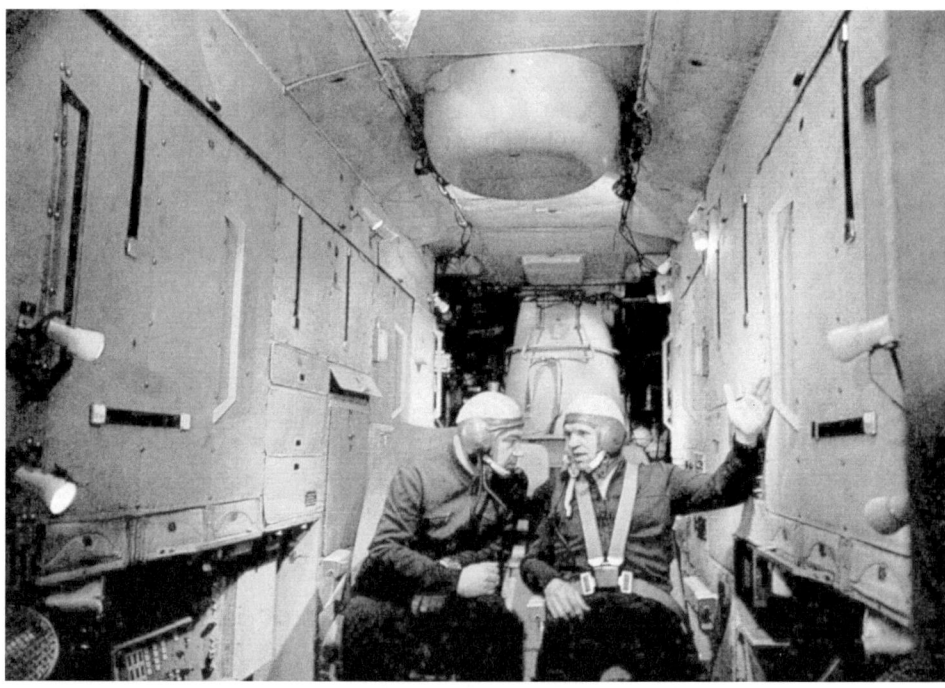

The 'third crew' for DOS-3: Grechko (left) and Gubaryev in the station simulator.

The 'fourth crew' for DOS-3: Sevastyanov (left) and Klimuk.

can mislead the attitude control system, and thereby increase the consumption of fuel. The DOS designers were aware of this, and decided to operate the thrusters at their weakest level in order to minimise the efflux. However, the disadvantage in using weak thrusters was that it would take a long time – possibly several hours – to achieve a major change in orientation, and the longer the time the greater the risk of the control system being misled. Unfortunately, because the system was new, it had not been tested in space to measure its susceptibility to thruster efflux. Nevertheless, the flight controllers were told to perform the ionic orientation as soon as the station was released into its initial orbit.

The last in the sequence of mistakes which led to the loss of DOS-3 was the weak organisation of the terrestrial NIP sites. While Mishin's team focused on testing the station and preparing it for launch, no real thought was given to the unique aspects of controlling it in flight. In fact, the greatest weakness of the Soviet system at that time was flight control – and not just for Salyut, for Soyuz too. On the one hand the designers failed to prepare the documentation in time to enable the flight controllers to appreciate the dynamical operations which DOS-3 would be required to perform. On the other hand the TsUP neglected to liaise with the experts that developed the control systems to draw up an effective plan for providing all the commands which the station would need during its hectic first few days in orbit. In the past, this kind of inadequate planning had been overcome by Pavel Agadzhanov, Boris Chertok, Yakov Tregub and Boris Raushenbakh, all of whom served on the Chief Operative and Control Group (GOGU). But of this group only Tregub was in the TsUP when DOS-3 was launched and, to make matters even worse, he was short of orientation system operators. General Agadzhanov, the head of the GOGU, was absent. He was represented by his assistant, Colonel Mikhail Pasternak. And, of course, the seven control stations across the Soviet Union were operated by the Army. As a result, the experts in telemetry and control who would require to coordinate closely in order to fly DOS-3 through its vital manoeuvres were isolated from each other. The flow of information through the system was slow, owing to the number of checks, protocols and certifications, and when flying a spacecraft through complex manoeuvres time is precious. Furthermore, as it had been accepted that it was impractical to continue to operate stations for months by communicating long lists of information passed by telephones and telegraphs, an automatic system for data processing was being tested at that time. So we see that the TsUP in Yevpatoriya was ill-prepared to swiftly and efficiently provide the commands which DOS-3 would require if it were to reach its operating orbit.

In fact, the leaders of the TsKBEM, Army and the Ministry of General Machine Building were aware of the difficulty of controlling manned spacecraft. Although a great deal had been done since 1966 to improve the system, it still suffered from the fact that the Army ran the ground stations and the technical communication systems and the civilian specialists were responsible for analysing the data and preparing the commands to be issued to the spacecraft. As yet, no one had attempted to unify the system in the manner that NASA had done a decade earlier by establishing Mission Control in Houston, Texas, and directly linking it to the global chain of tracking and communication stations.

Inspecting the DOS-3 simulator. Note the Soyuz docking probe on the right.

The first attempt to launch DOS-3 on 8 May 1973 had to be halted when a vent on one of the six oxidiser tanks of the first stage developed a leak 20 minutes prior to the scheduled time of lift-off. It prompted a major altercation between Mishin, who was the technical director of the DOS programme, and Chelomey, in charge of the rocket. Recalling that a launcher failure had been responsible for the loss of DOS-2, Mishin demanded that the station be transferred to a new rocket. Chelomey insisted that all that was required was to replace the vent. Chelomey prevailed, and the work was done at the pad. But Mishin persisted in demanding that the rocket be changed! Because this would impose a delay of at least a month his TsKBEM colleagues and members of the State Commission urged him to accept the rocket, so he reluctantly acceded. DOS-3 was successfully launched on 11 May 1973, just three days before the Americans launched Skylab.

The Proton delivered DOS-3 into the planned 155×215 km orbit without incident. The NIP-3 tracking facility at Sarishagan in Kazakhstan was the first to hear from the station and confirmed that the antennas and solar panels had deployed correctly. Twelve minutes into the flight, NIP-15 at Ussuriysk on Kamchatka, at the eastern end of the Soviet ground network, sent a command to activate the ionic orientation system. But despite the fact that the NIP-15 documentation specified that the thrusters were to be fired at minimum power, they were commanded to operate at their maximum! An investigation found that the order stating that the orientation engines must fire at full power was issued to NIP-15 by the TsUP in Yevpatoriya. A TsKBEM theorist who had modelled the performance of the thrusters in both regimes prior to going to the TsUP had discovered that if they were to be operated at their minimum power the slow pace of the orientation meant that there was a chance of the process halting during the station's second orbit. He therefore recommended that the orientation be conducted as rapidly as possible. This was forwarded to Tregub, who was the flight director. He accepted the reasoning, and ordered that a telegram be sent to NIP-15 to act accordingly. NIP-15 was in communication with the station for ten minutes, which was sufficient time to establish that the station had begun the orientation. But the only person present who was capable of doing so was isolated by the fact that all transmissions from the station had first to be registered by the Army's telemetric experts, who, after recording the data in their diaries, passed it to their superiors for further processing. When the TsKBEM's expert at NIP-15 received the data on the orientation he was shocked to see that the rate of rotation was ten times faster than the planned speed! Chertok later drew an analogy to convey what was happening to the station – it was like when a dog swings around suddenly to try to bite its tail. The thrusters were firing continuously at maximum power in an effort to stabilise a ship weighing 19 tonnes. The TsUP in Yevpatoriya should obviously have been notified immediately, but rather than just picking up the telephone, the operating procedure obliged that a telegram be written, signed by the appropriate senior officer and then entered into the NIP-15 log before being sent. Once the telegram reached the TsUP, it had to be printed out, logged and sealed before it could be delivered. In fact, the procedure was so time-consuming that meanwhile the station had completed its first orbit and entered the communication zone of NIP-16 in Yevpatoriya!

Because the TsUP controllers had expected that by this time the station would be correctly oriented to perform the first of the manoeuvres required to raise its orbit, they had everything ready to command this. But to their astonishment the data from the experimental automated data processing system indicated that it was not in the desired orientation, and that it had used a vast amount of fuel. The first thought was that the data processing system must not be working correctly; it was experimental, after all. But two young engineers, one an expert in the ionic orientation system and the other an expert in flight control, suspected that the data were correct. They ran to the room where the data was received, in order to examine the original tape, and confirmed that the orientation system had used so much fuel that if it continued to operate as it was doing then the tanks would soon run dry. Because the telephone in that room was not working they ran to the main control room and urged Tregub to command that the orientation system be switched off immediately – the station was still in communication with NIP-16, so this was feasible. But Tregub, who had rejected the plan to perform the orientation slowly and had directed that it be done rapidly, was reluctant to turn off the orientation system. He faced a dilemma. What would happen if he were to take the advice of the young engineers and it transpired that they had been wrong? Would it be possible to resume the orientation process in time to make the manoeuvre to increase the orbit? Unfortunately, he was unable to contact the TsKBEM leadership, as they were driving from Baykonur to the airport in order to fly to Yevpatoriya; they would not reach the TsUP for at least six hours. While Tregub pondered what he should do, the station flew out of range of NIP-16. It would not be able to be contacted again until it reached NIP-15 at Ussuriysk. All this time it continued to spin around 'hunting ions', consuming further fuel. Finally, Tregub decided that the best option would be to halt the orientation. He grabbed the telephone and ordered the NIP-15 operator to do this, but unfortunately the station had passed out of range two minutes earlier!

In the 40 minutes before DOS-3 flew back into range of Yevpatoriya, the experts at the TsUP analysed the available data and decided that the young engineers were right to have recommended immediately switching off the ionic orientation system. This was verified when contact was established and it was ascertained that the fuel was totally exhausted. If the orientation had been halted by NIP-16 at the end of the first orbit, it may have been possible to complete the task on the second orbit by firing the thrusters at their minimum level and then raise the orbit. But now it was lost! When the TsKBEM, Air Force, State Commission and MOM representatives reached Yevpatoriya they could not believe that the third space station in a row had been lost – all in a period of only ten months.

To disguise its identity, DOS-3 was announced by TASS as Cosmos 557; and for some reason its orbit was misquoted as 218 × 226 km. It re-entered the atmosphere on 22 May. Meanwhile, the Americans launched Skylab on 14 May. Although that station was damaged during its ascent through the atmosphere, its first crew of three took up residence on 25 May. They returned to Earth after 28 days, having beaten the record of the ill-fated Soyuz 11 cosmonauts. The second and third Skylab crews spent 59 and 84 days in space respectively, leaving the station 'mothballed'.

Flight director Yakov Tregub (left), cosmonaut Grechko (centre) and flight controller Vadim Kravets at the TsUP in Yevpatoriya.

THE DISMISSAL OF VASILIY MISHIN

Set against the tremendous success that the Americans had with Skylab, the dismal losses of DOS-2, OPS-1 and DOS-3 severely disappointed the Kremlin. The case of DOS-3 was unforgivable. A special investigating Commission was formed, chaired by Vyacheslav Kovtunenko, who was a Deputy Chief Designer at KB Yuzhnoye. Its members included experts in guidance and control – most notably Academician Nikolay Pilyugin, who was a colleague of Sergey Korolev, a legendary member of the Council of Chief Designers, and therefore had decades of experience in the development of rocket guidance. The KGB conducted a parallel investigation. What particularly caught the attention of the Commission was the change in the plan and the order to perform the orientation of the station by using the thrusters at their maximum level. Given that ionic sensors were in use, this sealed the fate of DOS-3. At an academic level, the question was why it had been decided to use the ionosphere, which is an extremely unstable part of the atmosphere, for such a crucial orientation process. A great deal of data on the operation of the sensor in such conditions should have been collected before attempting to use it in this manner. Finally, the Commission was confused by the fact that there was not a Chief Designer for guidance systems in the TsKBEM's structure. The last-minute proposal to change the plan by operating the thrusters at their full power ought to have been put to such a Chief Designer who, knowing the implications, would certainly have refused. Dozens of people who were in one way or another linked to the debacle were questioned, ranging from the TsKBEM managers to the people whose actions or inactions directly caused the loss of the station. The tempestuous outburst from the Kremlin that followed the Commission's report was of a nature never before seen in Soviet cosmonautics – not even after the deaths of cosmonauts.

The burden of blame fell on Yakov Tregub, the DOS-3 flight director. On being urged to leave the TsKBEM, he transferred to the design bureau which had built the Igla automatic docking system. Ex-cosmonaut Aleksey Yeliseyev was appointed in his place, and proceeded to completely revise the organisation and structure of the mission control operation. In addition to transferring the technical facilities from the Army to the TsKBEM, it was decided to create a new TsUP in Kaliningrad, not far from the TsKBEM.[8] After this became fully operational in early 1975, the facility in Yevpatoriya was used only for military space missions.

Also criticised was Boris Raushenbakh, who led the group that developed control and guidance systems. When one of his engineers said that modelling indicated that it would be better to perform the DOS-3 orientation process with the thrusters set at maximum power in order to complete the task as rapidly as possible, Raushenbakh had verbally agreed. When this engineer (whose identity remains unreported) made the suggestion to Tregub, he ordered the revision. Raushenbakh was relieved of his duties and replaced by Viktor Legostayev. Although Raushenbakh was retained as a consultant, he found this unacceptable and soon left the TsKBEM. His boss Boris Chertok was in charge of the general development of control and guidance systems, and received instructive admonition from both Minister Afanasyev of the MOM and the Communist Party organisation at the TsKBEM. Disciplinary measures were also taken against others involved in developing the ionic orientation system, as well as those from the TsKBEM and the Army who were at the TsUP and whose actions or inactions directly contributed to the loss of the station.

The DOS-3 debacle also highlighted weaknesses in the leadership structure at the TsKBEM. At the top was Vasiliy Mishin, who had regarded the DOS programme as a distraction. It had started because in late 1969 a group of his deputies and senior designers had, without his knowledge, put to Ustinov the idea that the Almaz which Chelomey was developing for the military could be made into a station for scientific research. Mishin had argued against the idea when he found out, but was told by the Kremlin to implement it. Wishing to concentrate on the N1-L3 lunar programme, in February 1971 Mishin had suggested to Ustinov that DOS should be handed over to Chelomey, but Ustinov, who did not like Chelomey, had refused to do this. Then in April 1972, during preparations to launch DOS-2, Mishin made an agreement with Chelomey that after four DOS were launched the programme would be transferred to the TsKBM, to enable the TsKBEM to concentrate on its N1-L3 work. The most important point of this Mishin-Chelomey 'contract' called for the production run of DOS stations to be limited to the four which were specified in February 1970 by the Central Committee of the Communist Party and the Council of Ministers. In a letter to Minister Afanasyev, Mishin and Chelomey recommended that future research in space intended to aid the national economy be done by the Almaz programme. This did not mean that Mishin was uninterested in space stations – he fully supported the TsKBEM's Multipurpose Orbital Complex (MOK). This was based on the Modular Space Base Station (MKBS) and would be launched by an N1 rocket. Pointing out

---

[8] The Kaliningrad mission control facility was designated TsUP-M, to distinguish it from TsUP-E at Yevpatoriya.

that the MOK would be larger than either the OPS or its DOS derivative, and hence would have greater requirements, Mishin and Chelomey suggested that the TKS be used to resupply it. Finally, they broached the subject of the joint mission with the Americans planned for 1975. One suggestion had been that an Apollo should dock with a DOS station, but Mishin and Chelomey rejected this, arguing instead that the docking should be between an Apollo and a Soyuz. Mishin and Chelomey sent their 'contract' to Minister Afanasyev, who gave it his endorsement.

Mishin had evidently not consulted his deputies prior to drawing up his agreement with Chelomey, for it provoked intense reactions in the TsKBEM. It was supported by those who sympathised with Mishin – most notably Yuriy Semyonov, a leading figure in the DOS programme,[9] and Sergey Okhapkin, one of Mishin's deputies for the N1 rocket. It was opposed by Konstantin Bushuyev, Boris Chertok and Dmitriy Kozlov. It was Bushuyev and Chertok who had recommended Mishin to supersede Korolev as Chief Designer in 1966. The critics also included Konstantin Feoktistov, who had led the conspiracy to approach Ustinov with the DOS proposal, and Sergey Kryukov, a close colleague of Korolev who had led the development of the R-7 missile and then been reduced in rank when Mishiin took over. In 1970 he moved to the Lavochkin Design Bureau, and became its manager in August 1971 after the death of Chief Designer Georgiy Babakin. The TsKBEM was therefore split into two factions, one of which favoured concentrating on the N1-L3 and the other wished to focus on space stations. As a result, the design, testing and preparations to launch DOS-3 occurred in a strained and unpleasant atmosphere. To the group centred on Bushuyev and Chertok, DOS was a more realistic project and of greater relevance to the nation. But to Mishin, DOS represented a distraction which he wished to rid himself of as soon as possible.

Afanasyev and Ustinov had for some time been concerned by the situation at the TsKBEM, and in February 1973 a working efficiency assessment conducted by the Ministry for General Machine Building criticised the TsKBEM's performance over the last several years. Deficiencies in the organisational structure directly influenced the entire organisation and had, in particular, resulted in the degradation of both the quality and the safety of its systems. Mishin was not mentioned by name, but the message was clear: the Kremlin was losing patience with his leadership of what was supposed to be the nation's principal space organisation. Soon after this assessment, Bushuyev, Chertok, Kozlov, Feoktistov and Kryukov, with the support of Ustinov, who as we have seen had rejected an earlier attempt by Mishin to offload the DOS project to Chelomey, sent a joint letter to the Central Committee of the Communist Party and the Council of Ministers in which they criticised both Mishin's work and the state of the TsKBEM, particularly expressing their dissatisfaction with both the manner in which Mishin ran projects and the fact that he ignored their criticism of his management. They concluded by demanding that Mishin be replaced.

---

[9] Although Semyonov was a leading figure in the DOS programme, he probably supported Mishin on this issue simply through loyalty to his boss. However, it is also possible that Semyonov realised that owing to the problems faced by the N1 the lunar programme was likely to be cancelled, whereupon the TsKBEM's only option would be the DOS programme.

Ustinov paid an unannounced personal visit to the TsKBEM. Such behaviour can be interpreted as being meant to signal to Mishin that the Politburo was concerned. As it was, when Mishin arrived Semyonov was showing Ustinov a scale model of DOS-3 and they were discussing the possibility of fitting a station with two docking ports. Of course, this idea was not new. The designers had been considering it since right after the first Salyut was launched in June 1971. It would enable an occupied station to be supplied with fuel, food, water and air. With regular servicing, a DOS would be able to be operated for years. The idea had been proposed by Semyonov, Feoktistov and Viktor Ovchinikov, an expert in spacecraft system development. But because Mishin was eager to hand the entire programme over to Chelomey he had refused to waste time on improvements beyond the DOS-3/4 configuration. Taking advantage of the moment, Semyonov asked if Ustinov would personally support the development of further DOS stations. Noting that Ustinov saw promise in the idea, Mishin figured that if he reversed his position and agreed to continue to build DOS stations, then he might gain Ustinov's support against those who had demanded his resignation. And that is how it turned out. Alone in Mishin's office, Ustinov pointed out that a station with two docking ports would have tremendous potential, and then he said in a friendly manner that Mishin should give some thought to his position at the TsKBEM. It was clear to Mishin that the only way in which he could remain as Chief Designer would be to support continued DOS development. This rendered the agreement with Chelomey obsolete. As Mishin's opponents had hoped, this behind the scenes manoeuvring ensured that the TsKBEM focused its efforts on operating space stations – which was just as well, because the N1-L3 lunar programme was in deep trouble from which it was destined never to recover. And, of course, by acting in this way Ustinov was able once again to frustrate Chelomey.

As a result, the TsKBEM directed its efforts towards designing a new generation of DOS with two docking ports, the first of which was launched in September 1977 as Salyut 6. It was manned by five long-term crews, four of which were able to set successive endurance records. By being supplied a dozen times by automated cargo ships and occupied for a total of 684 days, it was a spectacular demonstration of the soundness of the design.

Despite the appearance that Mishin had secured his position, he was undermined by the list of failures by the TsKBEM since his appointment as Chief Designer in January 1966:

- November 1966 – The first unmanned Soyuz (Cosmos 133) suffered a series of faults; it was deliberately destroyed during its return in order to prevent it landing in China.
- February 1967 – Although the second Soyuz (Cosmos 140) was better than the first, it also suffered various difficulties, and ended up on the floor of the Aral Sea.
- April 1967 – Despite two less than satisfactory unmanned test flights, it was decided to start manned flights. Soyuz 1 suffered serious problems early on,

and cosmonaut Vladimir Komarov was killed on impact after the parachute failed to deploy.

- October 1968 – Cosmonaut Georgiy Beregovoy failed to dock his Soyuz 3 with the unmanned Soyuz 2.
- January 1969 – The first launch of the N1 lunar rocket failed.
- July 1969 – The second N1 failed.
- October 1969 – The docking of Soyuz 8 with Soyuz 7 had to be cancelled in flight as a result of the failure of the Igla automated rendezvous system.
- November 1969 – The circumlunar L1 programme was abandoned without even one cosmonaut flying in the spacecraft.
- April 1971 – Soyuz 10 failed to completely dock with the first DOS space station owing to a technical failure.
- June 1971 – The third N1 failed.
- June 1971 – After spending a record time in space on board the first DOS space station, the Soyuz 11 crew died on their way home.
- July 1972 – The second DOS space station failed to reach orbit owing to a technical failure in the Proton launcher – although to be fair, this was not the fault of the TsKBEM.
- November 1972 – The fourth (and as events would prove, final) N1 failed.
- May 1973 – DOS-3 was lost soon after it achieved its initial orbit as a result of procedural errors.

As a result of losing DOS-2 and DOS-3, there were five Soyuz spacecraft sitting in storage. They could not be kept indefinitely, since their systems would gradually degrade to the degree that they would be unreliable. The State Commission decided that two would be flown unmanned and two would fly with crews on solo missions. On 15 June 1973, in the guise of Cosmos 573, a Soyuz spacecraft was launched into a 206×268 km orbit; it returned after two days. On 27 September 1973, more than two years after the Soyuz 11 tragedy, Soyuz 12 was launched. Aleksey Leonov and Valeriy Kubasov, veterans of the DOS-1 programme, had trained as the first crew for DOS-2, then for DOS-3, and immediately after DOS-3 was lost they were reassigned to the joint mission with the Americans that was to fly in 1975. The Soyuz 12 mission therefore went to Vasiliy Lazaryev and Oleg Makarov, who had trained as the second crew for both DOS-2 and DOS-3. On their two days in space they checked the Sokol-K pressure suit and the operation of all the revised systems. The spacecraft changed its orbital parameters several times. And, for the first time, NASA's Mission Control Centre played a role in controlling a Soviet mission, as an exercise in preparation for the joint mission.

On 30 November 1973 another Soyuz was launched to a 195×295 km orbit in the guise of Cosmos 613. This was the craft in which Leonov and Kubasov would have flown to DOS-2. It remained in orbit for two months to assess how well the systems stood up to prolonged exposure to the space environment, and then returned safely. DOS-2 had carried an Orion advanced astrophysical telescope, but DOS-3 had not, so it was decided to install this apparatus on a Soyuz by substituting it for the active docking system and make observations of Comet Kohoutek as this passed

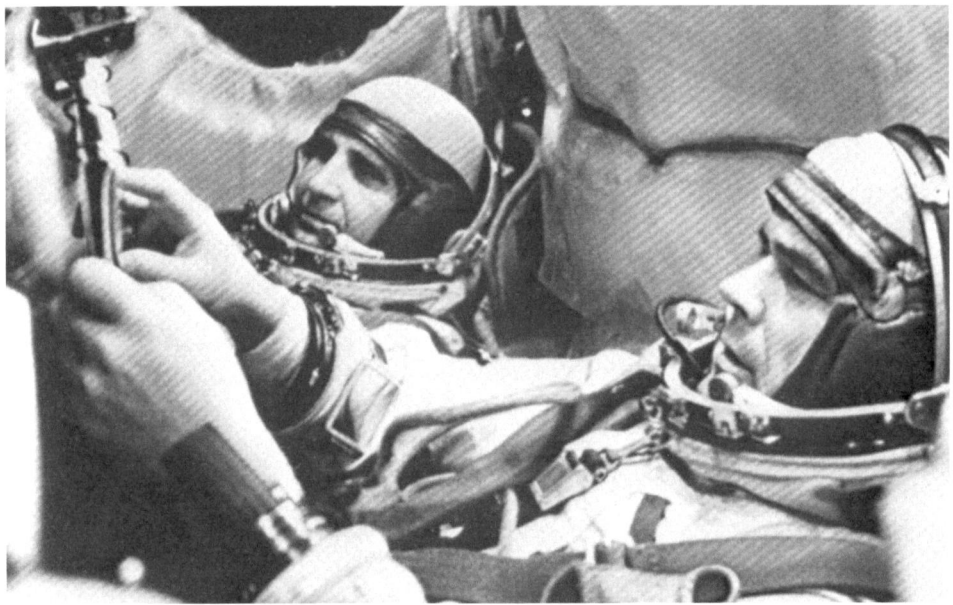

Soyuz 12, the first manned mission of modified Soyuz spacecraft, was flown by Makarov and Lazaryev (foreground).

Soyuz 13, the last mission before Vasiliy Mishin was dismissed as Chief Designer, was flown in December 1973 by Klimuk (left) and Lebedyev and was primarily to conduct astrophysical research.

In May 1974 Vasiliy Mishin was dismissed as the TsKBEM's Chief Designer.

On superseding Mishin, General Designer Valentin Glushko renamed the TsKBEM NPO Energiya.

near the Sun. In addition, solar panels were added to enable the spacecraft to remain in orbit for a week. Cosmonauts Pyotr Klimuk and Valentin Lebedyev flew this Soyuz 13 mission between 18 and 26 December 1973. The fifth spacecraft from the DOS-2 and DOS-3 stock was used for engineering tests of the special docking system made for the Apollo-Soyuz mission.

Soyuz 13 was the last manned mission to be launched under Mishin's leadership. His downfall came as no surprise to his TsKBEM colleagues – for many of whom it was long overdue. It would appear that after consulting Ustinov, Brezhnyev decided that Mishin would have to go, and Afanasyev, Mishin's protector, was powerless to intervene.

The formal decision was made at a meeting of the Politburo in mid-May 1974. As a result, Academician Pilyugin informed Chertok that Mishin was to be replaced by Valentin Pavlovich Glushko, the famous designer of rocket engines and, after Korolev, the most imposing figure in the early Soviet space programme. Chertok has written that it was clear from the behaviour of his colleagues that they knew what was going on, and yet no one wished to talk about it. In fact, Mishin must have been aware. On 22 May Afanasyev and Glushko arrived at the TsKBEM unannounced. Mishin was in hospital, but all of his deputies were convened. Afanasyev announced that the Politburo had decided to replace Mishin with Glushko. In shaking up the TsKBEM, Glushko merged it with his own bureau,[10] creating the Research and Production Association Energiya (NPO Energiya) with himself as Director and General Designer. This organisation became a veritable empire which addressed all areas of the manned space programme, from the development of motors and rockets, transport spacecraft, space stations, and even lunar bases. In mid-1974, therefore, a new era in the history of Soviet cosmonautics began.

Specific references
1. Chertok, B.Y., *Rockets and People – The Moon Race, Book 4*. Mashinostrenie, Moscow, 2002, pp. 422–434 (in Russian).
2. Afanasyev, I.B., Baturin, Y.M. and Belozerskiy, A.G., The World Manned Cosmonautics. RTSoft, Moscow, 2005, pp. 231–232 (in Russian).

[10] OKB-456, later to become Energomash.

15

Memories

For more than 36 years the ashes of cosmonauts Georgiy Dobrovolskiy, Vladislav Volkov and Viktor Patsayev have rested in niches in the Kremlin's wall. In addition to their families, they were mourned by hundreds of engineers, technicians, officers, cosmonauts and politicians. Despite the tragedy, there was a determination that the DOS programme must continue. The programme would never have come about if it were not for the support of Dmitriy Ustinov and Sergey Afanasyev, the so-called 'Space Minister'. They supported the proposal initiated by Boris Raushenbakh, Boris Chertok and Konstantin Feoktistov at the TsKBEM to modify the Almaz military reconnaissance station which was being developed by a rival bureau led by Vladimir Chelomey, to serve as a long-term station for scientific research. Although Vasiliy Mishin, in charge at the TsKBEM, was antagonistic, these men succeeded not only in getting the programme started but also in making it the dominant element of the Soviet space programme.

On the operational side, General Nikolay Kamanin managed the training of the cosmonauts. The cosmonauts whose lives were most affected by the early years of the DOS work were Vladimir Shatalov, Aleksey Yeliseyev, Nikolay Rukavishnikov, Aleksey Leonov, Valeriy Kubasov and Pyotr Kolodin. For months, together with Dobrovolskiy, Volkov and Patsayev, these men trained to operate the world's first space station, Salyut.

Let us conclude by reviewing the lives of the key people of the programme after its disastrous early years.

VASILIY PAVLOVICH MISHIN

Although Mishin's leadership of the TsKBEM was criticised in the aftermath of the Soyuz 11 tragedy, he retained his position owing to support by Sergey Afanasyev, the Minister of General Machine Building, and Andrey Kirilenko, who was a close colleague of Brezhnyev in the Politburo. Mishin's relationship with Ustinov is very interesting. At first sight it may appear that he was always backed by Ustinov (for how else could he have remained in post despite the deaths of four cosmonauts, the fiasco of the L1 circumlunar programme, the repeated failures of the N1 rocket for the N1-L3 lunar programme and the loss of two DOS stations before they could be visited) the relationship between the two men was actually much more complex. For instance, when asked in an interview with the eminent space journalist Vladimir Gubaryev about Ustinov's nomination to lead the Soviet rocket programme, Mishin said: "I am not sure that it was the best choice! It is hard to say whether he brought more harm or good."

During the eight years that Mishin ran the main Soviet space institution, he was a controversial figure. He was unfortunate in gaining leadership at a time that NASA accelerated its space programme and won the 'race' to be the first to land a man on the Moon. To understand how the Soviet Union lost this race it is necessary to analyse Mishin's leadership in the context of the roles of Afanasyev and Ustinov, and indeed of the input of Brezhnyev and Kosygin. However, in technical terms, the failures of Mishin's years in charge of the TsKBEM were, in large part, the result of decisions made by this organisation, initially by Korolev and later by himself.

In terms of Earth orbital flights, Mishin's period will be remembered for a series of failures, two of which concluded tragically for the crews – the only such losses to date in the programme. Even so, he retained the support of Afanasyev and Ustinov. He was replaced only after the cancellation of the N1-L3, the organisation of which was largely directed by Afanasyev and Ustinov!

The year 1971 marked a low point for the Soviet space programme, with the third launch of the giant N1 lunar rocket ending in failure, the Soyuz 11 tragedy and the deaths of three of the leading rocketry specialists: Aleksey Isayev, Georgiy Babakin and Mikhail Yangel. The disasters continued in 1972 with the loss of DOS-2 and the final N1, and into 1973 with the loss of DOS-3.[1] Although the design of the N1 was criticised by the leading designers at some of the other organisations (and indeed by some of the people in OKB-1/TsKBEM), Mishin continued to work on it, confident that it would soon become operational and enable cosmonauts to walk on the Moon. But the L3 concept was also criticised – if the manner in which the Americans had gone about landing on the Moon was extremely risky, the way that Mishin planned to do it seemed highly likely to result in the loss of the cosmonaut who attempted to execute it.

Mishin often did things in his own way. When dealing with issues about which he really ought to have consulted with his deputies, he made decisions on his own. An excellent example was his 'contract' with Chelomey – which marked the beginning of his downfall. Also, owing to his abrupt manner, his intolerance of criticism, and his frequent heavy drinking (sometimes at the TsUP during missions) the number of people whose respect he lost progressively grew. When he lost the support of some of his close colleagues, including Bushuyev and Chertok, this divided the TsKBEM into two factions, one wishing to push on with what was now really no more than a dream of a lunar programme and the other considering the DOS programme (which Mishin wished to discard) as the basis for a strong space programme. When Mishin ignored this 'mutiny' by his closest colleagues, the Kremlin stepped in and made its

---

[1] The loss of the OPS-1 station does not count in this context, because it was not a TsKBEM project.

"The gene of renunciation." During his 8 years in charge of the TsKBEM, Vasiliy
Mishin (third from the left), with the support of Minister Sergey Afanasyev (fourth from
the left), worked with the objective of reaching the Moon. After his dismissal in 1974,
Mishin (right photo) worked as a professor of space rocket technology at the Moscow
Aviation Institute.

dissatisfaction clear, and in 1974 he was replaced by his old rival Valentin Glushko.
To Ustinov, Mishin said: "I understand everything, except the reason for choosing
Glushko." Although Glushko's management had its critics, he successfully turned
the TsKBEM into an empire on a scale that Mishin could never have achieved.

Mishin was appointed as a professor of space rocket technology at his alma
mater, the Moscow Aviation Institute (MAI). In fact, since 1958 he had been
lecturing at Lomonosov University in Moscow, and he continued to do this in
parallel with MAI. One of his students was Valentin Lebedyev, who joined the
TsKBEM, trained as a cosmonaut, and flew as the flight engineer of Soyuz 13, which
was the last mission to be flown during Mishin's term as Chief Designer. While a
professor at the MAI, Mishin was able to supervise nine master's theses and eight
doctorates. Those who knew him in these years say he showed two different
personalities. At times he was rough, explosive, intolerant and brusque, just as he
had been when Chief Designer while speaking his mind in dealing with politicians
and generals. But the second personality on display at the MAI was much more
pleasant. As a teacher, he transmitted to generations of students his rich experience
in the design of rockets. He directed the Department for the Design and
Construction of Flying Vehicles at the MAI (later Department 601, Space Systems
and Rocket Design) until 1990, and in 2002 its laboratory was given Mishin's name.
He co-authored a number of study-books that are still in use today. In addition, he
directed a students' design bureau where, among other projects, the first Soviet non-
hermetic satellite was constructed.[2]

In the second half of the 1980s, after Mikhail Gorbachov had become the General
Secretary of the Communist Party of the Soviet Union, Mishin gave interviews and
published several works designed to vindicate his still controversial contribution to
cosmonautics. Although the CIA had been aware since the 1960s that a man named

---

[2] When launched, this was named Radio1.

Mishin was a key figure in the design of Soviet rockets, it was not until now that his identity was allowed to become public. In *Why Didn't We Fly to the Moon?* which was published in December 1990,[3] he described, for the first time, the Soviet lunar programme in detail. Always sharp and direct in his manner, he wrote:

> They accused me of not defeating the Americans. But everyone knew right from the beginning that the Americans would win. Our leaders did not listen. After the Americans had done it, we said that we were ready to do it better, but they would not let us try.

In conclusion, he wrote:

> Often the question arises: If Korolev had not died, what would have come of our space programme? It is my view that not even he, with all his authority, persistency and predisposition for achieving goals, could have dealt with all the processes that have caught all areas of activity in our society. It would have been difficult for him to work without directives, ... which followed an incomprehensible politics even during his lifetime. Without doubt, he would have achieved something. We could have had a landing on the Moon, ... but sadly not within the deadlines that were imposed on us for prestige over the USA. Too much time had been wasted, and so much money was needed, but the directives did not provide it.
>
> I do not wish readers to think that I am trying to avoid my responsibility as Chief Designer for some of the mistakes that were made in the course of the lunar programme – some by myself. He that does not do anything, does not make errors! We, the successors of Korolev, did everything that we could, but it was not enough.

Aleksey Leonov has strongly criticised Mishin for wasting the money available to the lunar programmes. Leonov firmly believes that in 1968 the Soviets could have beaten the Americans to a circumlunar flight. In fact, Leonov was to command the first L1 crew and, if the N1 rocket had worked and the N1-L3 programme had gone ahead, he would have been the first cosmonaut to attempt to land on the Moon. It is likely that Leonov's hostility towards Mishin originated with the cancellation of the L1 programme without even attempting a manned mission, and was then worsened by Mishin's order for Leonov's crew to stand down and let Dobrovolskiy's crew fly the Soyuz 11 mission.

Although Mishin persistently denied being directly responsible for the failures of the Soviet manned space programme in the years 1966 to 1973, when asked why he had been so antagonistic to the DOS programme he confessed: "I only understood it later on. In those years, I was not aware that I was making a mistake. The point is that 80 per cent of the tasks that were beneficial to the national economy could have been done by unmanned spacecraft."

Few people at the TsKBEM felt sorrow at Mishin's dismissal as Chief Designer.

---

[3] 'Почему мы не слетали на Луну?'

In writing his memoirs, Boris Chertok did not feel it appropriate to explain anything about Mishin's subsequent career.

After leaving the TsKBEM, Mishin left its work behind. Only twice did he cross the doorstep of NPO Energiya. His only real support was his family: his wife Nina Andreyevna, with whom he spent 63 years, and his daughters Yelena (who worked for Korolev and for her father for 40 years), Kira and Vera.

Vasiliy Mishin died on 10 October 2001, aged 84, and was buried five days later in Trekurovskoye Cemetery in Moscow.[4] During a ceremony on 18 January 2007 to mark the 90th anniversary of his birth, his eldest daughter, Yelena, said: "As time goes by, all the things which remind me of my father and link me to him become dearer to me. He did not have relatives in high positions or strong contacts with the top man. He had only his wife and three daughters. ... Yes, he always said what he thought. He never stepped back from anyone. He was wise, intellectual and a man of honour. It has been said that every scientist must have a gene of renunciation ... my father had such a gene."

VLADIMIR NIKOLAYEVICH CHELOMEY

The empire that Chelomey had spent many years building up began to decay when Ustinov became the Minister of Defence in 1976. As Ustinov did not wish to have two institutions working on manned space projects, Branch No. 1 of the TsKBM at Fili was transferred to NPO Energiya.[5] It was therefore ironic that whereas Mishin had sought to offload the DOS programme to Chelomey, Almaz was removed from Chelomey and handed to the TsKBEM's successor! However, later Fili became KB Salyut, and eventually joined the Khrunichev Centre.

Vladimir Chelomey. The DOS design was derived from his Almaz reconnaissance station.

---

[4] Mishin outlived his mentor, Sergey Afanasyev, by five months.
[5] It is now part of the Khrunichev Centre.

After Ustinov had left office, work began on an unmanned version of the Almaz, but progress was so protracted that Chelomey did not live to see its completion. He died on 8 December 1984, aged 70. He had been taken to hospital following a car accident, and during medication an artery became blocked. So ended the life of one of the Soviet Union's greatest designers of missiles and space rockets.

BORIS YEVSEYEVICH CHERTOK

As the last of the Pleiades of extraordinary members of Soviet rocketry, they called Chertok a patriarch of cosmonautics. For two decades (1946–1966) he worked with Sergey Korolev. He directed the department which developed guidance systems and their associated electronics. From 1966 until 1973 he was a member of the Chief Operative and Control Group at the TsUP in Yevpatoriya. He was also one of the men who in 1969 approached Ustinov behind Mishin's back and thereby started the DOS programme. Without Chertok's willingness to embark on such a major project without his boss's support, and to continue with it despite his boss's open antipathy, the history of the Soviet manned space programme would certainly have turned out very differently.

On Mishin's dismissal in 1974, Chertok was the first of the TsKBEM's senior people to meet the new director – doing so several days prior to Glushko's official

Boris Chertok, one of the leading figures in the Soviet rocket and space era.

appointment. On Glushko's death in 1989 Yuriy Semyonov took over, and in 1992 Chertok became an advisor to Semyonov. Although at the time of Semyonov's retirement Chertok was 95, he continued as the principal scientific consultant to Nikolay Sevastyanov, one of his former students, who took over the directorship of RKK Energiya (as NPO Energiya had become) in 2005.[6] Even after 60 years in the business he continued to work, and lectured at the N.E. Bauman University and at the Physics and Technical Institute.[7] His memoirs, in four volumes entitled *Rockets and People*,[8] provide a unique insight into the development of the Soviet rocket and space programmes.

BORIS VIKTOROVICH RAUSHENBAKH

As a result of the loss of DOS-3, Raushenbakh was dismissed from his post in charge of the development of systems for the guidance and orientation of vehicles in space, and soon thereafter left the TsKBEM to become a professor at Moscow's Physics and Technical Institute. This was a natural move, because while working at the TsKBEM he had been a part-time lecturer there. Raushenbakh was one of the most imposing senior personnel at the TsKBEM. In addition to being a theoretician and designer of one of the most complex aspects of rocketry (guidance systems) he was also an academician and a distinguished philosopher and student of religion. He had a friendly relationship with Korolev that started before the Second World War. In view of his German roots, he was committed to a concentration camp, as indeed was Korolev for a short period. After Stalin's death in 1953 Raushenbakh joined the Central Scientific Research Institute (TsNIIMash) created by Mstislav Keldysh. In 1955 he moved to OKB-1 to direct the development of guidance and orientation systems for rockets. He and his colleagues explained to the first cosmonauts how the Vostok spacecraft functioned. His genius is apparent from the fact that the systems that were developed under his direction were used for decades in Soviet spacecraft.

Boris Raushenbakh, who initiated the TsKBEM's rapid construction of the DOS space station.

---

[6] RKK stands for *Raketno-Kosmicheskaya Korporatsiya*, which means Space Rocket Corporation.
[7] Fizichesko-Tehnicheckiy Institut.
[8] 'Ракеты и люди'.

After leaving cosmonautics behind, Raushenbakh devoted his time to the analysis of philosophy, science, religion and art. He wrote two books on geometry in artistic paintings, two on the connections between science and early Russian iconography, and the last, which was published just before his death, on Russian science, Nazism and nationalism. He died on 27 March 2001 at the age of 87.

KONSTANTIN PETROVICH FEOKTISTOV

Great designer and famous cosmonaut Feoktistov played one of the most important roles in starting the DOS programme. In June 1974, soon after Mishin's dismissal, Glushko named Feoktistov as one of his deputies – a post he held until May 1990. In the summer of 1975 he worked as flight director for the second crew of Salyut 4, although only briefly. His principal task was the design of the 'Soyuz T' crew ferry and the automated 'Progress' cargo ship, but he also contributed to improved forms of the DOS, including Salyut 6 and the legendary Mir.

In October 1964 Feoktistov became the first space engineer to fly in space, when he was a member of the first Voskhod mission. Four years later he was a serious candidate for the one-man Soyuz 3 flight, but at that time the Air Force did not wish to allow civilians to pilot spacecraft. In the period May to October 1980 he trained to perform extensive maintenance on the thermal regulation system of Salyut 6 in order to extend the use of that station. He was to fly this Soyuz T-3 mission with Leonid Kizim (TsPK, commander) and Oleg Makarov (NPO Energiya). However, in October, less than a month before the scheduled date of launch, he was replaced by Gennadiy Strekalov. Although the official explanation was that Feoktistov had a medical problem, he insists otherwise: "It was the Air Force. I have battled them all the time. You see, I thought that those who knew most about cosmonautics should be the ones to fly. In fact, the point was reached at which the leader of the mission should have been a cosmonaut-engineer, not the spacecraft's commander. However, the soldiers did not like this idea." In October 1987, aged 62, he left the ranks of the cosmonauts. Yuriy Semyonov was once Feoktistov's boss on the DOS programme, but under Glushko was assigned to direct the development of the Buran space-plane. Feoktistov, who never held back in criticising the direction of the space programme, condemned this project. Semyonov never forgave him, and in May 1990, shortly after Semyonov was appointed head of NPO Energiya, Feoktistov drew his 35-year career as a spacecraft designer to an end and moved to Moscow's Higher Technical School (MVTU) Bauman. Many of the leading figures in Soviet rocketry and space technology came from Bauman – among them Feoktistov, who got his PhD there in 1967. He retired in 2005.

Feoktistov authored over 150 scientific papers and also several books. In *Seven Steps to the Sky*, published in 1984,[9] he wrote of a manned flight to Mars. As time went by he grew ever more critical of the space programme. Given that Feoktistov

---

[9] 'Семь шагов в небо'.

dedicated his best years to the development of space technology his autobiography, *Life Path*, published in 2000,[10] was written in a curious, sometimes sarcastic style.

Regarding the role of the International Space Station (ISS), whose lineage can be traced back to his own DOS work, and the future of manned space flight in general, he states:

> People should not work on this subject just now. There is nothing interesting at the ISS – or in space. There is no serious research. We and the Americans have both spent so much time and effort on manned fights and space stations, but the attainment of the main goal is not linked to these projects. However, the Hubble telescope has offered a great amount of new information. People should work in the areas where results can be obtained. The future belongs to

"There is nothing interesting at the ISS – or in space." Having devoted his career to the design of manned spacecraft, Konstantin Feoktistov (here between cosmonauts Makarov and Kizim) later became a critic of manned space flight.

[10] 'Траектория жизни'

automated stations. Manned cosmonautics lacks any practical sense and it will not have any meaning, not now, not in future times.

From three marriages Feoktistov has the largest family among all Soviet/Russian cosmonauts: comprising one daughter and three sons – one of whom was born in 1982 when Feoktistov was 56. He is the oldest of the still-living Soviet cosmonauts to have flown in space. A crater on the far side of the Moon, 19 km in diameter, was named in his honour. In February 2006 he celebrated his jubilee 80th birthday.

YURIY PAVLOVICH SEMYONOV

After leaving Mikhail Yangel's Design Bureau in 1964 to join Korolev, Semyonov rose steadily through the ranks. He started as an assistant to the main designer of the Soyuz spacecraft, but was then appointed the main designer for the L1 circumlunar variant, and finally the main designer for the DOS programme. During the period in which Glushko ran the company, Semyonov participated in improving the Soyuz to serve the second-generation Salyuts, and later Mir. He also directed the Interkosmos programme which trained cosmonauts from fraternal communist countries and gave them brief visits to Salyut 6. In 1981 he was appointed Glushko's principal deputy, and was placed in charge of the development of the Buran space-plane. During the Gorbachov era he tried politics, but was hindered by the fact that his wife was the daughter of Andrey Kirilenko, who was a senior man in the Politburo of Leonid Brezhnyev. After Glushko's death in 1989 Semyonov became General Designer of NPO Energiya. In 1991 he was made General Director, and played a key role in preserving the core of the national space programme. After the conglomerate was re-organised in early 1995 as Space Rocket Corporation (RKK) Energiya, he became its first president. He established strong links with the two leading Western space agencies: NASA and ESA. This collaboration prolonged the period of operation of the Mir space station. Nevertheless, once the American use of Mir ended there were no funds to continue to operate the station, and in March 2001 it was de-orbited. By this time, however, Russia was a partner in the International Space Station (ISS). Although RKK Energiya was playing a key role, a significant fraction of the work went to the M.V. Khrunichev Centre, which had become a commercial competitor in the space market. In an effort to improve RKK Energiya's income, Semyonov, against opposition from NASA, offered Soyuz 'tourist flights' to the ISS. Following the demise of the Soviet Union, Russia established a space agency to coordinate the national space programme. Even though the government owned 38 per cent of RKK Energiya, Semyonov tried to bypass the national space agency, thereby drawing criticism, and in May 2005, a month after his 70th birthday, the government told him to retire. He was superseded by 44-year-old Nikolay Sevastyanov. Semyonov holds a PhD in Technical Sciences and has authored over 200 scientific papers.

Yuriy Semyonov as General Designer at NPO Energiya.

General Nikolay Kamanin (left) and General Designer of NPO Energiya Valentin Glushko at the TsPK in Zvyozdniy in August 1974. (From the book *Hidden Space*, courtesy www.astronaut.ru)

NIKOLAY PETROVICH KAMANIN

Kamanin's involvement with the space programme began in the best possible way, with the selection and training of the first cosmonauts, including Yuriy Gagarin, the first man to orbit the Earth. It ended with the deaths of Dobrovolskiy, Volkov and Patsayev. However, during his 11 years as head of cosmonaut training there had been other deaths. In March 1961, shortly before Gagarin's launch, the young trainee cosmonaut Valentin Bondarenko lost his life as a result of burns from an accidental fire in a pressure chamber that was providing a pure-oxygen atmosphere. Although this occurred in the Institute for Space and Aviation Medicine, which was not under Kamanin's control, he directed the training regime. For many years Bondarenko's death remained a secret. And, of course, in 1967 Vladimir Komarov lost his life on impact with the ground after the failure of Soyuz 1's parachute system. Kamanin announced his attention to retire prior to the Soyuz 11 mission.[11] He was simply fed up of disputes with the leaders of the space programme – first with Korolev, then with Mishin. And it was not only with civilians that he clashed. His relationship with his new boss, Pavel Kutakhov, was far from ideal. Immediately upon taking office in 1969 Kutakhov had decided that Kamanin should be replaced, and it had been decided that cosmonaut Vladimir Shatalov would supersede him.

Although Kamanin became a Lieutenant-General at the end of the Second World War, aged 37, it was 22 years before he was promoted to Colonel-General in 1967.

---

[11] In fact, Kamanin's ceased to be in charge of cosmonaut training on 25 June 1971, a few days before Soyuz 11 was due to return to Earth.

Whereas cosmonauts gained medals and the Gold Star of Hero of the Soviet Union, he was persistently bypassed, even when space flights in which he played a key role were successful. But on his death he was laid to rest bearing the Hero of the Soviet Union which he received in 1934 for risking his life to save the passengers of the icebreaker *Chelyuskin*.

Kamanin was highly critical of the manner in which the Soviet space programme was directed. He particularly criticised the men who "pulled the strings" behind the scenes. In his celebrated diary, published as *Hidden Space*,[12] he sharply criticised Korolev for underestimating the role of cosmonauts in piloting a spacecraft, and for wasting time during the development of the Soyuz spacecraft automating its control and guidance systems. He also criticised the decision that cosmonauts should cease to wear pressure suits. Finally, he constantly battled with Mishin for the selection of crews. His relationship with the cosmonauts was sometimes very formal and harsh, but he displayed an affinity for some of the first group, especially Gherman Titov.

Kamanin's family life was also marked by tragedy – his son was lost in an aircraft crash. Although after retirement he played no further role in the space programme he sometimes visited Zvyozdniy, which he had done so much to create, to take part in celebrations involving the cosmonauts. He died on 11 March 1982 at the age of 74, and was buried in Novadevechye Cemetery in Moscow.

VLADIMIR ALEKSANDROVICH SHATALOV

On 26 April 1971, immediately after his return from the Soyuz 10 mission, Shatalov was promoted to Major-General. Two months later he superseded Kamanin. This appointment was largely the result of his close relationship with Marshal Kutakhov, who was his mentor prior to becoming a cosmonaut, his participation in organising the historic visit to Baykonur of President Charles De Gaulle in June 1966,[13] and his excellent management skills. His promotion coincided with the Soyuz 11 mission, and his first duty was to participate in the State Commission which investigated the loss of that crew. This recommended a thorough restructuring of the manned space programme. As part of this review, the training of cosmonauts was broadened and, among other things, they became more actively involved in the preparation of the experiments which they were to perform in space.

During Shatalov's 16 years as the head of cosmonaut training he was responsible for equipping the TsPK with simulators and other training facilities, the recruitment of new military cosmonauts, and the selection of crews for a succession of DOS and Almaz space stations. In the meantime, in April 1972 he defended a master's thesis at the Gagarin Military Air Force Academy, and in 1974 he was promoted to the rank of Lieutenant-General.

---

[12] 'Скрытый космос'.

[13] Interestingly, it was at this time that the Voskhod 3 mission was cancelled, and Shatalov had been a member of the backup crew.

Kamanin's successor, General Vladimir Shatalov, with Lebedyev (centre) and Klimuk in training for Soyuz 13.

When Gorbachov initiated his reforms of the Soviet Union in 1985, these affected the Army too. The structure of the Air Force began to change. Many older generals were retired, including Georgiy Beregovoy, the 65-year-old Director of the TsPK. After three possible successors had been rejected, it was suggested to Shatalov that he should take on this role in addition to his existing duties. As he was 58 years old at the time and according to the new law his retirement would take effect at age 60, he asked that he be allowed to remain in post for longer. On being promised that he would be able to serve until 65, he accepted. His task would be to organise training for the ambitious Mir and Buran programmes. For Shatalov, the direction of this unique training center, which at that time employed about 400 people, represented a special challenge. When it was put to him that the TsPK should be transferred from the Air Force to the Space Army (as the Strategic Rocket Forces, which managed the Baykonur cosmodrome, had become) he disagreed. When he was promised the rank of Colonel-General in return for his acquiescence, he refused. He also resisted Glushko's efforts to transfer the TsPK to NPO Energiya. His campaign to keep the

General Shatalov with the Apollo-Soyuz crewmembers at the TsPK. Left to right: Leonov, Shatalov, Stafford, Brand and Kubasov. Slayton is out of frame.

TsPK in the Air Force was not helped by the antipathy of the leading military figures. In the aftermath of the upheaval in the Soviet Union in 1991 he was transferred to the reserve corps of the Commander in Chief of the Air Force, and then ordered to retire in March 1992.

By September 1991 Shatalov had been a member of all State Commissions over two decades that had made the final decisions on launching manned space missions. He related his role in the space programme in his autobiography, published in 1978 as *The Hard Road to Space*.[14] He also co-authored the books *The Application of Computers in Spacecraft Guidance System* (1974), *People and Space* (1975), *The Soviet Cosmonauts* (1979) and *Space to Earth* (1981). For almost three decades he led a society which promoted friendly relations with Cuba. A 21-km-diameter crater on the Moon was named after him.

As regards retirement: "I have decided that after 65 years of life, of which almost 50 were devoted to the Army, I have some right to live for myself. I have not started attaching myself to political, religious or commercial organisations. ... They have no interest for me. Although I had many offers, especially in 1992, I have not gone into politics either – it was suggested that I become a member of one of the many parties and enter parliament." Ever since 1957 he has been interested in underwater fishing. The sea attracts him now, as once he was attracted to the sky and to space: "I cannot imagine the sea without submarine life. What is the point of lying on the sand to sunbathe, without wondering what might be found under water. No matter where I travel, I carry my mask, my fins and 8 kg of iron as ballast for diving. If the weather is suitable, I also like to ski. I love to spend time in my vacation house, to mow, and to tend my small garden. My wife is devoted to agronomy. Every spring we start with transplantation, and plant new sprouts. ... I want to spend what is left of my life peacefully with my family, children and grandchildren. Now, as a retired person, I analyse the events of my tempestuous life in cosmonautics."

ALEKSEY STANISLAVOVICH YELISEYEV

After his third and final space flight on Soyuz 10 in April 1971 Yeliseyev became one of Yakov Tregub's deputies, being responsible for the preparation and control of manned missions. He worked mainly on the development of the programmes for missions and the onboard instruction, the technical aspects of crew training, and the control of a flight. When Prime Minister Aleksey Kosygin and America's President Richard Nixon signed an agreement which called for the first joint space mission involving the two space-faring nations, Tregub suggested that Yeliseyev should fly, but Yeliseyev did not wish to make any more flights. "I had the feeling that Tregub saw me as a potential rival. Supposing that I would be interested in working with the Americans, he said several times that perhaps I should fly as the flight engineer, but I had already made up my mind and did not wish to change my decision. Very soon I

---

[14] 'Трудные дороги космоса'.

realised that my suspicion about Tregub was justified. When I was called by Minister Afanasyev, ... he asked me whether I would agree to run mission control for the Soviet-American flight. ... By rank, this offer ought to have been made to Tregub, so I concluded that they must have something against him. I accepted the offer." In this way Yeliseyev became the director of manned spaceflight. In parallel, he gained his doctorate in February 1973.

It was at this time that the new Mission Control Centre was built in Kaliningrad. It was more capable than the old facility in Yevpatoriya. In the next eight years (until 1981) thirty manned flights were conducted under Yeliseyev's technical direction. In the case of the historic Soyuz-Apollo mission he was given a delicate task. Once the docking had occurred, the intention was that two of the three astronauts would join the two cosmonauts in the orbital module of the Soyuz. Shortly beforehand, Yeliseyev was called to the office of the State Commission and was told by Ustinov that he was to transmit a congratulatory message by Leonid Brezhnyev. This would not be easy, as every minute of the joint activities had been meticulously planned. But the message had to be read. To avoid sending the message himself, Yeliseyev proposed that it should be done by a professional TV reporter. Ustinov agreed. Such a person was urgently delivered to the TsUP, given the text and instructed to read it, word by word, without omitting anything. Some minutes later the reporter came to Yeliseyev:

"I cannot do it."
"What can't you do?"
"I can't read this word by word."
"Why not?"
"It says here 'L. Brezhnyev'."
"So what?"
"I cannot say 'L. Brezhnyev'. I can say 'Leonid Brezhnyev', 'Leonid Ilich Brezhnyev', or simply 'Brezhnyev'. What should I say?"

Just in case, I decided to ask Ustinov. There was not much time left before the start of the session with the cosmonauts. Without hesitating, I sprinted to Ustinov and put to him the question posed by the reporter. Ustinov remained silent, pretending that he had not heard me. I realised that he did not wish to take the responsibility. I tried my best to assist: I suggested that the reporter should say 'Leonid Brezhnyev'. Still Ustinov said nothing. I started to sweat. There was now just one minute to the beginning of the session. Noticing my nervousness, Afanasyev volunteered, "I agree with the suggestion." Ustinov made a slight ambiguous nod that could be interpreted as his agreement, but also indicated that he considered the conversation over.

As the director of space flight, Yeliseyev sometimes had to make decisions on which depended the lives of cosmonauts in orbit: for example during the flight of Soyuz 25 in October 1977, which was unable to dock with Salyut 6 and had to be ordered home, and again when Nikolay Rukavishnikov was commanding Soyuz 33 in April 1979 and had a problem with his main engine (see below).

In December 1985 Yeliseyev officially left the cosmonaut-engineer group briefly

Flight director Aleksey Yeliseyev (insert) and the new Mission Control Centre in Kaliningrad, Moscow.

in order to serve as a deputy to the General Designer of NPO Energiya. Then, at the suggestion of the Minister of Education, for five and a half years he was the rector of Moscow's Higher Technical School (MVTU) Bauman. But this was not a happy experience because his proposal to restructure the school faced opposition. He gave up the rectorship and went to work for IBM, which had begun to make a presence in the Soviet and Asian markets, remaining with them until January 1996. Today he is the head of the Festo international fund. In Russia this fund promotes education and directs a department of the Moscow Energy Institute. He travels a lot. With his wife Larisa he visits historic places in Russia. He reads books on economics by Western authors. He also thinks about life and his contribution to the space programme: "On asking myself what I achieved in all those years, it does not seem very much when compared to what was being done around me. Obviously that is the way it ought to be. A life time is no more than a particle in the kaleidoscope that represents men's destinies – no more than a drop in the sea." This sentiment inspired the title of his autobiography, *Life – A Drop in the Sea*, which was published in 1998.[15]

NIKOLAY NIKOLAYEVICH RUKAVISHNIKOV

After the mission of Soyuz 10 Rukavishnikov was nominated as the flight engineer on Leonov's Soyuz 12 crew, which was to make the second visit to the Salyut space station, but this crew was stood down when it became necessary to revise the design of the spacecraft after the deaths of the Soyuz 11 crew. He made two further space flights. The first occasion was on Soyuz 16, which was a six-day test in December 1974 in preparation for the joint mission with the Americans the following summer. The spacecraft was commanded by Anatoliy Filipchenko, and Rukavishnikov was the flight engineer. He trained as the commander of the backup crew for Soyuz 28. This was the first flight of the Interkosmos programme, and took a Czechoslovakian cosmonaut to Salyut 6. When he flew Soyuz 33 in April 1979 he became the first civilian to command a Soviet ship. His passenger was Georgiy Ivanov (Kakalov) of Bulgaria. The Igla rendezvous system locked onto Salyut 6 and began to navigate towards it, but when the range reduced to 4 km and they saw the station for the first time a six-second firing of the main rocket engine was cut short after three seconds! Rukavishnikov manually restarted the engine, but there was a terrible noise and it cut off again. On board the station, Vladimir Lyakhov and Valeriy Ryumin reported to the TsUP that they had observed sparks from the spacecraft's propulsion module. The rendezvous had to be abandoned. This was the first and only failure of the main engine of a Soyuz. The crew were told to rest while the engineers on Earth decided what to do. The technical director of the flight was Yeliseyev, who precisely eight years earlier had flown with Rukavishnikov on Soyuz 10 in an attempt to dock with the original Salyut station!

Meanwhile in space, Rukavishnikov found it difficult simply to rest:

> Throughout the night I told myself that as commander I was responsible not only for myself and the ship but also for Georgiy. I had to analyse all of the variables and be ready to answer any queries from Earth, or to execute any directions they provided.
>
> As I was thinking, Georgiy asked: "Captain, shall we refresh a bit?"
>
> We carried Bulgarian foodstuffs as a gift for the Salyut 6 crew. "Let's get out the presents," I decided.
>
> "Can we?"
>
> "Now we can."
>
> We refreshed ourselves. I only had a little, but Georgiy really ate well.
>
> "Off to sleep," I told him. "We have to get good rest. Tomorrow we'll be busy."

---

[15] 'Жизнь – капля в море'.

Meanwhile, Yeliseyev called the engine designers and experts in ballistics to the TsUP and together they thoroughly analysed the situation. Luckily, the Soyuz had a reserve braking engine (DKD). Unlike the main engine this could be fired just once, for the braking manoeuvre. But there was some concern, because its propellant and electrical lines were located close to those of the main engine, which had evidently suffered a serious problem. It was to be hoped that the reserve engine had not been damaged. It was possible that the engine would start and then cut off prematurely. If it were to fire for less than 90 seconds, the crew would require to execute a series of firings of the small docking and orientation engines (DPO) to depart from orbit, but this would result in a return far from the planned landing site. At 6.46.49 p.m. on 12 April the reserve engine was activated to make a 188-second burn. Rukavishnikov inferred from the buzzing sound transmitted through the structure of the spacecraft that the engine was not operating at full power. After 188 seconds had elapsed and it failed to shut off automatically he took the decision to keep it running for another 25 seconds before he turned it off. As a result, the descent was steeper than normal, and followed a ballistic trajectory that subjected the occupants to peak deceleration load of 8 g. To everyone's relief, the descent module landed safely at 7.35 p.m. at a point 320 km southeast of Dzhezkazgan in Kazakhstan. When he reflected upon his second failed attempt to dock with a Salyut, he joked: "the stations did not wish me on board".

In April 1980 Rukavishnikov gained a master's degree at the Moscow Institute of Engineering and Physics (MIFI). Meanwhile, he was training as commander of the backup crew for Soyuz T-3. Initially, the objective of this flight was to undertake an extensive medical research programme on board Salyut 6, but this was altered to perform maintenance on the station to enable it to operate long enough to complete the Interkosmos programme. Undeterred, Rukavishnikov focused his hopes on the forthcoming Salyut 7, and from September 1983 to February 1984 trained as flight engineer for the mission that was to carry the first Indian cosmonaut. However, with just two months remaining to the launch date he caught the flu, and thereby lost not only his opportunity to visit a space station but also the chance to become one of the few Soviets to fly four times in space. His unsympathetic colleagues joked that all Salyuts had a built in "anti-Rukavishnikov device".

On leaving the cosmonaut group in July 1978 Rukavishnikov became a deputy to the director of one of the departments of NPO Energiya, then retired in November 1999. In 1981 he became president of the Soviet Cosmonautics Federation,[16] and in this role vigorously sought support from the Kremlin for a number of programmes. He also arranged for a medal to be given to an anonymous artist who had for many years painted artwork depicting the space programme. In addition, he led the radio show *On Space Orbits*. Although he gave the appearance of having a very serious personality, those who knew him well said he was vibrant and always interesting to be with.

---

[16] In 1999 this became the Russian Cosmonautics Federation.

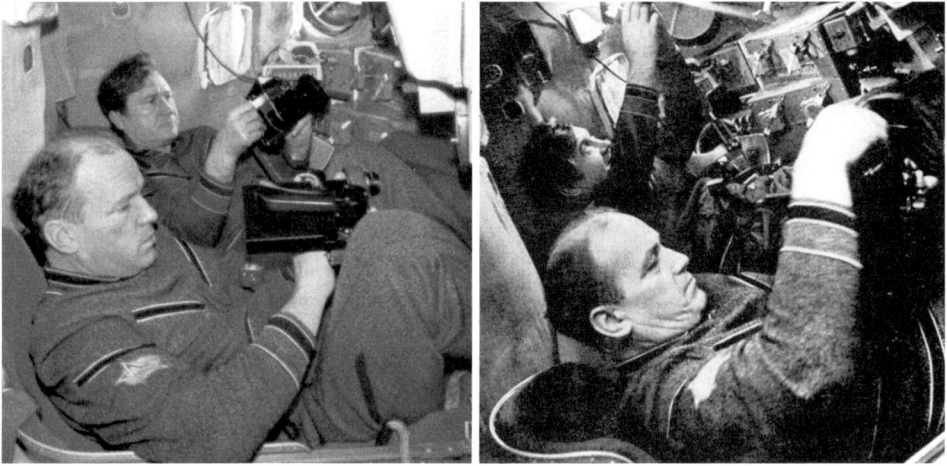

Two views of Nikolay Rukavishnikov (foreground) and Anatoliy Filipchenko in the Soyuz simulator.

"The stations did not wish me on board." Rukavishnikov (foreground) and the Bulgarian cosmonaut Georgiy Ivanov made a dramatic return after their Soyuz 33 spacecraft suffered a main engine failure on the way to the Salyut 6 space station.

Rukavishnikov stands in front of the Soyuz simulator with the prime and backup crewmembers for the Indian mission to Salyut 7, but a medical complaint caused his replacement 2 months before the launch.

In the space of six years Rukavishnikov's family suffered three tragedies. First his wife Nina died in 2000. Those closest to him gathered for his 70th birthday on 18 September 2002, but his memory was impaired by Alzheimer disease. Although he had survived one heart attack, the second was followed by pulmonary problems and he died in Burdenko hospital on 19 October 2002. He was buried in Ostankinsko Cemetery. Finally, in January 2006 his only child, Vladimir, succumbed to a severe illness and died aged only 41. He was buried alongside his father. Until the very end of his life Vladimir had unselfishly offered details of his father's life to anyone who expressed an interest.

ALEKSEY ARKHIPOVICH LEONOV

Following the loss of the DOS 3 station in May 1973 Leonov and Kubasov were reassigned to the Apollo-Soyuz programme. They lifted off in Soyuz 19 on 15 July 1975. The Apollo carried a special airlock on its nose and was the active partner in the docking. In the two days during which the two spacecraft were linked the crews visited each other several times. Half an hour after separation, the Soyuz played the

After almost 10 years of continuous training for circumlunar, lunar landing and space station missions, Aleksey Leonov (foreground) finally returned to space in July 1975 with flight engineer Valeriy Kubasov, for the Apollo-Soyuz mission.

Historic encounter in orbit. Leonov (in the middle) in the company of astronauts Thomas Stafford (above) and Donald Slayton in the Soyuz 19 orbital module.

As one of the chiefs at the TsPK, Leonov was responsible for selecting and training cosmonauts. On the left he takes the exams from Popov and Lebedyev in 1980 (for the Salyut 6 programme), and on the right from Manarov, Titov and Levchenko in 1987 (for the Mir programme).

active role and docked again. Three hours later they undocked and took spectacular photographs of one another prior to moving clear. Leonov and Kubasov returned to Earth on 21 July. As had been the case for the launch and docking, the landing was also broadcast 'live' on TV – marking a new degree of openness in the Soviet space programme. For this mission Leonov was awarded a second Gold Star of a Hero of the Soviet Union and was promoted to Major-General.

In March 1976 he was appointed commander of the cosmonauts at the TsPK. In March 1981 he was awarded a master's degree by the N. A. Zhukovskiy Air Force Engineering Academy. In January 1982 he was made first deputy to the chief of the TsPK, General Beregovoy. This marked the end of his 22-year cosmonaut career. In 1985 Beregovoy retired. Shatalov proposed Leonov as his successor, but Glushko at NPO Energiya objected so Leonov continued to run the training of the cosmonauts, their exams, and the selection of the crews. He played a key role in the preparations for the mission of Soyuz T-13 which brought the crippled Salyut 7 back to life in 1985. Leonov was retired in September 1991 at the age of 57. He got a letter from the Commander in Chief of the Air Force in which, together with an expression of thanks, he was praised as a "founder of the school for cosmonaut training" – which was not far from the truth, given that since the death of Gagarin in 1968 he had not only trained space flight candidates but had also prepared many of the crews.

Having left the TsPK, Leonov was made president of the Alfa-Capital Investment Fund, and also of the companies Bering-Vostok and Vostok-Capital. In July 1996 his family was struck a terrible blow when his elder daughter Victoria Alekseyevna died during a medical operation at the age of 35. In 1997 he became vice-president of the Alfa-Bank, and in December 2005 also became a consultant to the Sladko confectionery company. Despite his age, he is in excellent health and is very active in the Association of Space Flight Participants, often attending the meetings which deal with key events in the history of cosmonautics. He was one of the initiators of the project to make a movie about the life of Sergey Korolev, which was released in January 2007 to commemorate the 100th anniversary of the birth of this giant of the Soviet space programme. In 2005 Leonov became president of the Russian-Serbian

Society of Friendship. A 38-km-diameter crater on the Moon has been named after him. Having previously co-authored two books, in 2004 he co-authored *Two Sides of the Moon* with the American astronaut David R. Scott, in which they interleave their parallel stories in the rival space programmes as a unique dual autobiography. Although he travels a lot, Leonov frequently indulges his old passion by painting cosmic themes. So far, eight albums of paintings have been published with Leonov as author or co-author.

VALERIY NIKOLAYEVICH KUBASOV

After training for missions which never flew to the first three DOS stations, in May 1973 Kubasov was assigned with Leonov to the Apollo-Soyuz programme. After the two spacecraft were docked, Thomas Stafford and Donald Slayton transferred through the special airlock to the hatch of Soyuz 19, where the historic handshake between men of the rival space-faring nations occurred. Meanwhile, their colleague Vance Brand remained in the Apollo. The cosmonauts had prepared a surprise for their guests: "We knew that after the docking we would have lunch on our ship with the Americans, so we decided to entertain them. We had brought several samples of Stolichnaya vodka, and once in space we glued these to juice and soup tubes. When ready to eat, we put these 'rarities' on the table. After a moment of confusion, the astronauts started to cheer like kids! Of course, they realised that this was a Russian tradition ... And on trying it, they laughed heartily." After spending two days in the docked configuration, the spacecraft separated on 19 July and the Soyuz returned to Earth two days later.

In August 1977 Kubasov began to train for his third space flight, which was to be to deliver a foreign cosmonaut to Salyut 6 for the Interkosmos programme. Initially, he was commander of the backup crew for the Polish flight but in November 1978 he was given command of the first crew for the Hungarian flight, making him only the second civilian cosmonaut to be given command of a Soviet spacecraft (the first such assignment having gone to Rukavishnikov). The plan was that the Hungarian flight should be in May 1979, but when Rukavishnikov's mission in April ran into difficulties the Hungarian flight had to be cancelled to enable an unmanned Soyuz to be sent up to the station to replace the aging ferry which was docked there. As a result, Kubasov and Bertalan Farkash did not launch until 26 May 1980, and then it was on Soyuz 36. The next day Kubasov became the first cosmonaut-engineer to dock a spacecraft with a station. During their week-long visit to Leonid Popov and Valeriy Ryumin, he achieved his ten-year-old dream of working on board a Salyut station.

Kubasov had hoped to make further flights, but it was decided that henceforth the spacecraft commanders must be military cosmonauts. Having been a cosmonaut for 15 years and made three flights he argued that he could not accept flying under the command of an inexperienced military cosmonaut, and in July 1981 he declined to be a candidate for further flights. He managed the training of cosmonaut-engineers for ten years, then worked on the design of life support systems,

After the fiasco of DOS-2 and DOS-3, Valeriy Kubasov and Aleksey Leonov were nominated as the prime crew for the Soviet element of the Soyuz-Apollo mission.

On his third and final space flight, Kubasov was commander of Soviet-Hungarian crew with Bertalan Farkash. They spent a week on Salyut 6 in May 1981, thereby fulfilling Kubasov's dream of visiting a space station. Here they undergo water training.

biological-medical and thermal regulation equipment. He resigned as a cosmonaut in November 1993 but stayed at NPO Energiya for another four years as a scientific consultant. He has authored one book and co-authored two others. He is currently writing a book about the joint Soyuz-Apollo mission. He periodically goes hunting and although over 70 years of age still plays tennis very well.

PYOTR IVANOVICH KOLODIN

Fate was not very kind to Kolodin. Seven years after losing his chance to fly to the first Salyut in 1971 with Leonov and Kubasov he was named as flight engineer for Soyuz 27. It would be commanded by Lieutenant-Colonel Vladimir Dzhanibekov, who, like Kolodin, had not yet been in space. The objective of the mission, planned

Pyotr Kolodin, the eternal backup.

for launch on 28 January 1978, was to dock with Salyut 6 in order to exchange the ferry for the station's main crew. It would be a historic mission for the Soviet space programme because for the first time two spacecraft would be docked at a station. However, when the rookie crew of Soyuz 25 failed to dock on the inaugural mission to the station it was decided that in the future at least one cosmonaut of each crew must be experienced. Although Kolodin had been a member of the cosmonaut corps for 13 years he was replaced just two months before launch by Oleg Makarov, who had flight experience. In his autobiography, Kolodin used on 14 occasions phrases such as: "He was training...", "he was third backup", "second backup...", "was training as first backup..." and "member of the prime crew...". However he never flew in space. Among the cosmonauts, he was legendary as one on whom the stars did not shine. In April 1983 he left the Air Force's cosmonaut group but continued to work at the TsPK. In November 1986 he retired with the rank of Colonel, then worked as a principal engineer in the Mission Control Centre in Kaliningrad.

REMINISCENCE AND LEGACIES

But the real heroes of this outstanding epoch in the Soviet space programme are Georgiy Dobrovolskiy, Vladislav Volkov and Viktor Patsayev. In the towns of their births, Odessa, Moscow and Aktyubinsk, there are monuments to them at which colleagues, friends, relatives and ordinary people with a passion for space leave flowers. Every year, on the anniversary of their tragic deaths, members of the current cosmonaut corps pay their respects at the niches in the Kremlin's wall where the ashes of their fallen colleagues are interred.

Also, when the Apollo 15 astronauts David R. Scott and James B. Irwin landed on the Moon a month after the Soyuz 11 tragedy they left behind a plaque which bore the names of all the astronauts and cosmonauts then known to have died, including Dobrovolskiy, Volkov and Patsayev. In addition, they gave the name Salyut to one of the craters near their landing site, which was alongside the rim of Hadley Rille at the base of the Apennine mountain range. The International Astronomical Union has named craters on the far side of the Moon after the Soyuz 11 crew. The crater Dobrovolskiy is at 12.8°S, 129.7°E and is 39 km in diameter; Volkov is at 13.6°S, 131.7°E and is 35 km (or 40 km according to another source); and Patsayev is at 16.7°S, 133.4°E and is 55 km in diameter. They are all located near the large crater named after Konstantin Tsiolkovskiy, the 'father of cosmonautics', which was first seen in 1959 when a Soviet space probe took the first pictures of the far side of the Moon.

In June 1977 three small asteroids were named after the Soyuz 11 crew. Asteroid 1789 Dobrovolskiy (1966QC) was discovered in August 1966, orbits the Sun at an average distance of 1.79 AU and is 33 km in diameter.[17] Asteroid 1790 Volkov (1967ER) was discovered in March 1967, orbits at 2.01 AU and with a diameter of

---

[17] One Astronomical Unit (AU) is defined as the mean radius of the Earth's orbit of the Sun.

Cosmonauts place flowers at the niches in the wall of the Kremlin of their fallen colleagues.

42 km is the largest of the trio. Asteroid 1791 Patsayev (1967RE) was discovered in September 1967, orbits at 2.35 AU and has a diameter of 29 km.

Tracking ships of the scientific research fleet that communicated with spacecraft were also named for the Soyuz 11 crew.[18] *Cosmonaut Vladislav Volkov* was launched on 18 October 1977. It was joined by *Cosmonaut Georgiy Dobrovolskiy* on 14 October 1978 and *Cosmonaut Viktor Patsayev* on 19 June 1979. All three ships have a mass of 8,950 tonnes, are 122 metres in length, 17 metres wide, and are served by a crew of over 60 people. After the Soyuz 11 tragedy it was suggested that the old tracking ship *Bezhitsa* be renamed *Cosmonaut Georgiy Dobrovolskiy* but it was decided that a new vessel must be built. *Cosmonaut Viktor Patsayev* was based on a ship which was launched in 1968 and was modified in 1977/1978 to suit the new requirements. In addition to controlling automated satellites and interplanetary probes, these ships were used extensively in operating the Salyut 6 and Salyut 7 space stations. And in 1988 the Buran space-plane was controlled by *Cosmonaut Vladislav Volkov* in the Atlantic and *Cosmonaut Georgiy Dobrovolskiy* in the Pacific. *Cosmonaut Vladislav Volkov* made its 17th and final cruise in 1992, *Cosmonaut Georgiy Dobrovolskiy* its 17th in 1993 and *Cosmonaut Viktor Patsayev* its 14th in 1994. In 2001 *Cosmonaut Georgiy Dobrovolskiy* and *Cosmonaut Viktor Patsayev*, which were now owned by the Russian Space Agency, were docked at Kaliningrad on the Baltic and in 2005, against the opposition of veterans, *Cosmonaut Georgiy Dobrovolskiy* was sold to a foreign company and is still sailing the seas. *Cosmonaut*

---

[18] Such ships had already been named after Sergey Korolev, Vladimir Komarov, Yuriy Gagarin and Pavel Belyeyev.

Three craters near the large crater Tsiolkovskiy on the far side of the Moon have been assigned the names Dobrovolskiy, Volkov and Patsayev.

Tracking ships named after cosmonauts Volkov (left), Dobrovolskiy (top right) and Patsayev (bottom right).

Viktor Patsayev remained in Kaliningrad, and in April 2001 opened as a permanent Space Odyssey exhibition. Although it never sails, occasionally its antennas link the TsUP with the crew of the International Space Station. All the other tracking ships were discarded in the years following the collapse of the Soviet Union.[19]

Sports competitions are also associated with two of the Soyuz 11 crew. In 1972 a traditional cup in parachuting was named after Georgiy Dobrovolskiy. A year later, a cup in acrobatics was named in honour of Vladislav Volkov. Since 2002 this has been permanently celebrated at the town of Velikiy Novgorod, and it was recently opened to international participation. Interestingly, a solar observatory in Auckland in New Zealand was named Dobrovolskiy. A museum was opened in October 1974 on the grounds of the Aviation Institute in Moscow in honour of Vladislav Volkov, and a nearby street was named after him and a bust of him placed at its end. Streets in Vladivostok, Rostov-na-Donu and Mariupole in Ukraine were similarly named. Schools No. 10 in Odessa and No. 54 in Vladivostok were named in honour of Dobrovolskiy, as were streets in several Russian and Ukrainian cities and towns. A monument to him was placed in a square in Odessa, and a memorial plate was placed in the small town of Mogilyev.

Dobrovolskiy is an honoured citizen of Odessa, Volkov of Kaluga and Kirov, and Patsayev of Aktyubinsk. A monument to Patsayev was placed in the main square of Aktyubinsk and one of the main streets was named after him, as indeed was a street in Kaluga where Konstantin Tsiolkovskiy lived. The museum at Aktyubinsk has a collection of personal objects that belonged to Patsayev. His wife Vera donated the flight suit which he wore in preparing for his space flight. It bears

---

[19] The fate of *Cosmonaut Vladislav Volkov* is uncertain. It was very likely sold to a private company and scrapped. After the collapse of the Soviet Union, the largest tracking ships *Academician Sergey Korolev* and *Cosmonaut Yuriy Gagarin* were anchored in Odessa in Ukraine. Despite protests from Russia, both were sold to a private company that broke them up and sold the scrap to India in 1996.

the mission patch designed by Leonov. The museum also has two books from Patsayev's library, rare photographs, and a star map that Patsayev used on Salyut as the first man to work a telescope in space. A famous school for young pilots in Aktyubinsk was named in his honour. It became a tradition that at the conclusion of their studies the cadets were awarded their diplomas by Maria Sergeyevna, Patsayev's mother. She had been in Aktyubinsk paying her daughter a visit when the deaths of the Soyuz 11 crew were announced. She decided to spend the remainder of her life in the place where his monument was placed, and died there in August 2004 at the age of 91.

In the presence of Vera Patsayeva, a memorial was erected at the place where the Soyuz 11 capsule landed. Marina and Natalia Dobrovolskaya, Lyudmila Volkova and her granddaughter, and cosmonaut representatives visited it on the 20th anniversary. Marina recalls: "Curiously, it was a dry June, it had not rained in a month, and then as we began the ceremony, suddenly, and right on the spot of the landing, it started to rain."

However, perhaps the most commemorative tradition is the one that is performed at the Air Squadron in Sevastopol in which Georgiy Dobrovolskiy served as a pilot. Ever since September 1971 his name is called each evening during the roll call, as if he were still a member of the regiment. A pilot explains his absence: "Hero of the Soviet Union Lieutenant-Colonel Georgiy Timofeyevich Dobrovolskiy has lost his life in service to his country, obeying his duty in testing space technology." On 30 June 2006 Marina Dobrovolskaya marked the 35th anniversary of her father's death by attending this roll call.

The wives of the fallen cosmonauts never truly recovered from the shocking news of the accidental and tragic deaths of their husbands.

Lyudmila Dobrovolskaya died in 1986. Marina Dobrovolskaya is a teacher of literature at Moscow State University (MGU). She is married and has one daughter. Natalia Dobrovolskiy is in Moscow and looks after her family. Lyudmila Volkova is retired, while Vladimir Volkov works abroad as a TV camera operator.

As an expert in atmospheric physics and remote-sensing of the Earth from space, Vera Patsayeva had an opportunity to access original materials relating to the flight of Soyuz 11. In 1973, with the aid of a group of enthusiasts, colleagues and friends, she published the book *Salyut in Orbit*, drawing extracts from the diaries written by the crew. Svetlana Patsayeva says: "I can only now understand how much strength she needed to gain permission to publish the book, which presented information on the Soyuz and the Salyut station which, at that time, was highly confidential. At the same time, she was surviving her own tragic loss. Sadly, during those times it was impossible to write completely openly on the subject. The manuscript was censored and the most important part on the causes of the tragedy were deleted from the book. Later, she collected the designers' opinions, as well as the opinions of scientists and doctors, and interviewed eminent people whose names she could not reveal. She did this with the goal of determining whether the tragedy could have been avoided – and who was responsible for the accident. This is all in the manuscript, but it could not be published in the Soviet era. Later, unfortunately, her health deteriorated and she was unable to complete it."

Monuments to the memories of Vladislav Volkov (top left) and Viktor Patsayev (top right). Bottom: Svetlana and Dmitriy Patsayev (second from the right) with their sons visiting Red Square on 30 June 2007. (Copyright Svetlana Patsayeva)

As mentioned earlier, Vera Patsayeva had kept a diary during the Soyuz 11 flight. Of this, Svetlana said: "I dream about publishing mom's and dad's diaries in one book, as a day-by-day record of what was happening in space set alongside my mother's feelings about the things which she had to endure during that period."

Vera Patsayeva died in 2002. Viktor Patsayev's children have finished studies at Moscow State University. Both have followed the interests of their parents. Dmitriy Patsayev works at the Space Research Institute of the Russian Academy of Sciences on the development of instruments to investigate the planets Mars and Venus. He is married and has two sons. Svetlana Patsayeva is married and has one son. She is an assistant professor at the Physics Department of Moscow State University, holds a master's degree in science, and is an expert in the application of spectral analysis in ecology.

ZARYA AND ZVEZDA

When the Almaz and DOS programmes were initiated, no one could have predicted that such hardware would form the core of a space station at the turn of the century, but the Russian-built Zarya ('Dawn') and Zvezda ('Star') modules are key parts of the International Space Station. And certainly not even Sergey Korolev could have dreamed that his Soyuz spacecraft would still be in use ferrying crews to this station. This legacy is truly the best of monuments to the lost crew of the first space station.

One day, a space crew will depart from a space station to head once again for the Moon, as a stepping stone to the planets. These future space travellers will owe a tremendous debt of thanks to cosmonauts Georgiy Dobrovolskiy, Vladislav Volkov and Viktor Patsayev, whose names are by now written between the stars.

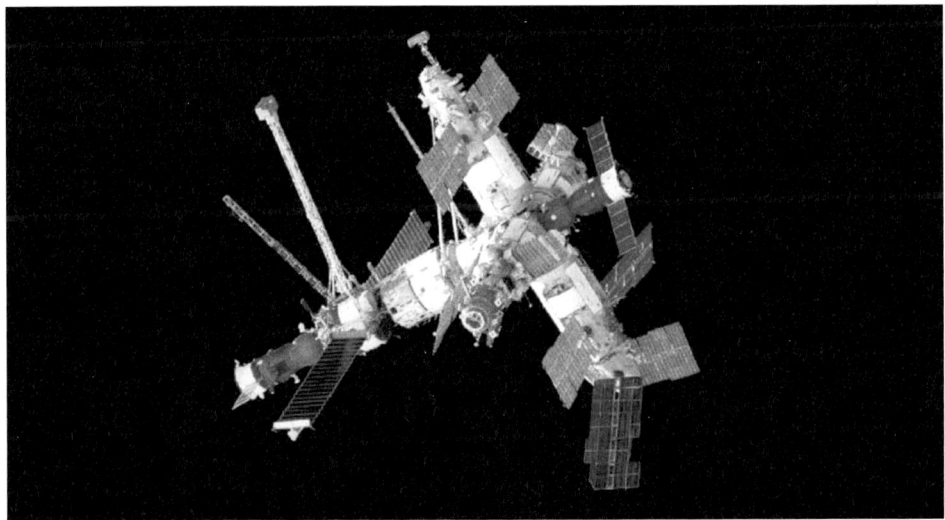

The Mir orbital complex. (Courtesy NASA)

Two Soyuz TMA spacecraft docked with the International Space Station. (Courtesy NASA)

The Zvezda and Zarya modules that form the core of the International Space Station are the direct legacy of Salyut and its heroic crew. Korolev's legacy is evident from the two Soyuz and one Progress spacecraft docked with the station. (Courtesy NASA)

Specific references

1. Mishin, V. P., *Why Didn't We Fly to the Moon?* Znaniye, 12/1990, Moscow, 1990 (in Russian).
2. Gubaryev, V.S., *Russian Space, Book 3*. Exmo, Algorithm, Moscow, 2006, pp. 390–412 (in Russian).
3. Loskutov, A., 'Tenable Gene' (Interview with Mishin's daughter), Daily News, Moscow, No. 8, 18 January 2007.
4. Novosti kosmonavtiki (in Russian)
   No. 12, 2002 (Eulogy for Nikolay Rukavishnikov)
   No. 3, 2003 (Necrology for Kerim Kerimov)
5. Molchanov, V.E., *About Those Who Did Not Reach Orbit*. Znaniye, Moscow 1990 (in Russian).
6. Soviet Cuban (Krasnodar), No. 29, 5 August 2005 (Interview with Konstantin Feoktistov).
7. Biographies of cosmonauts www.astronaut.ru
8. Tracking ships www.ski-omer.ru

Interviews by the author:

1. Marina Dobrovolskaya, 24 May 2007
2. Svetlana Patsayeva, 1 August 2007
3. Dmitry Patsayev, 5 September 2007

The immortal crew of the world's first space station – Viktor Patsayev, Georgiy Dobrovolskiy and Vladislav Volkov.

Glossary

Anna-III	Gamma-ray telescope (Salyut)
AN-SSSR	USSR Academy of Sciences
CK-KPSS	Central Committee of the Soviet Communist Party
DM	Descent module (Soyuz)
DMP	Solid-propellant braking rocket (Soyuz)
DOS	Long-duration orbital station programme
Era	Equipment for detection of high-energy ionospheric electrons (Salyut)
FEK-7	Photo-emulsion camera (Salyut)
FGB	Functional-cargo block
GOGU	Chief Operative and Control Group
IBMP	Institute for Biomedical Problems
Igla	Automatic rendezvous system
ISS	International Space Station
Kazbek-U	Soyuz crew couches
KIK	Command-measurement complex
KPSS	Communist Party of the Soviet Union
KTDU	Correction and braking engine system (Soyuz)
KTF	Complex for physical training
L1	Circumlunar manned programme (TsKBEM)
LK	Lunar ship (TsKBEM)
LK-1	Circumlunar manned project (TsKBM)
LK-700	Lunar landing manned programme (TsKBM)
MAI	Moscow Aviation Institute
MFTI	Moscow Institute for Physics and Technology
MGU	Moscow State University
MIFI	Moscow Institute of Engineering and Physics
MIK	Assembly-Test Building
MKBS	Multirole space base station
MOK	Multipurpose orbital complex
MOL	Manned orbiting laboratory
MOM	Ministry of General Machine Building

MVTU	Moscow Higher Technical School N.E. Bauman
N1	Lunar rocket (TsKBEM)
N1-L3	Lunar landing manned programme (TsKBEM)
NII	Scientific Research Institute
NIP	Ground-test polygon
NIS	Scientific exploration vessel
NPO	Scientific and Production Enterprise
NPO Energiya	Earlier OKB-1 and TsKBEM, now RKK Energiya named after S.P. Korolev
Oazis	Hydroponics chamber (Salyut)
ODNT	Lower-body negative-pressure apparatus (Salyut); also known as 'Veter' and 'Chibis'
OIS	Orbital exploration station
OKB	Special Design Bureau
OKB Zvezda	Design Bureau for the development of spacesuits, EVA airlock and ejection seats; established in 1952 and led until 1964 by Semyon M. Alekseyev, then by Gay I. Severin
OKB-1	OKB for the development of manned, unmanned spacecraft and interplanetary probes; established in 1956 and led until 1966 by Sergey Korolev; later TsKBEM and NPO Energiya and today RKK Energiya named after S.P. Korolev
OKB-2	OKB for the development of spacecraft engines; established in 1944 under the leadership of Aleksey Isayev; today KB Himmash
OKB-456	OKB for the development of rocket engines for missiles and space launchers; established in 1946 under the leadership of Valentin Glushko; today NPO EnergoMash named after V.P. Glushko
OKB-52	OKB for the development of military satellites, space launchers, military space stations and manned spacecraft; established in 1955 under the control of Vladimir Chelomey; later TsKBM; today NPO Mashinostroyenie
ONA	Main scientific equipment module (Salyut)
OPS	Orbital piloted station (known as Almaz or 'Military Salyut')
OST-1	Orbital solar telescope (Salyut)
Polynom	Medical equipment (Salyut)
RKK Energiya	Earlier OKB-1, TsKBEM and NPO Energiya from 1974 leaded by Valentin Glushko, Yuriy Semyonov, Nikolay Sevastyanov and today Vitaliy Lopota; full name RKK Energiya named after S.P. Korolev
Sokol-K	Soyuz pressure suit
Spirala	Soviet rocket-plane programme
TKS	Transport and supply ship
TNK	Training loading suit (known also as 'Athlete' and 'penguin')
TsKBEM	Central Design Bureau of Experimental Machine Building; between 1966 and 1974 leaded by Vasiliy Mishin; originally OKB-1; today

	Rocket and Space Corporation RKK Energiya named after S.P. Korolev
TsKBM	Central Design Bureau of Machine Building headed by Vladimir Chelomey; originally OKB-52; today NPO Mashinostroyenie
TsNII	Central Scientific Research Institute
TsNIIMash	Central Scientific-Research Institute for Machine Building. The institute is responsible for systems analysis, and for research and development of spacecraft and rocket technologies. Founded in 1946, it was called NI-88 until 1967
TsPK	Cosmonaut Training Centre
TsUP	Flight Control Centre
TsVNIAG	Central Air Force Scientific Research Hospital
USSR	Union of Soviet Socialist Republics
VPK	Military-Industrial Commission
VVS	Soviet Air Force
ZEM	Plant for Experimental Machine Building (TsKBEM)
ZIKh	M.V. Khrunichev Machine Building Plant

Personnel

Academy of Sciences

Keldysh, Mstislav V. (1911–1978)
President of the Soviet Academy of Sciences (1961–1975)

Petrov, Boris N. (1913–1980)
A senior academician in the Soviet Academy of Science involved in space research; the head of Interkosmos organisation

Yershov, Valentin G. (1928–1998)
Research-Cosmonaut of Academy of Sciences (1966–1974) involved in the testing of lunar landing equipment

Communist Party and MOM Senior Officials

Afanasyev, Sergey A. (1918–2001)
The first Minister of the Ministry of General Machine Building or Soviet Space Industry (1965–1980)

Brezhnyev, Leonid I. (1906–1982)
General Secretary of the Communist Party of the Soviet Union (1964–1982)

Gorbachov, Mikhail S. (b. 1931)
The last General Secretary of the Communist Party of the Soviet Union and the last head of state of the USSR (1985–1991)

Kerimov, Kerim A. (1917–2000)
Chairman of the State Commission on Manned Space Flights (1966–1991)

Khrushchov, Nikitha S. (1894–1971)
The First Secretary of the Communist Party of the Soviet Union from 1953 to 1964, and Chairman of the Council of Ministers from 1958 to 1964.

Kosygin, Aleksey N. (1904–1980)
Premier of the Soviet Union (1964–1980)

Kirilenko, Andrey P. (1906–1990)
Member of the Soviet Politburo (Presidium of the Communist Party), with particular attention to heavy industry

Podgorny, Nikolay V. (1903–1983)
Chairman of the Presidium of the Supreme Soviet of the USSR (1967–1977)

Serbin, Ivan D. (1910–1981)
The head of the defence and space program in the party's Central Committee

Smirnov, Leonid V. (1916–2001)
The head of the Military Industrial Commission (VPK) overseen development of the Soviet space programme

Ustinov, Dmitriy F. (1908–1984)
Since 1965 Secretary of the Central Committee, with oversight of the military, defence and space industry

Cosmonauts

1. Air Force (VVS) Cosmonauts

Alekseyev, Vladimir B. (b. 1933)
Cosmonaut of the 4th VVS Group (1967–1984). Member of DOS 2 cosmonaut team.

Artyukhin, Yuriy P. (1930–1998)
Cosmonaut of the 2nd VVS Group (1963–1982). Member of DOS 2 cosmonaut team. Flight engineer of the first Almaz/Salyut 3 crew (mission Soyuz 14, 1974)

Belyayev, Pavel I. (1925–1970)
Cosmonaut of the 1st VVS Group (1960–1970). Commander of Voskhod 2 mission (1965)

Beregovoy, Georgiy T. (1921–1995)
Cosmonaut of the 2nd VVS Group (1964–1982). Commander of Soyuz 3 mission (1968) and Director of Cosmonaut Training Centre (TsPK) in 1972–1986

Bondarenko, Valentin V. (1937–1961)
Cosmonaut of the 1st VVS Group (1960–1971). Died during the finish of training in the attitude chamber

Buynovskiy, Eduard I. (b. 1936)
Cosmonaut of the 2nd VVS Group (1963–1964)

Bykovskiy, Valeriy F. (b. 1934)
Cosmonaut of the 1st VVS Group (1960–1982). Commander of Vostok 5 (1963), Soyuz 22 (1976) and Soyuz 31-Salyut 6 (1978) missions, member of L1 and L3 cosmonaut teams

Dobrovolskiy, Georgiy T. (1928–1971)
Cosmonaut of the 2nd VVS Group (1963–1971). Commander of Soyuz 11 mission (1971) and first commander of first space station Salyut. Died during the final stage of Soyuz 11 mission

Dyomin, Lev S. (1926–1998)
Cosmonaut of the 2nd VVS Group (1963–1982). Member of the Almaz cosmonaut team and flight engineer of Soyuz 15 mission (1974) that failed to dock with Almaz 2/Salyut 3 space station

Dzhanibekov, Vladimir A. (b. 1942)
Cosmonaut of the 5th VVS Group (1970–1986). Commander of five space missions aboard space stations Salyut 6 and Salyut 7 between 1978 and 1985

Filipchenko, Anatoliy V. (b. 1928)
Cosmonaut of the 2nd VVS Group (1963–1982). Commander of Soyuz 7 (1969) and Soyuz 16 (1974) missions

Gagarin, Yuriy A. (1934–1968)
Cosmonaut of the 1st VVS Group (1960–1968). Commander of Vostok (1961) and the first man in space. In 1963–1968 Deputy Director of Cosmonaut Training Centre

Gorbatko, Viktor V. (b. 1934)
Cosmonaut of the 1st VVS Group (1960–1982) – Research-cosmonaut in mission Soyuz 7 (1969), and commander of Soyuz 24 – Almaz 3/Salyut 5 (1977) and Soyuz 37 – Salyut 6 (1980) missions

Gubaryev, Aleksey A. (b. 1930)
Cosmonaut of the 2nd VVS Group (1963–1981). Commander of Soyuz 17-Salyut 4 mission (1975) and Soyuz 28-Salyut 6 mission (1978)

Khrunov, Yevgeniy V. (1933–2000)
Cosmonaut of the 1st VVS Group (1960-1980). Research-cosmonaut in mission Soyuz 5/4 mission (1969)

Kizim, Leonid D. (b. 1941)
Cosmonaut of the 3rd VVS Group (1965–1987). Commander of Soyuz T-3 – Salyut 6 (1980), Soyuz T-10-Salyut 7 (1984) and Soyuz T-15 – Mir – Salyut 7 missions (1987)

Klimuk, Pyotr I. (b. 1942)
Cosmonaut of the 3rd VVS Group (1965–1987). Member of DOS-2 cosmonaut team, commander of Soyuz 13 (1973), Soyuz 18-Salyut 4 (1975) and Soyuz 30-Salyut 6 (1978) missions, then Director of Gagarin Cosmonaut Training Centre in 1991–2003

Kolodin, Pyotr I. (b. 1930)
Cosmonaut of the 2nd VVS Group (1963–1983). Member of the second DOS-1 crew (Soyuz 11, 1971) and original crew of Soyuz 27 mission (1978) replaced just before its start.

Komarov, Vladimir M. (1927–1967)
Cosmonaut of the 1st VVS Group (1960–1967). Commander of Voskhod (1964) and Soyuz 1 (1967) missions. Died during the Soyuz 1 landing.

Kuklin, Anatoliy P. (1932–2005)
Cosmonaut of the 2nd VVS Group (1963–1975)

Lazaryev, Vasíliy G. (1928–1990)
Cosmonaut of the 3rd VVS Group (1966–1985). Member of the second DOS-2 crew. Commander of Soyuz 12 mission (1973)

Leonov, Aleksey A. (b. 1934)
Cosmonaut of the 1st VVS Group (1960–1982). Pilot of Voskhod 2 mission and first man in open space (1965), commander of crews trained for L1, L3, DOS-1 and DOS-2 missions, commander of Soyuz 19 mission (1975), then in 1982-1991 one of Deputy Directors in Gagarin Cosmonaut Training Centre

Lyakhov, Vladimir A. (b. 1941)
Cosmonaut of the 4th VVS Group (1967–1994). Commander of Soyuz 32-Salyut 6 (1979), Soyuz T-9-Salyut 7 (1983) and Soyuz TM-6-Mir (1988).

Nikolayev, Andriyan G. (1929–2004)
Cosmonaut of the 1st VVS Group (1960–1982). Commander of Vostok 3 (1962) and Soyuz 9 (1970) missions; in 1968–1992 one of Deputy Directors in Gagarin Cosmonaut Training Centre

Popov, Leonid I. (b. 1945)
Cosmonaut of the 5th VVS Group (1970–1987). Commander of three space missions aboard space stations Salyut 6 and Salyut 7 between 1980 and 1982

Popovich, Pavel R. (b. 1930)
Cosmonaut of the 1st VVS Group (1960-1982). Commander of Vostok 4 (1962) and Soyuz 14-Almaz 2/Salyut 3 (1974) missions; in 1978–1989 one of Deputy Directors in Gagarin Cosmonaut Training Centre

Preobrazhenskiy, Vladimir Y. (1939–1993)
Cosmonaut of the 3rd VVS Group (1965–1980)

Rozhdestvenskiy, Valeriy I. (b. 1939)
Cosmonaut of the 3rd VVS Group (1965–1986). Flight engineer of Soyuz 23 mission (1976) that failed to dock with Almaz 3/Salyut 5 space station

Sarafanov, Gennadiy V. (1942–2005)
Cosmonaut of the 3rd VVS Group (1965–1986). Commander of Soyuz 15 mission (1974) that failed to dock with Almaz 2/Salyut 3 space station

Shatalov, Vladimir A. (b. 1927)
Cosmonaut of the 2nd VVS Group (1963–1971). Commander of Soyuz 4 (1969), Soyuz 8 (1969) and Soyuz 10 mission (1971). In 1971–1987 succeeded Kamanin served as Commander-in-Chief's Aide of Air Force for space and in 1987–1991 as Commander of Gagarin Cosmonaut Training Centre

Shonin, Georgiy S. (1935–1997)
Cosmonaut of the 1st VVS Group (1960–1979). Commander of Soyuz 6 (1969) and the original first crew of DOS-1 space station, but failed to fly due to problems with alcohol

Teryeshkova, Valentina V. (b. 1937)
Cosmonaut of the first VVS Woman Group (1962–1997). Commander of Vostok 6 (1963) and first woman in space

Titov, Gherman S. (1935–2000)
Cosmonaut of the 1st VVS Group (1960–1970). Commander of Vostok 2 (1961) and second man in orbit

Voloshin, Valeriy A. (b. 1942)
Cosmonaut of the 3rd VVS Group (1965–1969)

Volynov, Boris V. (b. 1934)
Cosmonaut of the 1st VVS Group (1960–1990). Commander of Soyuz 5 (1969) and Soyuz 21-Almaz 3/Salyut 5 missions (1976). In 1983–1990 commander of the group of Air Force cosmonauts

Vorobyev, Lev V. (b. 1931)
Cosmonaut of the 2nd VVS Group (1963–1974)

Voronov, Anatoliy F. (1930–1993)
Cosmonaut of the 2nd VVS Group (1963–1979). Member of DOS-1 and originally DOS-2 cosmonaut team.

Zholobov, Vitaliy M. (b. 1937)
Cosmonaut of the 2nd VVS Group (1963–1981). Flight engineer of Soyuz 21-Almaz 3/Salyut 5 mission (1976)

Zudov, Vyacheslav D. (b. 1942)
Cosmonaut of the 3rd VVS Group (1965–1987). Commander of Soyuz 23 mission (1976) that failed to dock with Almaz 3/Salyut 5 space station

2. TsKBEM Cosmonauts

Anyokhin, Sergey N. (1910–1986)
Cosmonaut of the 1st TsKBEM Group (1966–1968) and commander of civilian cosmonaut group at TsKBEM

Bugrov, Vladimir Y. (b. 1933)
Cosmonaut of the 1st TsKBEM Group (1966–1968)

Dolgopolov, Gennadiy A. (b. 1935)
Cosmonaut of the 1st TsKBEM Group (1966–1967)

Grechko, Georgiy M. (b. 1930)
Cosmonaut of the 1st TsKBEM Group (1966–1986), member of DOS-2 cosmonaut team, flight engineer of space missions Soyuz 17 – Salyut 4 (1975), Soyuz 26 – Salyut 6 (1977/78) and Soyuz T14 – Salyut 7 (1985)

Kubasov, Valeriy N. (b. 1935)
Cosmonaut of the 1st TsKBEM Group (1966–1993), member of DOS-1 and DOS-2 cosmonaut teams, flight engineer of space missions Soyuz 6 (1969) and Soyuz 19 (1975) and commander of Soyuz 36 – Salyut 6 (1980) flight

Lebedyev, Valentin V. (b. 1942)
Cosmonaut of the 3rd TsKBEM Group (1972–1989), flight engineer of space missions Soyuz 13 (1973) and Soyuz T-5 – Salyut 7 (1982)

Makarov, Oleg G. (1933–2003)
Cosmonaut of the 1st TsKBEM Group (1966–1986), member of L1, L3 and DOS-2 cosmonaut teams, flight engineer of space missions Soyuz 12 (1973), Soyuz 27 – Salyut 6 (1978) and Soyuz T-3 – Salyut 6 (1980)

Nikitskiy, Vladimir P. (b. 1939)
Cosmonaut of the 2nd TsKBEM Group (1967–1968)

Patsayev, Viktor I. (1933–1971)
Cosmonaut of the 2nd TsKBEM Group (1967–1971), research – cosmonaut of Soyuz 11 – Salyut mission (1971), died during the re-entry stage of Soyuz 11 flight

Rukavishnikov, Nikolay N. (1932–2002),
Cosmonaut of the 1st TsKBEM Group (1967–1987), member of L1, L3, DOS-1 and
DOS-2 cosmonaut teams, research – cosmonaut of Soyuz 10 flight (1971), flight
engineer of Soyuz 16 (1974) and commander of Soyuz 33 mission (1979)

Ryumin, Valeriy V. (b. 1939)
Cosmonaut of the 4th TsKBGEM Group (1973–1987), involved in construction of
DOS space station, then flight engineer of four space missions aboard Soyuz and
Space Shuttle spacecrafts and Salyut 6 and Mir space stations and director of
mission control during Salyut 7 and Mir flights

Sevastyanov, Vitaliy I. (b. 1935)
Cosmonaut of the 1st TsKBEM Group (1967–1993), flight engineer of Soyuz 9
mission (1970), member of DOS-1 and DOS-2 cosmonaut teams and flight engineer
of Soyuz 18 – Salyut 4 mission (1975)

Strekalov, Gennadiy M. (1940–2004)
Cosmonaut of the 4th TsKBEM Group (1973–1995), flight engineer of five space
missions Soyuz T-3 – Salyut 6 (1980), Soyuz T-8 (1983), Soyuz T-11 – Salyut 7
(1984), Soyuz TM-10 – Mir (1990) and Soyuz TM-21 – Mir (1995)

Volkov, Vladislav N. (1935–1971)
Cosmonaut of the 1st TsKBEM Group (1966–1971), flight engineer of Soyuz 7
(1969) and Soyuz 11 – Salyut mission (1971), died during the re-entry stage of Soyuz
11 flight

Yeliseyev, Aleksey S. (b. 1934)
Cosmonaut of the 1st TsKBEM Group (1966-1985), flight engineer of Soyuz 5/4
(1969), Soyuz 8 (1969) and Soyuz 10 (1971) missions (1971); in 1974–1981 director of
mission control during solo flights of Soyuz spacecrafts and missions aboard Salyut
4 and Salyut 6 space stations

Cosmonauts' Family Members

Dobrovolskaya (Kamenchuk), Mariya A. (1907–?)
Mother of Georgiy Dobrovolskiy

Dobrovolskaya (Steblyova), Lyudmila T. (1938–1986)
Wife of Georgiy Dobrovolskiy

Dobrovolskaya, Marina G. (b. 1959)
Daughter of Georgiy Dobrovolskiy

Dobrovolskaya, Nataliya G. (b. 1967)
Daughter of Georgiy Dobrovolskiy

Dobrovolskiy, Aleksandar T. (b. 1946)
Step-brother of Georgiy Dobrovolskiy

Dobrovolskiy, Timofey T. (1908–?)
Father of Georgiy Dobrovolskiy

Kubasov, Dmitriy V. (b. 1971)
Son of Valeriy Kubasov

Kuraytis, Stanislav A. (1905–1978)
Father of Aleksey Yeliseyev

Leonova, Victoria A. (1961–1996)
Daughter of Aleksey Leonov

Mikheyev, Mikhail G. (1889–1962)
Step-father of Nikolay Rukavishnikov

Patsayev, Dmitriy V. (b. 1957)
Son of Viktor Patsayev

Patsayev, Ivan P. (1910–1941)
Father of Viktor Patsayev

Patsayeva (Koltsova), Mariya S. (1913–2004)
Mother of Viktor Patsayev

Patsayeva (Kryazheva), Vera A. (1931–2002)
Wife of Viktor Patsayev

Patsayeva, Galina I. (b. 1937)
Sister of Viktor Patsayev

Patsayeva, Svetlana V. (b. 1962)
Daughter of Viktor Patsayev

Rukavishnikov, Vladimir N. (1965–2006)
Son of Nikolay Rukavishnikov

Rukavishnikova (Mikheyeva), Galina I. (1910–1982)
Mother of Nikolay Rukavishnikov

Rukavishnikova (Pavlova), Nina V. (1939–2000)
Wife of Nikolay Rukavishnikov

Shatalov, Aleksandr B. (1890–1970)
Father of Vladimir Shatalov

Shatalov, Igor V. (b. 1952)
Son of Vladimir Shatalov

Shatalova (Tolubeyeva), Yelena V. (b. 1958)
Daughter of Vladimir Shatalov

Shatalova (Yonova), Muza A. (b. 1928)
Wife of Vladimir Shatalov

Volkov, Ivan I. (?)
Step-father of Viktor Patsayev

Volkov, Nikolay G. (1914–?)
Father of Vladislav Volkov

Volkov, Vladimir V. (b. 1958)
Son of Vladislav Volkov

Volkova (Birykova), Lyudmila A. (b. 1937)
Wife of Vladislav Volkov

Volkova (Kotova), Olga M. (1912–?)
Mother of Vladislav Volkov

Yeliseyeva (Komarova), Larisa I. (b. 1934)
Second wife of Aleksey Yeliseyev

Yeliseyeva (Shpalikova), Valentina P. (b. 1935)
First wife of Aleksey Yeliseyev

Yeliseyeva, Valentina I. (b. 1909)
Mother of Aleksey Yeliseyev

Yeliseyeva, Yelena A. (b. 1960)
Daughter of Aleksey Yeliseyev

Defence Ministry, Air Forces and Strategic Rocket Forces

Agadzhanov, Pavel A. (1923–2001)
Led flight control for Soyuz space missions at Yevpatoriya

Bolshoy, Amos A.
Led flight control teams for early piloted space missions

Borisenko, Ivan G. (1921–2004)
'Sport Commissar' responsible for records of manned space mission data

Fadeyev, Nikolay G.
Led L1/Zond flight control teams

Goreglyad, Leonid I. (1915–1986)
General Staff representative at the TsPK and an aide to Kamanin

Grechko, Andrey A. (1903–1976)
Defence Minister of the Soviet Union (1967–1976)

Kamanin, Nikolay P. (1909–1982)
Deputy Chief of General Staff for space (1958–1966) and an aide to Air Force
Commander in 1966–1971 oversaw cosmonaut training at the TsPK

Karas, Andrey G. (1918–1979)
Commander of the Control Directorate of Space Assets (today Russian Military
Space Forces) and member of State Commission for Almaz space stations

Kutakhov, Pavel S. (1914–1984)
Commander of the Soviet Air Force (1969–1984)

Kuznyetsov, Nikolay F. (1916–2000)
Commander of Cosmonaut Training Centre (TsPK) in 1963–1972

Malinovskiy, Rodion Y. (1898–1967)
Defence Minister of the Soviet Union (1957–1967)

Pasternak, Mikhail
Member of the flight control team, an aide to Agadzhanov

Ponomaryev, Boris N.
Kutakhov's Deputy

Uglyanskiy, ?,
One of the Air Force members of the recovery team

Vershinin, Konstantin A. (1900–1973)
Commander of the Soviet Air Force (1946–1949 and 1957–1969)

Medical Officials

Burnazyan, Avetik I. (1906–?)
Deputy Minister of Health

Gazenko, Oleg G. (1917–2007)
Director of IBMP 1969–1988

Genin, Abram M. (b. 1922)
Director Chief at Institute of Aviation and Space Medicine in 1964–1975

Gurovskiy, Nikolay N.
Doctor at Institute of Aviation and Space Medicine and later Deputy Director at IBMP

Lebedyev, Anatoliy A.
Leading Air Force doctor in the Soyuz recovery team

Stezhadze, Levan
Member of the Soyuz recovery team

Vorobyev, Dr. Yevgeniy I.
Doctor at Institute of Aviation and Space Medicine

Yegorov, Boris B. (1937–1994)
Doctor at Institute of Aviation and Space Medicine and first cosmonaut-doctor (1964)

Officials from other Design Bureaus

Babakin, Georgiy N. (1914–1971)
Chief Designer at OKB Lavochkin in 1965–1971, led work on interplanetary and lunar probes

Barmin, Vladimir P. (1909–1993),
Chief Designer at Design Bureau SpetsMash in 1941–1993, led work on launch complexes

Darevskiy, Sergey G.
Chief Designer at Design Bureau SOKB of Gromov LII Institute in 1965–1975, led work on simulators and cockpit consoles

Glushko, Valentin P. (1908–1989)
General Designer of OKB-456 in 1946–1989, led work of development of rocket engines; in 1974–1989 led NPO Energiya (earlier TsKBEM)

Isayev, Aleksey M. (1908–1971)
Chief Designer of OKB-2 in 1947–1971, led work of development of spacecraft engines

Kovtunenko, Vyacheslav M. (1921–1995)
Chief Designer at NPO Lavochkin in 1977–1995, led work on interplanetary and lunar probes

Kryukov, Sergey S. (1918–2005)
Deputy Chief Designer at OKB-1 in the early 1960s, then General Designer at NPO Lavochkin

Kuznetsov, Viktor I. (1913–1991)
Chief Designer at NII-10 and NII-944 worked on missile and spacecraft gyros

Lobanov, Nikolay A. (1909–1978)
Chief Designer in 1968–1977 for parachutes and spacecraft landing systems

Mnatsakanyan, Armen S. (1918–1992)
Chief Designer at NII-648 worked on spacecraft telemetry and radar systems

Myasishchev, Vladimir P. (1902–1978)
Chief Designer of OKB-23 in 1951–1960, led work of development of space-plan

Pilyugin, Nikolay A. (1908–1982)
Chief Designer at NII-885 in 1948–1982 worked on missile and spacecraft guidance

Ryazanskiy, Mikhail S. (1909–1987)
Chief Designer at NII-885 in 1946–1951 and 1955–1987 worked on missile and spacecraft radio guidance

Severin, Gay I. (1926–2008)
Chief Designer at OKB Zvezda from 1964 worked on spacesuits and EVA airlocks

Yangel, Mikhail K. (1911–1971)
Chief Designer at OKB-586 in 1954–1971, led work on missiles and robot spacecrafts

TsKBEM – Senior Officials

Bushuyev, Konstantin D. (1914–1978)
Deputy Chief Designer in 1954–1972 at OKB-1 and TsKBEM led piloted space projects

Chertok, Boris Y. (b. 1912)
Deputy Chief Designer in 1956–1991 at OKB-1, TsKBEM and NPO (RKK) Energiya led development of rocket and spacecraft guidance systems

Feoktistov, Konstantin P. (b. 1926)
Department Chief at OKB-1 and TsKBEM, worked on development of Vostok, Voskhod and Soyuz spacecrafts and Salyut space stations. Cosmonaut of the 2nd TsKBEM Group (1968–1987) and the first cosmonaut-engineer in space (mission Voskhod in 1964)

Korolev, Sergey P. (1907–1966)
Chief Designer of Soviet rockets and spacecrafts in 1946–1966 at NII-3 and OKB-1, founder of the Soviet space programme

Kozlov, Dmitriy I. (b. 1919)
Head of OKB-1/TsKBEM branch No 3 from 1959, led development of reconnaissance spacecrafts

Legostayev, Viktor P.
Replaced Raushenbakh in 1973 as the head of development of spacecraft guidance systems

Mishin ,Vasiliy P. (1917–2001)
Chief Designer of TsKBEM in 1966–1974; as Korolev's successor led works on development of Soyuz, L1, N1-L3 and DOS programmes

Okhapkin, Sergey O. (1910–1980)
Deputy Chief Designer in 1952–1976 at Korolev's OKB, Mishin's First Deputy, led work on development of N1 lunar rocket

Raushenbakh, Boris V. (1915–2001)
Department Chief at OKB-1 and TsKBEM in 1960–1973, worked on rocket and spacecraft guidance systems

Semyonov, Yuriy P. (b. 1935)
The lead designer of Soyuz, L1/Zond spacecrafts and DOS space stations at OKB-1 and TsKBEM. In 1989–2005 General Designer at RKK Energiya

Sevastyanov, Nikolay N. (b. 1964)
Director of RKK Energiya in 2005–2007

Shabarov, Yevgeniy V. (b. 1922)
Deputy Chief Designer at OKB-1/TsKBEM, led the testing of manned spacecrafts

Tikhonravov, Mikhail K. (1900–1974)
One of Soviet rocket pioneers, designer at NII-3, NII-4 and OKB-1, worked on development of Sputnik and Vostok spacecrafts

Tregub, Yakov I. (b. 1918)
Deputy Chief Designer at OKB-1/TsKBEM in 1964–1973, led cosmonaut team at TsKBEM and flight control for manned space missions

Tsybin, Pavel V. (1905–1992)
Deputy Chief Designer at OKB-1/TsKBEM, involved in development of early space plans, L1/Zond rescue systems and Soyuz transport spacecrafts

TsKBEM – DOS Designers and Officials

Abramov, Aleksey P. (1919–1998)
Worked the Soyuz/DOS launch preparations at the cosmodrome

Bashkin, Yevgeniy
Worked on cosmonaut training

Bezverby, Vitaliy K.
Led MKBS project

Bugrova, Skella A.
Worked in flight control team

Gorshkov, Leonid A.
Bushuyev's Deputy worked on design of DOS space station

Inelaur, Viktor T.
Semyonov's Deputy in programme DOS responsible for development of the guidance apparatus

Isayev, Gennadiy F.
Worked on cosmonaut training

Karashtin, Vladimir M.
Worked the Soyuz/DOS launch preparations at the cosmodrome

Krapivina, Eleonora
Worked on cosmonaut training

Kuzmin, Viktor P.
Worked on Soyuz and DOS docking systems

Ostashev, Arkadiy I. (1925–1998)
Worked on the final Soyuz and DOS testings on cosmodrome

Ovchinikov, Viktor S.
Designer of DOS systems

Pallo, Arvid V. (b. 1912–2001)
One of Semyonov Deputies in programme DOS

Pravetskiy, Vladimir N.
Worked on development of DOS life support systems

Semyonov, Gherman Y.
Led the final commissioning of the DOS space station

Slesarev, Dmitriy A.
Semyonov's deputy in programme DOS responsible for development of the 7K-T
Soyuz transport spacecraft

Surgachov, Oleg V.
Worked on Soyuz and DOS thermal-regulation systems

Syromyatnikov, Vladimir S. (1936–2006)
Worked on Soyuz and DOS docking systems

Varshavskiy, Viktor P.
Worked on spacecraft on-board technical documentation

Vilnitskiy, Lev B.
Worked on Soyuz and DOS docking systems

Yurasov, Igor Y.
Chertok's Deputy in programme DOS responsible for development of the guidance
apparatus

Zelenshchikov, Boris I.
Tregub's Deputy in programme DOS responsible for testing of the systems,
cosmonaut training and mission control

Zhivoglotov, Vsevolod N.
Worked on Soyuz and DOS docking systems

TsKBM – Senior Officials

Bugayskiy, Viktor N. (b. 1922)
The head of TsKBEM Branch No1 in 1960–1973, worked on development of rockets, spacecrafts and DOS and Almaz space stations

Chelomey, Vladimir N. (1914–1984)
General Designer at OKB-52/TsKBM IN 1954–1984 led work on cruise missiles, Proton rockets, military and lunar manned spacecrafts and Almaz military space stations

Nodelyman, Yakov
Led development of TKS for OPS space station

Pallo, Vladimir V.
Bugayskiy's Deputy at TsKBM for development of DOS and Almaz space stations

Others Personnel

Berry, Dr. Charles A.
Space medicine researcher and the leading medical official at NASA during Mercury, Gemini and Apollo space programmes

Borman, Frank (b. 1928)
NASA astronaut; commander of Gemini 7 (1965) and Apollo 8 (1968)

Brand, Vance D. (b. 1931)
NASA astronaut; pilot of Apollo 18 (1975) and commander of space shuttle missions STS-5 (1982), STS-41B (1984) and STS-35 (1990)

Chou En-lai (1989–1976)
Premier of the People's Republic of China (1949–1976)

De Gaulle, Charles (1890–1970)
President of France (1958–1969)

Farkash, Bertalan (b. 1949)
First Hungarian cosmonaut – Soyuz 36 mission (1980)

Fokin, Yuriy
The commentator of the Central USSR TV

Frolov, Yevgeniy
The commentator of the Central USSR TV

Gandhi, Indira (1917–1984)
Prime Minister of India (1966–1977 and 1980–1984)

Gatland, Kenneth W.,
Space writer

Golovanov, Yaroslav K. (1932–2003)
Space Journalist

Gubaryev, Vladimir S. (b. 1938)
Space Journalist

Irwin, James B. (1930–1991)
Astronaut; lunar module pilot of the Apollo 15 (1971)

Ivanov, Georgiy (b. 1941) (real surname Kakalov)
First Bulgarian cosmonaut – Soyuz 33 mission (1979)

Johnson, Lyndon B. (1908–1973)
President of the United States (1963–1969)

Jones, Dr. Walton L.
Space medicine researcher at NASA during Apollo programme

Kennedy, John F. (1917–1963)
President of the United States (1961–1963)

Kibalchich, Nikolay I. (1853–1881)
A Russian revolutionary and a rocket pioneer

Low, George M. (1926–1984)
NASA Deputy Administrator (1969–1976)

Moore, Patrick (b. 1923)
Space writer

Nixon, Richard M. (1913–1994)
President of the United States (1969–1974)

Perry, Geoffrey E. (1927–2000)
Leader of a group of amateurs involved in tracking Soviet spacecraft

Pompidou, Georges (1911–1974)
President of France (1969–1974)

Pope Paul VI (1897–1978)
Pope of the Catholic Church (1963–1978)

Queen Elizabeth II (b. 1926)
Queen of the United Kingdom and the other Commonwealth realms

Reagan, Ronald W. (1911–2004)
President of the United States (1981–1989)

Rebrov, Mikhail F. (b. 1931)
Space Journalist

Romanov, Alexandar P.
Space Journalist

Scott, David R. (b. 1932)
NASA Astronaut; pilot of Gemini 8 (1966) and Apollo 9 (1969), and commander of
Apollo 15 (1971)

Shepard, Alan B. (1923–1998)
NASA Astronaut; pilot of Mercury-Redstone 3 (1961) and commander of Apollo 14
(1971)

Simonov, Konstantin M. (1915–1979)
Poet

Slayton, Donald K. (1924–1993)
NASA Astronaut; pilot of Apollo 18 (1975)

Stafford, Thomas P. (b. 1930)
NASA Astronaut; pilot of Gemini 6 (1965) and commander of Gemini 9 (1966),
Apollo 10 (1969) and Apollo 18 (1975)

Tsander, Fridrich A. (1887–1933)
A rocket scientist and pioneer of cosmonautic theory, who designed the first Soviet
liquid-propellant rocket (1933)

Tsiolkovskiy, Konstantin E. (1857–1935)
"Father of cosmonautics" – a rocket scientist and pioneer of cosmonautic theory

Verne, Jules (1828–1905)
Pioneer of the science-fiction genre

Yevtushenko, Yevgeniy A. (b. 1933)
Poet

Bibliography

Russian Language Books

1. Волков Вл., Шагаем в небо. – М.: Молодая гвардия, 1971
   Volkov Vl., We Pace into the Sky, Molodaya Gvardiya, Moscow, 1971
2. Колтовой Б. И., Коновалов Б. П., Мост в космос. – М.: Известия, 1971
   Koltovoy B.I., Konovalov B.P., The Bridge in Space, Izvestiya, Moscow, 1973
3. Василев М. П., ред. кол., Салют на орбите. – М.: Машиностроение, 1973
   Vasilev M.P. (a pseudonym of Vasiliy P. Mishin), ed, Salyut on Orbit,
   Mashinostroenie, Moscow, 1973
4. Борисенко И. Г., На космических стартах и финишах, – М.: Знание, 1975
   Borisenko I.G., On space starts and finishes Znaniye, Moscow, 1975
5. Пацаева Г. И., Отвага исканий. – М.: Молодая гвардия, 1976
   Patsayeva G.I., Boldness of Aspiration, Molodaya Gvardiya, Moscow, 1976
6. Береговой Г.Т., Тищенко А.А., Шибанов Г. П., Ярополов В. И.
   Безопасность космических полетов, – М.: Машиностроение, 1977
   Beregovoy G.T., Tishchenko A.A., Shibanov G.P., Yaropolov V.I.
   Bezopasnost kosmicheskih poletov, Mashinostroenie, Moscow,1977
7. Шаталов В. Трудные дороги космоса. – М: Молодая гвардия, 1978
   Shatalov V., Hard Roads to Space, Molodaya Gvardiya, Moscow, 1978
8. Филипченко А.В., Надежная орбита. – М.: Изд – во ДОСААФ СССР, 1978
   Filipchenko A.V., Reliable Orbit, DOSAAF USSR, Moscow, 1978
9. Лебедев Л., Лукьянов Б., Романов А., Сыны голубой планеты 1961–1981
   – М.: Политиздат, 1981
   Lebedyev L., Lukyanov B., Romanov A., Sons of the Blue Planet 1961–1981
   Politizdat, Moscow, 1981
10. Глушко В. П., ред. кол., Космонавтика, Энциклопедия
    М.: Машиностроение, 1985
    Glushko V.P., ed, Cosmonautics, Encyclopaedia,
    Mashinostrenie, Moscow, 1985
11. Мишин В. П., Почему мы не слетали на Луну?, – М.: Знание, 12/1990
    Mishin V.P., Why Didn't We Fly to the Moon?
    Znaniye 12/1990, Moscow, 1990

12. Молчанов В. Е., О тех, кто не вышел на орбиты. – М.: Знание, 1990.
Molchanov V.E., About Those Who Did Not Reach Orbit
Znaniye, Moscow 1990

13. Афанасьев И. Б., Неизвестные корабли. М.: Знание, 12/1991
Afanasyev I.B., Unknown Spacecrafts, Znaniye, 12/1991

14. Ракетно- космическая корпорация "Энергия" им. С. П. Королева/Под. ред. Ю. П. Семенова, 1996
Semyonov Y.P., ed, Rocket and Space Corporation Energiya named after S.P. Korolev, 1996

15. Ребров М., Космические катастрофы: Странички из секретного досье. - М.: ЭКСПРИНТ НВ, 1996
Rebrov M., Space Catastrophes: Notes from the Secret Dossier
EKSPRINT NV, Moscow, 1996

16. Елисеев А. С., Жизнь – капля в море.
– М.: ИД « Авиация и космонавтика », 1998
Yeliseyev A.S., Life – A Drop in the Sea
ID Aviatsiya and kosmonavtika, Moscow, 1998

17. Черток Б. Е., Ракеты и люди. Книга 1, – М.: Машиностроение, 1999.
Chertok B.Y., Rockets and People, Book 1, Mashinostrenie, Moscow, 1999

18. Феоктистов К. П., Траектория жизни. Между вчера и завтра.
– М.: ВАГРИУС, 2000.
Feoktistov K.P., Trajectory of Life. Between Yesterday and Tomorrow
VAGRIUS, Moscow, 2000

19. Давыдов И. В. Триумф и традегия советской космонавтики.
– М.: Глобус, 2000.
Davidov I.V., Triumph and Tragedies of Soviet Cosmonautics
Globus, Moscow, 2000

20. Каманин Н. П., Скрытый космос. Книга 4.
– М.: ООО ИИД « Новости космонавтики », 2001.
Kamanin N.P., Hidden Space, Book 4, Novosti kosmonavtiki, 2001

21. Черток Б. Е., Ракеты и люди. Лунная гонка. Книга 4.
– М.: Машиностроение, 2002.
Chertok B.Y., Rockets and People – The Moon Race, Book 4,
Mashinostrenie, Moscow, 2002

22. Сыромятников В. С.
100 рассказов о стыковке и других приключениях в космосе и на Земле. Часть 1: 20 лет назад. – М.: Логос, 2003.
Siromyatnikov V.S., A Hundred Stories About Docking and Other Connections in Space and on the Earth, Part 1: Twenty Years Ago
Logos, Moscow, 2003

23. Буйновский Э. И., Повседневная жизнь первых российских ракетчиков и космонавтов – М.: Молодая гвардия, 2004.
Buynovskiy E.I., Daily Life of the first Russian Rocket men and Cosmonauts
Molodaya Gvardiya, Moscow, 2004

24. Железняков А., Тайны ракетных катастроф. – М.: Яуза, 2004.
Zheleznyakov A., Secrets of Rocket Catastrophes, Yauza, Moscow, 2004

25. Афанасьев И. Б., Батурин Ю. М., Белозерский А. Г.
Мировая пилотируемая космонавтика. – М.: "РТСофт", 2005
Afanasyev I.B., Baturin Y.M., Belozerskiy A.G.
The World Manned Cosmonautics, – RTSoft, Moscow, 2005

26. Феоктистов К. П., Зато мы делали ракеты. – М.: Время, 2005.
Feoktistov K.P., It Is Why We Made Rockets, Vremya, Moscow, 2005

27. Караш Ю. Ю., Тайны лунной гонки. СССР и США: сотрудничество в космосе, – М.: ОЛМА-ПРЕСС. Инвест, 2005
Karash Y.Y., Secrets of the Moon Race. USSR and USA: Cooperation in Space, OLMA-PRESS, Moscow, 2005

28. Губарев В. С., Русский космос, III, Эксмо, Алгоритам, Москва 2006
Gubaryev V.S., Russian Space, Book 3,
Exmo, Algorithm, Moscow, 2006

English Language Books

1. Oberg, James E.
Red Star in Orbit – The inside story of the Soviet Space programme.
Harrap, London, 1981

2. Clark, Phillip
The Soviet Manned Space Programme.
Salamander Books, London, 1988

3. Cassutt, Michael
Who's Who in Space.
Macmillan Publishing Company, New York, 1993

4. Harvey, Brian
The New Russian Space Programme.
Wiley-Praxis, UK, 1996

5. Shayler, David J.
Disasters and Accidents in Manned Spaceflight. – Springer-Praxis, UK, 2000

6. Stafford, Thomas P. with Cassutt Michael.
We Have Capture – Tom Stafford and the Space Race.
Smithsonian Institution Press, 2002

7. Godwin, Robert, ed.
Rocket and Space Corporation Energia – The Legacy of S.P. Korolev
Apogee Books, Burlington, 2002

8. Burgess, Colin and Doolan, Kate (with Vis, Bert)
Fallen Astronauts, Heroes who died Reaching for the Moon.
University of Nebraska, 2003

9. Siddiqi, Asif A.
The Soviet Space Race with Apollo.
University Press of Florida, 2003

10. Hall, Rex, Shayler, David J.
    Soyuz – A Universal Spacecraft.
    Springer-Praxis, UK, 2003
11. Scott David and Leonov Alexei.
    Two Sides of the Moon – Our Story of the Cold War Space Race.
    Simon & Schuster, UK, 2004
12. Hall, Rex, Shayler, David J., Vis, Bert
    Russia's Cosmonauts.
    Springer-Praxis, UK, 2005

Articles

1. The Guardian, Russia plans first station in space, 16/03/1971
2. New York Times, U.S.S.R. may be building a station in space, 20/04/1971
3. U.P.I. Soyuz crew begin new stage in space exploration, 23/04/1971
4. Daily Telegraph, Soyuz-10 likely to attempt link with space station, 24/04/1971
5. Scottish Sunday Express, House in space a stage nearer, 25/04/1971
6. Observer, Soyuz-10 catches up with Salyut, 25/04/1971
7. Daily Telegraph, Surprise end to Soyuz mission after difficult flight, 25/04/1971
8. New York Times, Space links continue, says Russia, 26/04/1971
9. Daily Telegraph, Cosmonauts describe Salyut space station, 26/04/1971
10. The Guardian, Spaceman was not ill, 27/04/1971
11. The Guardian, Breeze saved Soyuz, 27/04/1971
12. Soviet Weekly, The Space Station Flies On, 1/05/1971
13. The Guardian, "Space" boys scoop Tass, 07/06/1971
14. The Irish Times, U.S. space plan stands despite Soyuz deaths, 1/07/1971
15. The Times, Soyuz systems under suspicion, 01/07/1971
16. The Times, Little sing of injury to spacemen, 02/07/1971
17. Politika, Ostvarili su podvig, Beograd, 02/07/1971
    Politika, They Made Accomplishment, Belgrade, 2 July 1971
18. Politika, Sahranjeni sovjetski kosmonauti, Beograd, 03/07/1971
    Politika, Soviet Cosmonauts Buried, Belgrade, 3 July 1971
19. The Times, Russian leader justifies space programme as cosmonauts are buried, 03/07/1971
20. The Sunday Times, Breathless clue to Soyuz space deaths, 4/07/1971
21. The Times, Moscow to go ahead with plans for manned space stations despite Soyuz disaster, 05/07/1971
22. The Times, Ominous Development, 12/07/1971
23. The Washington Post, Cosmonauts died because valve was forced open, 29/10/1973
24. The Guardian, Vasily Mishin Soviet space boss scapegoat for failure to put a man on the moon, 01/11/2001

25. Экспресс-К (Алматы) N145, Уходят из жизни ..., 30.7.2004
    Express-K (Alma-Ata) N145, Walkout from the life ..., 30/07/2004
26. Grahn, Sven, "Salyut-1, its origin, flights to it and radio tracking thereof", web
    site, www.svengrahn.pp.se/trackind/salyut1/salyut1.html
27. Таррасов А., Полеты во сне и на яву, "Правда", 20/10/1989.
    Tarasov A., Missions in dreams and Reality, Pravda, 20 October 1989
    (Interview with Vasiliy Mishin)
28. Салахутдинов Г. М., Ещё раз о космосе, Огонёк, No. 34, 1990.
    Salahutdinov G.M., Once More about Space, Aganyok, No. 34, 1990
    (Interview with Vasiliy Mishin)
29. Лоскутов А., Ген одержимости – Интервью с дочерью Мишина,
    Ежедневнвые новости-Подмосковье (Москва), No. 8, 18.1.2007
    Loskutov A., Tenable Gene (Interview with Mishin's daughter)
    Daily News, Moscow, No. 8, 18 January 2007
30. Марина Добровольская, – Пусть тебе приснится солнце ...
    Одесская газета, 07/1988
    Marina Dobrovolskaya, – Let to you dream the sun ...
    Odesskaya gazeta, July 1988

Interviews by the author

1. Marina Dobrovolskaya, 24 May 2007
2. Svetlana Patsayeva, 1 August 2007
3. Dmitriy Patsayev, 5 September 2007

Periodicals

1. Soviet Weekly, Space Station Gets Down to its Real Work, 26/06/1971
2. "Новости космонавтики"
   Novosti kosmonavtiki (Cosmonautics' News), Russian monthly magazine
       No. 10, 1996
       No. 11, 1996
       No. 12/13, 1996
       No. 7, 1997
       No. 4, 2002 (Interview with Vladimir Shatalov)
       No. 6, 2002 (Interview with Aleksey Yeliseyev)
       No. 12, 2002 (Necrology for Nikolay Rukavishnikov)
       No. 3, 2003 (Necrology for Kerim Kerimov)
       No. 3, 2005 (Interview with Valeriy Kubasov)
3. Spaceflight, No. 1, 2000, British Interplanetary Society
4. Советская Кубань (Краснодар), No. 29, 5.8.2005, Яблони на Марсе не нужны
   – интервью с Феоктистовым
   Soviet Cuban (Krasnodar), No. 29, 5 August 2005 (Interview with Konstantin
   Feoktistov)

5. Российский космос, Rossiyskiy kosmos (Russian Space), Russian monthly magazine
   No. 6, 2006 – First Space Home
   No. 6, 2006 – Fragments from Notes of Salyut Crew
   No. 9, 2006 – Finish of Martian Expedition

Web sites

1 www.astronaut.ru
2 www.cosmoworld.ru
3 www.russianspaceweb.com
4 www.astronautix.com
5 www.svengrahn.pp.se
6 www.spacefacts.de
7 www.sff.net
8 www.ski-omer.ru

Index

Printing: Mercedes-Druck, Berlin
Binding: Stein+Lehmann, Berlin